SOUND WAVES IN SOLIDS
H.F. Pollard

Applied physics series

Series editor H.J.Goldsmid

Series editors H.J.Goldsmid and D.W.G.Ballentyne

SOUND WAVES IN SOLIDS
H.F. Pollard

repl

6476 - 025X

PHYSICS

Published in the USA by

p *Methuen, Inc.* ury Park, London NW2 5JN

733 Third Avenue,
New York, New York 10017

© 197... ...on Limited

ISBN 0 85086 053 9

Printed in Great Britain

Preface

Many interesting practical processes and devices depend on a knowledge of sound wave effects in solid materials. Any study of the mechanical properties of solids requires at least a nodding acquaintance with basic elasticity theory, so a brief account of it is presented in the first chapter. A study of the behaviour of mechanical waves, both in a very large elastic medium and a limited medium with surfaces, then leads to a chapter on acoustic waveguides. At this stage there is a chapter on experimental techniques, most of which rely on a recognition of the type and characteristics of the wave motion employed.

The problem of energy losses associated with wave propagation in real materials is the subject of chapters 6 and 7, in which the role played by dislocations in solids is given considerable attention. Finally, in chapter 8, high-frequency techniques are discussed with particular application to biomedical problems and acoustical holography.

An attempt has been made to make the text self-consistent. Theory is introduced that is necessary for the development of the subject and not just for its own sake. The book has been designed to be useful in senior undergraduate and graduate level courses, including current awareness courses in this field. It should also serve as a background text in the theory and methods used in ultrasonic nondestructive testing and for those who are faced with making measurements of mechanical properties of materials.

A special feature of the book is the inclusion of numerous synopses of published experiments. These are intended to be illustrative either of typical investigations or of interesting related phenomena. Original data and diagrams are included. Each chapter is followed by a number of questions which are based directly on the indicated section of the text. These questions are designed as a simple test of absorption of the material by the reader. The problems vary in standard and often act as an extension of the text material. Full solutions of all problems are provided. Among special topics discussed in the book are higher-order elastic constants, acoustic amplification in piezoelectric semiconductors, acoustic emission, acoustic holography, and biomedical scanning methods.

Acknowledgements

I would like to thank the Executive Officer and staff of the Australian Institute for Nuclear Science and Engineering for providing accommodation and facilities needed during the writing of this book. Special thanks are also due to Michael Benton for his willing cooperation in locating and copying source material. To my wife, Margaret, and my family I am indebted for a sterling display of patience and tolerance during the apparently endless process of writing.

I also wish to express my gratitude to the authors quoted in the text for the permissions received to reproduce material from their papers and books, and also to the following publishers and organisations for allowing me to use their copyright material:

Academic Press, Inc. (London) Ltd., London (Wells: *Physical Principles of Ultrasonic Engineering*)

Academic Press, Inc., New York (Mason and Thurstons, Editors: *Physical Acoustics;* Truell, Elbaum, and Chick: *Ultrasonic Methods in Solid State Physics; Solid State Physics*)

The American Institute of Physics, New York (*Journal of the Acoustical Society of America; Journal of the Optical Society of America; Journal of Applied Physics; Review of Scientific Instruments*)

The American Physical Society, New York (*The Physical Review; Physical Review Letters*)

Bendix Research Laboratories, Southfields, Michigan, USA (Kock: *Seeing Sound*)

Chapman and Hall, London (*Journal of Materials Science*)

C.S.I.R.O., Melbourne (*Australian Journal of Applied Science; Australian Journal of Physics*)

Elsevier–Sequoia S.A., Lausanne (*Materials Science and Engineering*)

Hirzel Verlag, Stuttgart (*Acoustica*)

The Institute of Electrical and Electronic Engineers, Inc., New York (*IEEE Transactions, Sonics and Ultrasonics*)

The Institute of Physics, London (*Journal of Physics, British Journal of Applied Physics, Proceedings of the Physical Society, Reports on Progress in Physics*)

The Institution of Radio and Electronics Engineers of Australia (*Proceedings of IREE, Australia*)

I.P.C. Science and Technology Press, Guildford, Surrey, UK (*Optics and Laser Technology; Ultrasonics*)

McGraw-Hill Book Co., New York (Hirth and Lothe: *Theory of Dislocations*)

North Holland Publishing Co., Amsterdam (Van Bueren: *Imperfections in Crystals; Progress in Solid Mechanics*)

Pergamon Publishing Co., New York (*Acta Metallurgica*)

Taylor and Francis, Ltd., London (*Philosophical Magazine*)

John Wiley and Sons, New York (Hayden, Moffatt, and Wulff: *The Structure and Properties of Materials,* Vol.3; Kock: *Seeing Sound*)

Contents

List of symbols

A	area, anisotropy factor, amplitude, dislocation mass per unit length
B	amplitude, magnetic induction, dislocation damping force per unit length
C	capacitance, dislocation line tension
C_a	acoustic compliance
C_ϵ	specific heat at constant elastic strain
C_m	mechanical compliance
C_σ	specific heat at constant stress
D	thermal diffusivity
D_i	electric displacement vector
D_n	electronic diffusion constant
D^p	coefficient of pipe diffusion
E	Young's modulus, energy
E_{ad}	adiabatic Young's modulus
E_d	atomic displacement energy
E_i	energy flow vector, electric field strength
E_T	isothermal Young's modulus
F	elastic free energy (Helmholtz function)
F_i	force vector
$F(i\omega)$	input spectrum
G	Gibbs function
G	Lamé constant, shear modulus
$G(i\omega)$	output spectrum
$H(i\omega)$	transfer function
I	intensity
J	current density
K	bulk modulus, radius of gyration of cross-section of a rod, stress intensity factor, electro-mechanical coupling constant
K_{ik}	thermal conductivity tensor
L	inductance, average dislocation loop length
L_N	network loop length
L_c	intermediate loop length
M	mass
M_a	acoustic mass
M_n	modified mode mass
M_R	relaxed elastic modulus
M_U	unrelaxed elastic modulus
M^*	complex modulus
P_i	external surface force vector, polarisation per unit area
Q	quantity of heat, quality factor, dislocation–impurity interaction energy
R	reflection coefficient, electrical resistance
R_a	acoustic resistance
R_m	mechanical resistance

R_n	modified mode resistance
$R(\sigma)$	stress reflection coefficient
$R(v)$	velocity reflection coefficient
$R_{11}(\tau)$	auto-correlation coefficient
$R_{12}(\tau)$	cross-correlation coefficient
S	entropy, area, strength of impulse
$S(x)$	sampling function
$S_{11}(\tau)$	power spectral density
T	absolute temperature, transmission coefficient
U	internal energy, volume flow, activation energy, light wave displacement
V	potential energy, volume, velocity amplitude, voltage
W	work
X	electrical reactance
Y	electrical admittance
Y_n	mode admittance
$Y(t)$	normalised modulus
Z	wave impedance, electrical impedance
Z_m	mechanical impedance
Z_n	mode impedance
Z_o	characteristic impedance
Z_T	terminating impedance
$Z(t)$	normalised decrement
Z_{11}	input impedance
Z_{12}	transfer impedance
a	radius
$\boldsymbol{b}$	Burgers vector
c, c_p	phase velocity
c_g	group velocity
c_i	defect concentration
c_ϱ	phase velocity of longitudinal wave in infinite medium
c_n	mode compliance
c_t	phase velocity of transverse wave in infinite medium
c_F	phase velocity of flexural waves
c_L	phase velocity of longitudinal wave in a rod or bar
c_T	phase velocity of torsional wave in a rod
c_{iklm}	elastic constant or stiffness tensor
d	cross-sectional dimension, spacing, thickness
d_{ijk}	piezoelectric moduli, piezoelectric constant tensor
e	voltage, charge on an electron
f	frequency, force
$f(t)$	input function
g_i	gravity acceleration vector
$g(t)$	output function

h	thickness, Planck constant
$h(t)$	impulse response
$\hbar$	modified Planck constant $(= h/2\pi)$
i	electric current
i_o	current amplitude
k	Boltzmann's constant
k	complex wave vector
k_i	wave vector
l	distance
m	density ratio, mode number
m_n	mode mass
n	wave velocity ratio, charge carrier density
n	unit normal
n_d	pinning point density
p	hydrostatic pressure, acoustic pressure
q	electric charge
q	phonon wave vector
q_n	mode constant
r	relative Rayleigh wave velocity, radial displacement
r_n	mode resistance
r_o	separation between atoms
s	ratio of transverse to longitudinal wave velocity
s_i	slowness vector
s_{iklm}	elastic compliance tensor
t	time
t	unit tangent vector
t_{ik}	thermodynamic tension components
u_i	displacement vector
$u(t)$	unit step function
v	particle velocity
x_i	Cartesian coordinates
z	number of atoms per unit cell
$\mathscr{E}$	kinetic energy
$\mathscr{F}[\ ]$	Fourier transform
$\mathscr{F}^{-1}[\ ]$	inverse Fourier transform
Δ	decrement
Δ_0	relaxation strength
Δ_t	total decrement of composite system
Δ_H	amplitude–dependent decrement
Δ_I	amplitude – independent decrement
Λ	dislocation density
Π	radiation pressure
Φ_i	body force vector
Ω	orientation factor

α	angle of incidence, attenuation coefficient
α_i	direction cosines of particle displacements
$\alpha(t)$	potential variable
α_{cr}	critical angle of incidence
β	angle of reflection (longitudinal wave), linear thermal expansion coefficient
β'	angle of refraction
$\beta(t)$	flow variable
γ	angle of reflection (transverse wave), propagation constant, phase shift, coefficient of surface tension, drift parameter
γ_{ik}	temperature coefficients of stress
δ	phase difference between strain and stress, pulse round-trip delay time, damping coefficient
δ_{ik}	unit tensor
$\delta(t)$	delta function
$\delta'(t)$	unit doublet
ϵ	permittivity
ϵ_0	strain amplitude
ϵ_{ik}	strain tensor (second order)
ϵ_r	relative permittivity (dielectric constant)
ζ	displacement in z direction
η	loss factor, shear viscosity coefficient, Cottrell's misfit parameter
η_{ik}	material strain components
η_{iklm}	viscosity tensor
θ	dilatation, angle
$\theta(\omega)$	phase characteristic
κ_{ik}	dielectric tensor
κ_n	excitation constant
λ	Lamé constant, wavelength
λ_{im}	Christoffel's tensor
μ	Poisson's ratio, electronic mobility
ν	frequency
ν_c	composite system resonant frequency
ν_0	waveguide cut-off frequency, relaxation frequency
ξ	normal coordinate
ρ	volume density
σ_c	electrical conductivity
σ_0	stress wave amplitude
σ_{cr}	critical resolved shear stress
σ_{ij}	stress tensor
τ	relaxation time
ϕ	scalar potential, phase difference
χ	compressional viscosity coefficient
ψ_i	vector potential

ω	circular frequency
ω_D	electron diffusion frequency
ω_c	conductivity relaxation frequency
ω_d	damped circular frequency
ω_{ij}	rotation tensor

Linear elasticity

1.1 Introduction

When external forces are applied to a solid body, the general motion of the body involves linear translation, rotation, and deformation. In the linear theory of elasticity the only deformation that is considered is that in which stress is linearly related to strain. Further theories have been developed to allow for higher order strains, thermal strain, viscous flow, etc. For many purposes a continuum theory is satisfactory in explaining large scale elastic phenomena. An atomic-based theory of elasticity only becomes necessary when deformation effects must be examined very close to the atoms concerned (within one or two atomic spacings).

In figure 1.1 are shown graphs of potential energy, V, versus the separation, r, between atoms in a solid. As may be observed, as r is decreased there is initially an attractive long-range potential (and hence an attractive force since $F = -\partial V/\partial r$), but as the atoms are pushed closer together a strong short-range repulsive force is set up when the atomic orbitals begin to overlap. At the distance r_0 there is a balance between attractive and repulsive forces resulting in a minimum in the potential energy curve. It may be shown (see, for instance, Cottrell, 1964) that, within the limits of the linear theory of elasticity, the elastic constants of the material depend on the curvature of the $V = f(r)$ curve at the equilibrium point r_0. Only small excursions on either side of r_0 are permissible if the curvature is not to change appreciably.

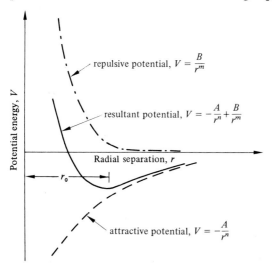

repulsive potential, $V = \dfrac{B}{r^m}$

resultant potential, $V = -\dfrac{A}{r^n} + \dfrac{B}{r^m}$

Potential energy, V

Radial separation, r

r_0

attractive potential, $V = -\dfrac{A}{r^n}$

Figure 1.1. Schematic representation of the relationship between potential energy and radial separation between atoms in a solid. r_0 is the equilibrium separation corresponding to minimum resultant potential energy.

In the following sections an outline will be given of the main concepts and results of the linear theory of elasticity. For conciseness in stating the sets of equations, cartesian tensor notation is used. The essential definitions and properties of tensors involved may be found in standard texts such as Jeffreys (1931) and Nye (1960). A summary of tensor properties is given in Appendix 1.1.

1.2 Stress

Consider a small volume of a solid body bounded by a surface. In general, two types of mechanical force may act on the body: (i) body forces for which the force is proportional to the volume of the body, (ii) surface forces for which the force is proportional to the area of the surface. Gravity is an example of the first type of force which involves the force acting on each particle of the body. The stress applied to an elastic body is an example of the second type of force in which the force experienced by the surface particles is transmitted to particles in the interior.

Stress may be defined broadly as the force divided by the area on which the force acts. The stress is therefore not only proportional to the direction in which the force acts but also to the orientation of the surface element. In figure 1.2 a unidirectional force F is shown acting on a number of unit areas of surface with different orientations of their respective unit normals.

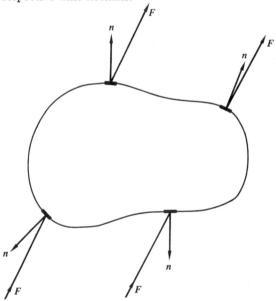

Figure 1.2. A unidirectional force F acts on the surface of a small portion of a solid body. n is the unit normal drawn from each of the unit areas shown. The stress on the surface depends on both the direction of the force and the orientation of the area on which it acts.

Consider one of the unit areas shown in figure 1.2. If we set up cartesian coordinates with respect to this area, then there are three components of the stress, denoted by σ_i, where $i = 1, 2, 3$, one component along the direction of the normal with unit vector n_1 and two parallel to the surface in the directions of unit vectors n_2 and n_3. The vector equation for the stress on this area is

$$\sigma = \sigma_1 n_1 + \sigma_2 n_2 + \sigma_3 n_3 \,. \tag{1.1}$$

However, in order to define the state of stress at a point in a continuous body, it is necessary in principle to find the force acting on every unit area that includes the point in question. In practice it is sufficient to choose three unit areas at the point, one normal to each of three reference coordinate axes.

We may therefore examine the forces acting on the faces of an infinitesimal rectangular parallelepiped within a stressed body, as shown in figure 1.3. The applied force is assumed to be unidirectional, the body is assumed to be in static equilibrium, and body forces are ignored. There are three stress components on each of the three faces shown giving a total of nine stress components. A double subscript notation is used to identify the components, the first subscript denoting the direction of the stress component and the second subscript denoting the direction of the normal to the plane across which the component acts. Since the stress is uniform, the components across the other three faces of the parallelepiped must be equal and opposite to those shown.

σ_{11}, σ_{22}, and σ_{33} are called **normal components** of stress; σ_{12}, σ_{13}, σ_{23}, etc. are called **shear components** of stress. It is readily shown (see, for instance, Nye, 1960; Long, 1961; Landau and Lifshitz, 1970) that the set of components σ_{ik}, where i and k have the values 1, 2, 3, are the components of a second rank tensor called the **stress tensor**.

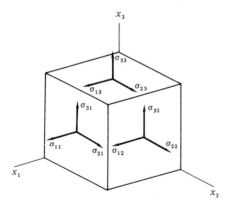

Figure 1.3. The stress components acting on each face of an elementary parallelepiped when a uniform force acts on the body.

By taking moments about one of the coordinate axes it may be shown that the stress tensor is symmetrical, that is

$$\sigma_{ik} = \sigma_{ki} . \tag{1.2}$$

Thus there are only six independent components of stress. The nine possible stress components may be written in the form of a matrix

$$\sigma_{ik} = \begin{bmatrix} \sigma_{11} & \sigma_{12} & \sigma_{13} \\ \sigma_{21} & \sigma_{22} & \sigma_{23} \\ \sigma_{31} & \sigma_{32} & \sigma_{33} \end{bmatrix} . \tag{1.3}$$

Hydrostatic pressure is a special case of the stress tensor in which only the normal stresses are finite and are all equal. Thus, the pressure, p, can be written in matrix form as

$$\begin{bmatrix} -p & 0 & 0 \\ 0 & -p & 0 \\ 0 & 0 & -p \end{bmatrix} , \tag{1.4}$$

or, as a tensor, by introducing the **unit tensor**, δ_{ik}, so that

$$\sigma_{ik} = -p\delta_{ik} , \tag{1.5}$$

where $\delta_{ik} = 1$ when $i = k$, and $\delta_{ik} = 0$ when $i \neq k$. The minus sign is used because of the convention that a stress component is taken as positive if the stress is tensile and negative if the stress results in a compression.

1.3 Strain
A deformable body is one in which relative motions may occur between its constituent parts giving rise to a change in shape or volume or both. In general, the changes produced by the applied forces may be (a) longitudinal —involving a change in length of a specimen, (b) tangential or shear— involving a relative displacement in parallel layers of the material, (c) bulk compressional—involving a change in volume without change of shape. The deformation is said to be elastic if the body regains its original size and shape after the forces are removed. If the applied forces exceed a defined elastic limit, the body will retain a permanent set on removal of the forces. If the forces are large enough, the body may exhibit plastic flow or may fracture.

In order to derive a general expression for the strain induced in a solid body consider a point P within the body whose position coordinates are x_i with respect to a set of cartesian axes. If P is displaced to a new position P$'$ with coordinates x_i', the **displacement vector**, u_i, is defined as

$$u_i = x_i' - x_i . \tag{1.6}$$

Consider a second point Q adjacent to P with coordinates $x_i + dx_i$, as shown in figure 1.4. If the body is deformed, then the element PQ becomes the element P'Q'. While P moves to P', Q moves to Q' where the coordinates of Q' are $x_i' + dx_i'$. Then the displacement of Q is

$$u_i' = (x_i' + dx_i') - (x_i + dx_i) = (x_i' - x_i) + (dx_i' - dx_i) = u_i + du_i . \qquad (1.7)$$

Thus, du_i is a measure of the deformation of the element, since, if $du_i = 0$, PQ is displaced without deformation.

Using Taylor's theorem and ignoring higher order terms we may write

$$u_i' = u_i + du_i = u_i + \frac{\partial u_i}{\partial x_k} dx_k + \dots$$

$$= u_i + \frac{1}{2}\left(\frac{\partial u_i}{\partial x_k} + \frac{\partial u_k}{\partial x_i}\right) dx_k - \frac{1}{2}\left(\frac{\partial u_k}{\partial x_i} - \frac{\partial u_i}{\partial x_k}\right) dx_k$$

$$= u_i + \epsilon_{ik} \, dx_k - \omega_{ik} \, dx_k . \qquad (1.8)$$

Each of the terms in equation (1.8) denotes a vector while ϵ_{ik} and ω_{ik} denote second rank tensors.

The tensor

$$\epsilon_{ik} = \frac{1}{2}\left(\frac{\partial u_i}{\partial x_k} + \frac{\partial u_k}{\partial x_i}\right) \qquad (1.9)$$

is called the **strain tensor**. From equation (1.9) it follows that ϵ_{ik} is symmetrical, that is

$$\epsilon_{ik} = \epsilon_{ki} . \qquad (1.10)$$

Thus, of the possible nine components of the strain tensor, only six components are independent.

The tensor ω_{ik} is antisymmetrical ($\omega_{ik} = -\omega_{ki}$) with three independent components and can be identified with a pure rotation of the element

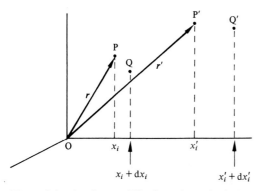

Figure 1.4. An element PQ of an elastic body becomes the element P'Q' after deformation.

(Nye, 1960). The result expressed in equation (1.8) is referred to as **Helmholtz's theorem**. In words, this theorem states that any arbitrary displacement of an element PQ of a body can be resolved into (a) a pure translation of the element, u_i, (b) a deformation $\epsilon_{ik} dx_k$, and (c) a pure rotation $-\omega_{ik} dx_k$.

It is not possible to measure individually the components of the strain tensor except under circumstances in which only the diagonal components ϵ_{11}, ϵ_{22}, ϵ_{33} have finite values. Landau and Lifshitz (1970) show that, if the distance, dl', between P' and Q' after deformation is computed and then ϵ_{ik} is diagonalised,

$$(dl')^2 = (1 + 2\epsilon_{11})dx_1^2 + (1 + 2\epsilon_{22})dx_2^2 + (1 + 2\epsilon_{33})dx_3^2 . \tag{1.11}$$

Thus, an arbitrary strain may be represented by three independent strains in the directions of the principal axes of the strain tensor. It is easy to show that each of these strains is a simple extension or compression. For instance, the length dx_1 along the first principal axis becomes

$$dx_1' = (1 + 2\epsilon_{11})^{\frac{1}{2}} dx_1 \approx (1 + \epsilon_{11})dx_1 .$$

The relative extension, or **longitudinal strain**, is then

$$\frac{dx_1' - dx}{dx_1} = \epsilon_{11} . \tag{1.12}$$

Similar expressions hold for ϵ_{22} and ϵ_{33}.

The volume strain, or **dilatation**, θ, may be found by making use of the above result. Consider an infinitesimal rectangular parallelepiped of volume dV which after deformation has a volume dV'. Take the principal axes of the strain tensor as coordinate axes. Then,

$$dV' = dx_1' dx_2' dx_3' = dV(1 + \epsilon_{11})(1 + \epsilon_{22})(1 + \epsilon_{33}) ,$$

where $dV = dx_1 dx_2 dx_3$. Neglecting higher order terms, we have

$$dV' = dV(1 + \epsilon_{11} + \epsilon_{22} + \epsilon_{33}) = dV(1 + \epsilon_{ii}) ,$$

where ϵ_{ii} is the sum of the principal values of the strain tensor (an invariant)[1].

The relative volume change, or dilatation, is

$$\theta = \frac{dV' - dV}{dV} = \epsilon_{ii} . \tag{1.13}$$

Note that ϵ_{ii} is a scalar since it is formed by contraction of a second rank tensor. Also, from equation (1.9) it follows that $\epsilon_{ii} = \partial u_i / \partial x_i$ which is the divergence of u_i. In alternative symbols, $\theta = \text{div}\, u_i$.

---

[1] The Einstein summation convention is used throughout, that is repeated suffixes denote summation.

1.4 Equations of motion and equilibrium

Consider a body of given volume subject to a stress σ_{ik} applied to the surface enclosing the body. If F_i denotes the resultant internal force per unit volume, the resultant force acting on a volume element of the body, dV, is $F_i\,dV$. The resultant force on the whole body is $\int_V F_i\,dV$. The volume forces may be related to the surface forces with the aid of Gauss's divergence theorem,

$$\int_V F_i\,dV = \int_S \sigma_{ik}\,dA_k \ ,$$

where dA_k is an element of surface area. In words, this equation states that the integral of a vector over an arbitrary volume may be transformed into the integral of a second-rank tensor over the surface, provided the vector is the divergence of the tensor, that is, provided

$$F_i = \frac{\partial \sigma_{ik}}{\partial x_k} \ . \tag{1.14}$$

Note that, because of the summation convention, for each value of i there is a sum of three terms.

The **equations of motion** are obtained by equating the resultant force due to the internal stresses to the product of acceleration and mass per unit volume:

$$\frac{\partial \sigma_{ik}}{\partial x_k} = \rho \ddot{u}_i \ , \tag{1.15}$$

where ρ is the volume density and u_i is the displacement vector. The notation may be abbreviated further by writing $\partial_k \equiv \partial/\partial x_k$ so that the equations of motion become

$$\partial_k \sigma_{ik} = \rho \ddot{u}_i \ . \tag{1.16}$$

If body forces, Φ_i, are included,

$$\partial_k \sigma_{ik} + \Phi_i = \rho \ddot{u}_i \ . \tag{1.17}$$

The most common body force is gravity for which $\Phi_i = \rho g_i$, where g_i is the gravity acceleration vector.

The **equations of equilibrium** follow simply, since, in equilibrium, the resultant of the internal forces must be zero, that is

$$\partial_k \sigma_{ik} = 0 \ , \tag{1.18}$$

or, in the presence of body forces,

$$\partial_k \sigma_{ik} + \Phi_i = 0 \ . \tag{1.19}$$

1.5 Thermodynamics of deformation

By considering the strain produced in an infinitesimal volume by the action of forces applied to the surrounding surface, an expression can be found (see, for instance, Nye, 1960, or Landau and Lifshitz, 1970) for the work done, dW, per unit volume by the external forces:

$$dW = \sigma_{ik} d\epsilon_{ik} . \tag{1.20}$$

If the elastic deformation is assumed to occur slowly so that the process may be considered to be thermodynamically reversible, then by the first law of thermodynamics

$$dU = dW + dQ = \sigma_{ik} d\epsilon_{ik} + dQ , \tag{1.21}$$

where dU is the increase in internal energy and dQ is the heat flow into the volume. By the second law of thermodynamics

$$dQ = T dS , \tag{1.22}$$

where T is the absolute temperature and dS is the increase in entropy. Combining equations (1.21) and (1.22) gives

$$dU = \sigma_{ik} d\epsilon_{ik} + T dS . \tag{1.23}$$

In order to derive the stress or strain components from the energy it is convenient to use two further functions. The **elastic free energy**, F (or Helmholtz function), is defined as

$$F = U - TS . \tag{1.24}$$

Differentiating and combining with equation (1.23) gives

$$dF = \sigma_{ik} d\epsilon_{ik} - S dT . \tag{1.25}$$

The stress components can therefore be found from

$$\sigma_{ik} = \left(\frac{\partial U}{\partial \epsilon_{ik}} \right)_S = \left(\frac{\partial F}{\partial \epsilon_{ik}} \right)_T . \tag{1.26}$$

Similarly, by introducing the Gibbs function, G,

$$G = U - TS - \sigma_{ik} \epsilon_{ik} , \tag{1.27}$$

then differentiating and combining with equation (1.23), we obtain

$$dG = -S dT - \epsilon_{ik} d\sigma_{ik} . \tag{1.28}$$

The strain components can therefore be found from

$$\epsilon_{ik} = -\left(\frac{\partial G}{\partial \sigma_{ik}} \right)_T . \tag{1.29}$$

1.6 The perfectly elastic body

Hooke's law may be generalised to state that, for a perfectly elastic body, each component of stress is linearly related to each component of strain. Since there are six independent components of stress and six independent components of strain, for the most anisotropic body with no symmetry relations there will be a set of six equations each with six terms and therefore a total of 36 coefficients. These equations may be written concisely with the use of tensor notation (repeated suffixes denote summation):

$$\sigma_{ik} = c_{iklm}\epsilon_{lm} , \qquad (1.30)$$

where c_{iklm} is a fourth rank tensor called the **elastic constant tensor** or **stiffness tensor**. In the most general case c_{iklm} has 81 components. Fortunately it is nearly always possible to reduce this number to manageable proportions by use of symmetry conditions or by choosing the relative orientations in an actual experiment. For instance, since σ_{ik} is symmetrical, then

$$c_{iklm} = c_{kilm} ,$$

and since ϵ_{lm} is symmetrical,

$$c_{iklm} = c_{ikml} .$$

These two conditions reduce the number of independent constants to 36.

A further reduction may be made from energy considerations. If the process is thermodynamically reversible and takes place under **isothermal conditions**, the elastic free energy can be shown (Nye, 1960; Landau and Lifshitz, 1970) to be a quadratic function of the strain:

$$F = \tfrac{1}{2}c_{iklm}\epsilon_{ik}\epsilon_{lm} . \qquad (1.31)$$

It may be observed that the product $\epsilon_{ik}\epsilon_{lm}$ is unchanged not only when the pairs i, k, and l, m are interchanged but also when i, l, and k, m are interchanged. That is

$$c_{iklm} = c_{kilm} = c_{ikml} = c_{lmik} . \qquad (1.32)$$

The symmetry condition (1.32) has the effect of reducing the number of independent constants for the most general anisotropic body to 21.

Other conditions than isothermal may be accommodated in the theory by retaining the temperature and entropy term in the free energy. **Adiabatic deformation** is of importance and can be investigated by adding a term representing thermal strain. In place of equation (1.30) we have

$$\sigma_{ik} = c_{iklm}\epsilon_{lm} - \gamma_{ik}\Delta T , \qquad (1.33)$$

where $\gamma_{ik} = (\partial\sigma_{ik}/\partial T)_\epsilon$ are the components of a second rank tensor representing the **temperature coefficients of stress** (at constant elastic strain ϵ). In addition we need the relation

$$\Delta S = \gamma_{ik}\epsilon_{ik} + \frac{C_\epsilon}{T}\Delta T = 0 , \qquad (1.34)$$

where C_ϵ is the specific heat at constant elastic strain. Eliminating ΔT from equations (1.33) and (1.34) we obtain

$$c_{iklm}^S = c_{iklm}^T - \frac{\gamma_{ik}\gamma_{lm}T}{C_\epsilon} . \qquad (1.35)$$

Equation (1.35) gives the relationship between the **adiabatic elastic constants** c_{iklm}^S and the **isothermal constants** c_{iklm}^T. Equations of the same form may therefore be used for elastic problems involving either isothermal or adiabatic conditions, provided the appropriate constants are employed.

1.7 Elastic constants for crystal systems

When dealing with materials such as crystals which possess some degree of symmetry we can apply **Voigt's principle**. This principle states that the symmetry of the physical process is superimposed on the symmetry of the crystal. The number of independent elastic constants is therefore dramatically reduced as the degree of symmetry increases. A detailed derivation of the number of independent constants for the different crystal systems is given in texts such as Landau and Lifshitz (1970) and Nye (1960). The following list summarises the numbers of independent constants (the isotropic case is included for comparison):

triclinic	21
monoclinic	13
orthorhombic	9
tetragonal and trigonal	7 or 6 (depending on class)
hexagonal	5
cubic	3
isotropic	2

The independent constants for a triclinic crystal may be written in the form of a half-matrix where mixed tensor notation has been used for compactness; the full matrix is symmetrical about the diagonal shown:

$$
\begin{array}{cccccc}
c_{11}^{11} & c_{22}^{11} & c_{33}^{11} & c_{23}^{11} & c_{31}^{11} & c_{12}^{11} \\
& c_{22}^{22} & c_{33}^{22} & c_{23}^{22} & c_{31}^{22} & c_{12}^{22} \\
& & c_{33}^{33} & c_{23}^{33} & c_{31}^{33} & c_{12}^{33} \\
& & & c_{23}^{23} & c_{31}^{23} & c_{12}^{23} \\
& & & & c_{31}^{31} & c_{12}^{31} \\
& & & & & c_{12}^{12}
\end{array}
$$

In many texts and papers it is customary to use an abbreviated notation (sometimes referred to as matrix notation) in place of the full tensor notation. The following scheme is adopted:

Tensor notation: 11   22   33   23,32   31,13   12,21
Matrix notation: 1   2   3   4   5   6

Thus, the matrix of constants for a triclinic crystal may be written:

Triclinic

$$
\begin{matrix}
c_{11} & c_{12} & c_{13} & c_{14} & c_{15} & c_{16} \\
 & c_{22} & c_{23} & c_{24} & c_{25} & c_{26} \\
 & & c_{33} & c_{34} & c_{35} & c_{36} \\
 & & & c_{44} & c_{45} & c_{46} \\
 & & & & c_{55} & c_{56} \\
 & & & & & c_{66}
\end{matrix}
$$

The elastic constant matrices for other crystal systems are as follows:

Monoclinic (all classes): 13 constants

Diad parallel to x_2 Diad parallel to x_3
(standard orientation)

$$
\begin{matrix}
c_{11} & c_{12} & c_{13} & 0 & c_{15} & 0 \\
 & c_{22} & c_{23} & 0 & c_{25} & 0 \\
 & & c_{33} & 0 & c_{35} & 0 \\
 & & & c_{44} & 0 & c_{46} \\
 & & & & c_{55} & 0 \\
 & & & & & c_{66}
\end{matrix}
\qquad
\begin{matrix}
c_{11} & c_{12} & c_{13} & 0 & 0 & c_{16} \\
 & c_{22} & c_{23} & 0 & 0 & c_{26} \\
 & & c_{33} & 0 & 0 & c_{36} \\
 & & & c_{44} & c_{45} & 0 \\
 & & & & c_{55} & 0 \\
 & & & & & c_{66}
\end{matrix}
$$

Orthorhombic (all classes): 9 constants

$$
\begin{matrix}
c_{11} & c_{12} & c_{13} & 0 & 0 & 0 \\
 & c_{22} & c_{23} & 0 & 0 & 0 \\
 & & c_{33} & 0 & 0 & 0 \\
 & & & c_{44} & 0 & 0 \\
 & & & & c_{55} & 0 \\
 & & & & & c_{66}
\end{matrix}
$$

Tetragonal

Classes 4, $\bar{4}$, $4/m$:
7 constants

$$\begin{array}{cccccc} c_{11} & c_{12} & c_{13} & 0 & 0 & c_{16} \\ & c_{11} & c_{13} & 0 & 0 & -c_{16} \\ & & c_{33} & 0 & 0 & 0 \\ & & & c_{44} & 0 & 0 \\ & & & & c_{44} & 0 \\ & & & & & c_{66} \end{array}$$

Classes $4mm$, $\bar{4}2m$, 422, $4/mmm$:
6 constants

$$\begin{array}{cccccc} c_{11} & c_{12} & c_{13} & 0 & 0 & 0 \\ & c_{11} & c_{13} & 0 & 0 & 0 \\ & & c_{33} & 0 & 0 & 0 \\ & & & c_{44} & 0 & 0 \\ & & & & c_{44} & 0 \\ & & & & & c_{66} \end{array}$$

Trigonal

Classes 3, $\bar{3}$:
7 constants

$$\begin{array}{cccccc} c_{11} & c_{12} & c_{13} & c_{14} & -c_{25} & 0 \\ & c_{11} & c_{13} & -c_{14} & c_{25} & 0 \\ & & c_{33} & 0 & 0 & 0 \\ & & & c_{44} & 0 & c_{25} \\ & & & & c_{44} & c_{14} \\ & & & & & \tfrac{1}{2}(c_{11}-c_{12}) \end{array}$$

Classes 32, $\bar{3}m$, $3m$:
6 constants

$$\begin{array}{cccccc} c_{11} & c_{12} & c_{13} & c_{14} & 0 & 0 \\ & c_{11} & c_{13} & -c_{14} & 0 & 0 \\ & & c_{33} & 0 & 0 & 0 \\ & & & c_{44} & 0 & 0 \\ & & & & c_{44} & c_{14} \\ & & & & & \tfrac{1}{2}(c_{11}-c_{12}) \end{array}$$

Hexagonal (all classes): 5 constants

$$\begin{array}{cccccc} c_{11} & c_{12} & c_{13} & 0 & 0 & 0 \\ & c_{11} & c_{13} & 0 & 0 & 0 \\ & & c_{33} & 0 & 0 & 0 \\ & & & c_{44} & 0 & 0 \\ & & & & c_{44} & 0 \\ & & & & & \tfrac{1}{2}(c_{11}-c_{12}) \end{array}$$

Cubic (all classes): 3 constants

$$\begin{array}{cccccc} c_{11} & c_{12} & c_{12} & 0 & 0 & 0 \\ & c_{11} & c_{12} & 0 & 0 & 0 \\ & & c_{11} & 0 & 0 & 0 \\ & & & c_{44} & 0 & 0 \\ & & & & c_{44} & 0 \\ & & & & & c_{44} \end{array}$$

Isotropic: 2 constants

$$
\begin{array}{cccccc}
c_{11} & c_{12} & c_{12} & 0 & 0 & 0 \\
 & c_{11} & c_{12} & 0 & 0 & 0 \\
 & & c_{11} & 0 & 0 & 0 \\
 & & & \tfrac{1}{2}(c_{11}-c_{12}) & 0 & 0 \\
 & & & & \tfrac{1}{2}(c_{11}-c_{12}) & 0 \\
 & & & & & \tfrac{1}{2}(c_{11}-c_{12})
\end{array}
$$

For further details concerning crystal symmetry see Nye (1960).

Expressions may also be derived from equation (1.31) for the elastic free energy (sometimes called the elastic strain energy) for each crystal system. For instance, the elastic free energy for a cubic crystal is (in tensor notation):

$$F = \tfrac{1}{2}c_{11}^{11}(\epsilon_{11}^2 + \epsilon_{22}^2 + \epsilon_{33}^2) + c_{22}^{11}(\epsilon_{11}\epsilon_{22} + \epsilon_{11}\epsilon_{33} + \epsilon_{22}\epsilon_{33})$$
$$+ 2c_{23}^{23}(\epsilon_{12}^2 + \epsilon_{13}^2 + \epsilon_{23}^2) .$$

Note that, in such energy equations, all finite constants (and equivalent constants according to symmetry conditions) must be included, that is the complete matrix of constants must be used, not just the half matrix of independent constants.

1.8 The isotropic body

An **isotropic body** is one for which the physical properties do not depend on the orientation of the body. Many polycrystalline materials behave as if they were isotropic although they are actually composed of a more or less random arrangement of small crystals or grains. In polycrystalline metals produced by drawing or rolling preferred orientation of the grains often leads to anisotropic behaviour.

The elastic properties of an isotropic body may be deduced from equation (1.30) by converting the fourth rank tensor c_{iklm} into an isotropic tensor, as described by Jeffreys (1931). See also Appendix 1.1. The only independent constants for an isotropic body are found to be c_{11}^{11} and c_{22}^{11} so that Hooke's law reduces to

$$\sigma_{ik} = (c_{11}^{11} - c_{22}^{11})\epsilon_{ik} + c_{22}^{11}\theta\delta_{ik} ,$$

where the dilatation $\theta = \epsilon_{11} + \epsilon_{22} + \epsilon_{33}$.

In the historical development of the theory of elasticity an alternative argument based on symmetry considerations leads to a similar equation but with two different constants, λ and G, where $\lambda = c_{22}^{11}$ and $G = \tfrac{1}{2}(c_{11}^{11} - c_{22}^{11})$. The constants λ and G were introduced by Lamé and are frequently referred to as the **Lamé constants** (Lamé used the Greek letter μ in place of G, but in this text we are following current IUPAP recommendations). In terms of these constants Hooke's law for an isotropic material may be written

$$\sigma_{ik} = 2G\epsilon_{ik} + \lambda\theta\delta_{ik} . \tag{1.36}$$

Compared with the elastic constants of a cubic crystal, the elastic constants for an isotropic body are subject to the additional condition

$$c_{11}^{11} = c_{22}^{11} + 2c_{23}^{23} .$$

In discussions of the properties of materials it is useful to define an **anisotropy factor**, A:

$$A = \frac{2c_{23}^{23}}{c_{11}^{11} - c_{22}^{11}} ;$$

in matrix notation

$$A = \frac{2c_{44}}{c_{11} - c_{12}} .$$

If $A = 1$, then the material is elastically isotropic. The degree of anisotropy of certain cubic crystals is shown by the following values of A: iron $2 \cdot 4$; aluminium $1 \cdot 2$; copper $3 \cdot 2$; lead $4 \cdot 0$; NaCl $0 \cdot 7$; KCl $0 \cdot 36$; LiF $1 \cdot 6$.

The elastic free energy of an isotropic body that obeys Hooke's law is

$$F = G\epsilon_{ik}^2 + \tfrac{1}{2}\lambda\theta^2 . \tag{1.37}$$

The Lamé constants are useful in theoretical discussions but λ, in particular, has no direct physical meaning. It is customary therefore to introduce further elastic parameters, called moduli, that have a more direct physical meaning.

1.9 Elastic moduli for isotropic bodies
1.9.1 Shear modulus, G
From equation (1.36), it follows that when $i \neq k$, $\sigma_{ik} = 2G\epsilon_{ik}$; that is $2G$ is the ratio of shear stress to shear strain. G is called the **shear modulus** or rigidity modulus.

1.9.2 Bulk modulus, K
The **bulk modulus** is defined as the ratio of hydrostatic pressure to the fractional change in volume this pressure produces when it is applied to an isotropic body. By combining equation (1.5) with Hooke's law, equation (1.36), it may be shown that

$$K \equiv -\frac{\Delta p}{\Delta\theta} = \lambda + \tfrac{2}{3}G . \tag{1.38}$$

1.9.3 Young's modulus, E

Consider a perfectly elastic isotropic body in the form of a cylinder subjected to a uniform longitudinal deformation. Let the longitudinal axis be in the x_1 direction and let a uniform force be applied to one end of the cylinder so as to produce a longitudinal stress σ_{11}. Application of equation (1.36) shows that the following three equations are relevant:

$$\sigma_{11} = 2G\epsilon_{11} + \lambda\theta ,$$
$$0 = 2G\epsilon_{22} + \lambda\theta ,$$
$$0 = 2G\epsilon_{33} + \lambda\theta .$$

Young's modulus is defined as the ratio of longitudinal stress to longitudinal strain, so that

$$E \equiv \frac{\sigma_{11}}{\epsilon_{11}} = \frac{G(3\lambda + 2G)}{\lambda + G} . \tag{1.39}$$

1.9.4 Poisson's ratio, μ

The ratio of lateral strain to longitudinal strain is called **Poisson's ratio**:

$$\mu \equiv -\frac{\epsilon_{22}}{\epsilon_{11}} = \frac{\lambda}{2(\lambda + G)} . \tag{1.40}$$

1.9.5 Relationships between moduli

It is clear from equations (1.38) to (1.40) that the various moduli are related, since

$$\lambda = \frac{E\mu}{(1 + \mu)(1 - 2\mu)} \tag{1.41}$$

and

$$G = \frac{E}{2(1 + \mu)} . \tag{1.42}$$

It is of interest to note that the ratio of the Lamé constants is

$$\frac{\lambda}{G} = \frac{2\mu}{1 - 2\mu}$$

and therefore depends only on Poisson's ratio.

Hooke's law may be expressed in terms of the moduli:

$$\sigma_{ik} = 2G\epsilon_{ik} + (K - \tfrac{2}{3}G)\theta\delta_{ik} , \tag{1.43}$$

or

$$\sigma_{ik} = \frac{E}{1 + \mu}\left(\epsilon_{ik} + \frac{\mu}{1 - 2\mu}\theta\delta_{ik}\right) . \tag{1.44}$$

B

1.10 Equations of motion and equilibrium for an isotropic elastic body

Combining the general equations of motion, (1.17), and Hooke's law, equation (1.36), we obtain

$$\rho \ddot{u}_i = (\lambda + G)\partial_i \theta + G\partial_k \partial_k u_i + \Phi_i , \tag{1.45}$$

or, using conventional vector terminology,

$$\rho \ddot{u}_i = (\lambda + G)\operatorname{grad}\operatorname{div} u_i + G\nabla^2 u_i + \Phi_i . \tag{1.46}$$

The equations of equilibrium then follow:

$$(\lambda + G)\operatorname{grad}\operatorname{div} u_i + G\nabla^2 u_i + \Phi_i = 0 . \tag{1.47}$$

1.11 Surface boundary conditions

If P_i represents the external force acting on unit area of the surface of a body, then the force acting on a surface element of area dA is $P_i\,dA$. At equilibrium this force must be balanced by the force due to the internal stresses $-\sigma_{ik}\,dA_k$, so

$$P_i\,dA - \sigma_{ik}\,dA_k = 0 .$$

If we write $dA_k = n_k\,dA$, where n_k is the unit vector along the outward normal to the surface, then

$$\sigma_{ik} n_k = P_i \tag{1.48}$$

must be satisfied at the surface.

If the surface is free, then the condition becomes

$$\sigma_{ik} n_k = 0 . \tag{1.49}$$

An example of a free surface is that between a solid and a vacuum.

1.12 Atomic limits for linear elasticity theory

It is of interest to consider under what conditions the continuum theory of elasticity is applicable to problems involving crystal lattices, especially when point defects or dislocations are present. Concepts such as strain, stress, and elastic constants must be re-examined in view of the need to express them in terms of atomic displacements and potential functions.

Homogeneous strain in an ideal crystal may be written in terms of atomic displacements as

$$u_i(l, p) = e_{ik} x_k(l, p) + u_i'(p) , \tag{1.50}$$

where $u_i(l, p)$ is the displacement from its position $x_k(l, p)$ of atom p in unit cell l in the perfect lattice; $e_{ik} = \partial u_k / \partial x_i$ are the components of the distortion tensor, the symmetric part of which is the strain tensor $\epsilon_{ik} = \frac{1}{2}(u_{ik} + u_{ki})$, and the antisymmetric part is the rotation tensor ω_{ik} (Nye, 1960). $u_i'(p)$ is the displacement within each unit cell.

Continuum theory under these conditions is only valid if the displacements $u_i(l, p)$ do not vary appreciably between adjacent cells of the lattice so that they can be approximated by a continuous smooth function of position.

The state of stress in a crystal may involve the interaction of a large number of different atoms so that it is not possible to identify the forces required for the classical definition of stress. It is, however, possible to use the definition of stress in terms of the elastic free energy, as given by equation (1.26). The stress so obtained is the average over many atoms and assumes that the strain function does not vary too rapidly.

The elastic constants for a crystal lattice may be deduced from the force constants $\partial^2 V / \partial x_i(l, p) \partial x_k(m, q)$, where V is the total crystal potential energy.

Near crystal defects the continuum theory breaks down, because of the high strain fields and the fact that both stress and strain vary rapidly with position. A method that has been widely used for such problems involves setting up a two-region model. A central region, called region I and containing a few hundred atoms, is considered in which individual atomic displacements are computed. The defect under consideration is then situated at the centre of this region. Surrounding this region, region II is defined as consisting of the remainder of the crystal in which it is assumed that the continuum theory of elasticity applies. In this method, the positions of atoms in region I are adjusted until their total potential energy is a minimum. Adjustable parameters for region II may also be included in this process. An example of this method in relation to an edge dislocation is described by Sinclair (1971).

References

Cottrell, A. H., 1964, *The Mechanical Properties of Matter* (John Wiley, New York), p.82.
Jeffreys, H., 1931, *Cartesian Tensors* (Cambridge University Press, Cambridge), chapters 1 and 8.
Landau, L. D., Lifshitz, E. M., 1970, *Theory of Elasticity* (Pergamon Press, Oxford).
Long, R. R., 1961, *Mechanics of Solids and Fluids* (Prentice-Hall, Englewood Cliffs, NJ).
Nye, J. F., 1960, *Physical Properties of Crystals* (Oxford University Press, Oxford).
Sinclair, J. E., 1971, *J. Appl. Phys.*, **42**, 5321.

REVIEW QUESTIONS

1.1   How is the elastic strain tensor defined? What is meant by the statement that
      the elastic strain tensor is symmetrical? (§1.3)
1.2   Define dilatation. Find the relationship between the dilatation and the
      components of the strain tensor. (§1.3)
1.3   Show how the stress components may be obtained from the elastic free energy,
      F. (§1.5)
1.4   Discuss the form of Hooke's law for a perfectly elastic anisotropic solid. (§1.6)
1.5   Explain how the number of independent elastic constants may be reduced for a
      cubic crystal. (§1.7)
1.6   Derive the equations of motion for an elastic isotropic body. (§1.10)

PROBLEMS

1.7   Show that the elastic free energy for a cubic crystal is given by

$$F = \tfrac{1}{2}c_{11}^{11}(\epsilon_{11}^2 + \epsilon_{22}^2 + \epsilon_{33}^2) + c_{22}^{11}(\epsilon_{11}\epsilon_{22} + \epsilon_{22}\epsilon_{33} + \epsilon_{11}\epsilon_{33}) + 2c_{23}^{23}(\epsilon_{23}^2 + \epsilon_{13}^2 + \epsilon_{12}^2) .$$

1.8   Show that the elastic free energy for an orthorhombic crystal is given by

$$F = \tfrac{1}{2}c_{11}^{11}\epsilon_{11}^2 + \tfrac{1}{2}c_{22}^{22}\epsilon_{22}^2 + \tfrac{1}{2}c_{33}^{33}\epsilon_{33}^2 + c_{22}^{11}\epsilon_{11}\epsilon_{22} + c_{33}^{22}\epsilon_{22}\epsilon_{33} + c_{33}^{11}\epsilon_{11}\epsilon_{33}$$
$$+ 2c_{23}^{23}\epsilon_{23}^2 + 2c_{13}^{13}\epsilon_{13}^2 + 2c_{12}^{12}\epsilon_{12}^2 .$$

1.9   Show that, for an isotropic body, the only independent components of c_{iklm} are
      c_{1111} and c_{1122}.
1.10 (a) Show that for an isotropic body Hooke's law may be written

$$\sigma_{ik} = 2G\epsilon_{ik} + (K - \tfrac{2}{3}G)\epsilon_{ll}\delta_{ik} ,$$

      where K is the bulk modulus and G is the shear modulus.
      (b) By using tensor contraction, show that the relative change in volume in any
      deformation of an isotropic body depends only on the sum of the principal
      values of the stress tensor, that is $\epsilon_{ii} = \sigma_{ii}/3K$.
1.11 Derive equation (1.44). Hence show that the equations of equilibrium for an
      isotropic elastic body (excluding body forces) may be written

$$\frac{E}{2(1+\mu)}\partial_k\partial_k u_i + \frac{E}{2(1+\mu)(1-2\mu)}\partial_i\partial_l u_l = 0 .$$

1.12 Find an expression for Young's modulus of a cubic crystal as a function of
      direction.

Appendix 1.1 Summary of tensor properties

"It has been said that vector equations are like a pocket map, and it has been replied that a pocket map has to be taken out of the pocket and unfolded before it is of any use. The same applies to the tensor method, and for the same reason; but it has the great advantage that it is not a new notation, but a concise way of writing the ordinary notation, so that the unfolding can be carried out more conveniently when occasion arises" (from the preface to H. Jeffreys, 1931, *Cartesian Tensors*, Cambridge University Press, Cambridge).

A vector is a directed quantity which may be specified in polar form in terms of its magnitude and the cosines of the angles subtended by the vector with a convenient set of axes.

In component form, a vector is specified by a set of three numbers which give the resolved components of the vector with respect to a set of cartesian axes. For instance, particle displacement may be written as a directed quantity u, or as the set of cartesian components (u_x, u_y, u_z). The latter may be written more compactly as u_i, where $i = 1, 2, 3$. In this system the directions of the three cartesian axes are simply written as x_i. A vector quantity should retain its identity if the reference axes are changed.

Consider two sets of rectangular axes with the same origin. A point P has coordinates (x_1, x_2, x_3) with respect to one set of axes and coordinates (x'_1, x'_2, x'_3) with respect to the second set which is rotated relative to the first set.

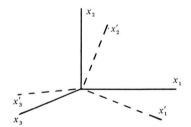

Then, the **axes transformation law** is

$$x'_1 = a_{11}x_1 + a_{12}x_2 + a_{13}x_3 ,$$
$$x'_2 = a_{21}x_1 + a_{22}x_2 + a_{23}x_3 ,$$
$$x'_3 = a_{31}x_1 + a_{32}x_2 + a_{33}x_3 ,$$

or, more briefly, it may be written

$$x'_i = \sum_{j=1,2,3} a_{ij} x_j ,$$

where a_{ij} is the cosine of the angle between x'_i and x_j, that is, $a_{ij} = \cos(x'_i, x_j)$. In each case, the first subscript refers to the new axes, the second to the old. The transformation law may also be written in matrix form

$$
\begin{bmatrix} x'_1 \\ x'_2 \\ x'_3 \end{bmatrix}
=
\begin{bmatrix} a_{11} & a_{12} & a_{13} \\ a_{21} & a_{22} & a_{23} \\ a_{31} & a_{32} & a_{33} \end{bmatrix}
\cdot
\begin{bmatrix} x_1 \\ x_2 \\ x_3 \end{bmatrix} .
$$

The **inverse transformation** (old coordinates in terms of new) is

$$x_i = \sum_{j=1,2,3} a_{ji} x'_j .$$

If the Einstein summation convention is adopted, the notation may be further shortened. According to this convention, a repeated suffix denotes summation over the values 1 to 3. Thus, the transformation equations may be written

$$x'_i = a_{ij}x_j \, ,$$

$$x_i = a_{ji}x'_j \, .$$

Transformation of components of a vector from one set of reference axes to another follows similar laws. If u_j denotes the components of a vector with respect to axes x_j, and u'_i the components with respect to axes x'_i, then the relationships between the two sets of components are:

$$u'_i = a_{ij}u_j \quad \text{(new in terms of old)},$$

$$u_i = a_{ji}u'_j \quad \text{(old in terms of new)}.$$

Because of the Einstein convention each of these relationships represents a set of three equations. The repeated suffix, j in this case, is sometimes called a **dummy suffix** since any letter may be used for the purpose. Note that in the transformation law for new components in terms of the old, similar subscripts are adjacent.

A **vector** is often defined as a set of three components which satisfy the above transformation relationships. A vector belongs to a more general class of quantities called tensors and is classified as a **tensor of first rank** (or order).

A **scalar** is a single quantity that has the same value for all sets of axes. A scalar is sometimes called a **tensor of zero rank**.

Multiplication of a vector u_j **by a scalar** m obeys the relationship:

$$mu'_i = a_{ij}(mu_j) \, .$$

Thus, mu_j acts like a tensor of the first rank. Note the summation with respect to j.

The sum of two vectors u_i and v_i transforms according to:

$$u'_i + v'_i = a_{ij}(u_j + v_j) \, .$$

Two vectors u_j **and** v_l are multiplied by multiplying each component of u_j by each component of v_l, thus generating a set of nine quantities, which may be written compactly as

$$u_j v_l \, .$$

On transformation to a new set of axes the following condition holds:

$$u'_i v'_k = (a_{ij}u_j)(a_{kl}v_l) = a_{ij}a_{kl}u_j v_l \, .$$

This is a set of nine equations, each with nine terms on the right-hand side (there is summation with respect to both j and l).

A **second rank tensor** is defined as a set of nine quantities, denoted by T_{ik}, which transform from one set of axes to another set by the transformation law

$$T'_{ik} = a_{ij}a_{kl}T_{jl} \, .$$

The components of a **second-rank tensor** may be arranged **in matrix form** as follows

$$\begin{bmatrix} T_{11} & T_{12} & T_{13} \\ T_{21} & T_{22} & T_{23} \\ T_{31} & T_{32} & T_{33} \end{bmatrix} \, .$$

A **symmetrical tensor** is unaltered if i and k are interchanged, i.e. $T_{ik} = T_{ki}$. Note that $(T_{ik} + T_{ki})$ is symmetrical.

In an **antisymmetrical tensor** all components reverse in sign when i and k are interchanged, i.e. $T_{ik} = -T_{ki}$. $(T_{ik} - T_{ki})$ is an antisymmetrical tensor whose components are

$$\begin{bmatrix} 0 & T_{12} & T_{13} \\ -T_{12} & 0 & T_{23} \\ -T_{13} & -T_{23} & 0 \end{bmatrix} .$$

There are only three independent components and therefore an antisymmetrical tensor can be considered as equivalent to a vector.

Theorem. Any second-rank tensor can be considered as the sum of symmetrical and antisymmetrical parts.

$$T_{ik} = \tfrac{1}{2}(T_{ik} + T_{ki}) + \tfrac{1}{2}(T_{ik} - T_{ki}) .$$

Tensors of third, fourth and higher rank may be formed by extending the previous definition. A **third-rank tensor**, T_{ijk}, is a set of 27 quantities that transforms according to the law

$$T'_{lmn} = a_{li}a_{mj}a_{nk}T_{ijk} .$$

Similarly, a **fourth-rank tensor**, T_{ijkl}, is a set of 81 quantities that transforms according to

$$T'_{mnop} = a_{mi}a_{nj}a_{ok}a_{pl}T_{ijkl} .$$

A vector quantity may be derived from a scalar ϕ by forming **the gradient** $\partial\phi/\partial x_j$. The gradient transforms according to the vector law

$$\frac{\partial\phi}{\partial x_i} = \frac{\partial x_j}{\partial x_i}\frac{\partial\phi}{\partial x_j} = a_{ij}\frac{\partial\phi}{\partial x_j} .$$

In a similar way, a tensor of rank two may be derived from a vector by forming the gradient

$$\frac{\partial u'_i}{\partial x'_k} = \frac{\partial x_l}{\partial x'_k}\frac{\partial u'_i}{\partial x_l} = a_{kl}\frac{\partial}{\partial x_l}(a_{ij}u_j) = a_{ij}a_{kl}\frac{\partial u_j}{\partial x_l} .$$

Contraction consists of making two suffixes in a tensor equal and then carrying out summation, when a new tensor is formed of order two less than that of the original tensor. For example
(a) $T_{ii} = T_{11} + T_{22} + T_{33}$ is a scalar (zero-rank tensor).
(b) $\partial u_i/\partial x_k$ on contraction gives

$$\frac{\partial u_i}{\partial x_i} = \frac{\partial u_1}{\partial x_1} + \frac{\partial u_2}{\partial x_2} + \frac{\partial u_3}{\partial x_3} ,$$

which is a scalar called the **divergence** of u_i.

The **unit tensor** δ_{ik} is a set of quantities such that if $i = k$, the components have value 1, if $i \neq k$, the components are zero:

$$\delta_{ik} = \begin{bmatrix} 1 & 0 & 0 \\ 0 & 1 & 0 \\ 0 & 0 & 1 \end{bmatrix} .$$

δ_{ik} is also called the **substitution tensor** because of the following property: We have noted that multiplication of two vectors (first-rank tensors) produces a second-rank tensor. Similarly, multiplication of a second-rank tensor by a vector produces a third-rank tensor, and so on. Thus, the quantity $\delta_{ik}u_m$ is a third-rank tensor. If we now put $m = k$ and then add for all values of k (repeated suffix), we obtain

$$\delta_{ik}u_k = u_i.$$

The rank of the product has been reduced by two and the suffix on the vector has been replaced by i. When used for this purpose, δ_{ik} is referred to as the substitution tensor.

An **isotropic tensor** is one whose components remain unaltered whatever the orientation of the axes. There are no isotropic tensors of the first rank. The only isotropic second-rank tensor is a scalar multiple of δ_{ik}. The only isotropic third-rank tensor is a scalar multiple of ϵ_{ikm} (the **alternating tensor**).

$$\epsilon_{ikm}\begin{cases}= 0 \text{ if any two of } i, k, m \text{ are equal} \\ = 1 \text{ if } i, k, m \text{ are all unequal and in cyclic order} \\ = -1 \text{ if } i, k, m \text{ are all unequal and not in cyclic order}\end{cases}$$

e.g. $\epsilon_{112} = 0$, $\epsilon_{231} = 1$, and $\epsilon_{321} = -1$.

There are three independent isotropic fourth-rank tensors: $\delta_{ik}\delta_{lm}$, $\delta_{il}\delta_{km} + \delta_{im}\delta_{kl}$, $\delta_{il}\delta_{km} - \delta_{im}\delta_{kl}$.

The most general isotropic fourth-rank tensor is therefore

$$T_{iklm} = \lambda\delta_{ik}\delta_{lm} + \mu(\delta_{il}\delta_{km} + \delta_{im}\delta_{kl}) + \nu(\delta_{il}\delta_{km} - \delta_{im}\delta_{kl}) ,$$

where λ, μ, and ν are scalars. This equation is useful when considering the elastic constants for an isotropic elastic solid.

An **invariant** has the same value in any coordinate system. For instance, the following are invariants:

T_{ii} is the sum of the diagonal elements of T_{ij} (called the trace of T_{ij}),

$\epsilon_{ijk}T_{i1}T_{j2}T_{k3}$.

A symmetrical second-rank tensor may be **diagonalised**. That is, three axes may be chosen so that only the diagonal components of the tensor (T_{11}, T_{22}, and T_{33}) have finite values, all other components being zero. The chosen axes are then called **principal axes** of the tensor. The diagonal components are called **principal values** of the tensor.

References
Jeffreys, H., 1931, *Cartesian Tensors* (Cambridge University Press, Cambridge).
Long, R. R., 1961, *Mechanics of Solids and Fluids* (Prentice-Hall, Englewood Cliffs, NJ).
Nye, J. F., 1960, *Physical Properties of Crystals* (Oxford University Press, Oxford).

Waves in an infinite elastic solid

2.1 Introduction
One of the most important applications of high-frequency ultrasonic techniques is in the measurement of the elastic constants of crystals. Depending on factors such as the dimensions of the crystal and the frequency of the waves, propagation may occur under free-field conditions or under waveguide conditions. In this chapter free-field propagation will be discussed and appropriate equations deduced for the determination of elastic constants. Waveguide propagation will be discussed in chapter 4.

When an arbitrary disturbance takes place in an infinite solid, two types of bulk wave are found to be possible. One type corresponds to longitudinal wave motion and involves volume changes of the medium. The second type does not involve any change in volume but does involve shearing motions and can be identified with a transverse wave motion. Both of these types of wave are well-known in seismology where they are referred to as P waves and S waves (primary and secondary waves). The velocity of P waves is appreciably higher than the velocity of S waves. When a solid is bounded by surfaces, further wave types become possible, such as Rayleigh waves, Love waves, Lamb waves, etc.

2.2 Infinite isotropic solid
The velocity equations will now be derived for the two wave types that are found in an infinite elastic isotropic solid. The equations of motion (1.46) for an isotropic elastic solid (neglecting body forces) were found to be

$$\rho \ddot{u}_i = (\lambda + G)\,\mathrm{grad}\,\mathrm{div}\,u_i + G\nabla^2 u_i . \tag{2.1}$$

It will be assumed in this chapter that adiabatic conditions are satisfied and hence that the appropriate elastic constants are involved.

Making use of the vector relationship

$$\nabla^2 u_i = \mathrm{grad}\,\mathrm{div}\,u_i - \mathrm{curl}\,\mathrm{curl}\,u_i$$

we can rewrite equation (2.1) as follows

$$\rho \ddot{u}_i = (\lambda + G)\,\mathrm{grad}\,\mathrm{div}\,u_i + G(\mathrm{grad}\,\mathrm{div}\,u_i - \mathrm{curl}\,\mathrm{curl}\,u_i)$$

$$= (\lambda + 2G)\,\mathrm{grad}\,\mathrm{div}\,u_i - G\,\mathrm{curl}\,\mathrm{curl}\,u_i . \tag{2.2}$$

This equation of motion may be separated into two parts:
(i) a *dilatational* part for which $\theta \equiv \mathrm{div}\,u_i$ is finite but for which $\mathrm{curl}\,u_i = 0$,
(ii) a *rotational* part for which $\mathrm{div}\,u_i = 0$ but $\mathrm{curl}\,u_i \neq 0$.

Thus, in the first case, when $\mathrm{curl}\,u_i = 0$, equation (2.2) simplifies to

$$\rho \ddot{u}_i = (\lambda + 2G)\,\mathrm{grad}\,\mathrm{div}\,u_i = (\lambda + 2G)\nabla^2 u_i ;$$

that is

$$\nabla^2 u_i = \frac{1}{c_1^2}\ddot{u}_i \, , \tag{2.3}$$

where

$$c_1^2 = \frac{\lambda + 2G}{\rho} \, . \tag{2.4}$$

Equation (2.3) may be recognised as the familiar general wave equation with c_1 having the dimensions of a velocity. By taking the divergence of each side we find

$$\nabla^2\theta = \frac{1}{c_1^2}\ddot{\theta}. \tag{2.5}$$

This is a wave equation in terms of the dilatation and shows that, when $\mathrm{curl}\, u_i = 0$, a wave type is obtained which involves changes in the volume of the medium. This type of wave is therefore called a **dilatational** or **compressional** wave.

In the second case, when $\mathrm{div}\, u_i = 0$,

$$\rho\ddot{u}_i = -G\,\mathrm{curl}\,\mathrm{curl}\,u_i = G\nabla^2 u_i \, ;$$

that is

$$\nabla^2 u_i = \frac{1}{c_2^2}\ddot{u}_i \, , \tag{2.6}$$

where

$$c_2^2 = \frac{G}{\rho} \, . \tag{2.7}$$

By taking the curl of each side of equation (2.6) we find

$$\nabla^2(\mathrm{curl}\,u_i) = \frac{1}{c_2^2}\frac{\partial^2}{\partial t^2}(\mathrm{curl}\,u_i) \, ,$$

or

$$\nabla^2\omega_i = \frac{1}{c_2^2}\ddot{\omega}_i \, , \tag{2.8}$$

where $\omega_i = \frac{1}{2}\mathrm{curl}\,u_i$ is the rotation vector.

This last equation shows that, when $\mathrm{div}\, u_i = 0$, a wave equation is obtained in terms of the rotation vector only, and does not involve any volume changes. Thus, in general, two types of wave propagate in an isotropic solid, a dilatational wave and a rotational wave. Solutions of equations (2.5) and (2.8) may be found for both plane and spherical waves.

We shall now investigate the relationships between the direction of propagation in each case and the particle motion.

2.2.1 Dilatational waves:

$\mathrm{div}\, u_i$ finite; $\mathrm{curl}\, u_i = 0$.

Consider plane wave solutions of equation (2.3) for which u_i depends only on x_1 and t and not on x_2 and x_3; then the following three equations hold:

$$\rho \ddot{u}_i = (\lambda + 2G)\frac{\partial^2 u_1}{\partial x_1^2} \,,$$

$$\rho \ddot{u}_2 = (\lambda + 2G)\frac{\partial^2 u_2}{\partial x_1^2} \,, \tag{2.9}$$

$$\rho \ddot{u}_3 = (\lambda + 2G)\frac{\partial^2 u_3}{\partial x_1^2} \,.$$

In addition, since $\operatorname{curl} u_i = 0$, each of its components will be zero, that is

$$\frac{\partial u_3}{\partial x_2} = \frac{\partial u_2}{\partial x_3} \,, \qquad \frac{\partial u_1}{\partial x_3} = \frac{\partial u_3}{\partial x_1} \,, \qquad \frac{\partial u_2}{\partial x_1} = \frac{\partial u_1}{\partial x_2} \,. \tag{2.10}$$

But u_i is a function of x_1 and t only; hence all terms in equations (2.10) are zero, and therefore only the first equation in (2.9) is valid. That is, the only finite displacement is u_1, which is in the direction of propagation. Such a plane wave is often termed a **longitudinal** wave and its velocity, c_ϱ, is

$$c_\varrho = \left(\frac{\lambda + 2G}{\rho}\right)^{\frac{1}{2}} . \tag{2.11}$$

From equation (1.38), c_1 may be expressed in terms of the bulk modulus:

$$c_\varrho = \left(\frac{K + \frac{4}{3}G}{\rho}\right)^{\frac{1}{2}} . \tag{2.12}$$

In an ideal fluid $G = 0$; hence

$$c_\varrho = \left(\frac{K}{\rho}\right)^{\frac{1}{2}} .$$

Longitudinal waves are also known as dilatational, compressional, irrotational, primary, or P waves.

2.2.2 Rotational waves

$\operatorname{div} u_i = 0$; $\operatorname{curl} u_i$ finite.

We shall now consider equation (2.6) for plane waves as depending only on x_1 and t; then three equations hold:

$$\rho \ddot{u}_1 = G\frac{\partial^2 u_1}{\partial x_1^2} \,, \qquad \rho \ddot{u}_2 = G\frac{\partial^2 u_2}{\partial x_1^2} \,, \qquad \rho \ddot{u}_3 = G\frac{\partial^2 u_3}{\partial x_1^2} \,. \tag{2.13}$$

Since $\operatorname{div} u_i \equiv \partial u_i/\partial x_i = 0$, we have $\partial u_1/\partial x_1 = 0$ and only the last two of equations (2.13) are relevant. We therefore conclude that two plane waves propagate in the x_1 direction, each with the same velocity but with displacements in the x_2 and x_3 directions, that is, perpendicular to the direction of propagation. This type of plane wave is termed a **transverse wave** or **shear wave**. A plane transverse wave propagating in the x_1 direction will be plane polarised in either the $x_1 x_2$ or $x_1 x_3$ plane. The velocity, c_t, of a transverse wave is then

$$c_t = \left(\frac{G}{\rho}\right)^{\frac{1}{2}} . \tag{2.14}$$

It is interesting to note that in an ideal fluid, for which $G = 0$, a transverse or shear wave cannot propagate. Transverse waves are also known as rotational, solenoidal, shear, secondary, or S waves.

2.3 Infinite anisotropic solid

In an anisotropic crystal there is often no simple relation between the direction of propagation of a wave and the direction of the particle displacement. In other words, in general the waves are not pure longitudinal or transverse types but are some form of mixed type. There are, however, certain directions in which pure longitudinal and transverse waves may propagate. In this section the wave equation for waves in an anisotropic material will be derived and specific solutions will be examined that correspond to pure wave types.

The equations of motion for an anisotropic body may be found by combining Newton's law (1.16)

$$\partial_k \sigma_{ik} = \rho \ddot{u}_i$$

with Hooke's law (1.30)

$$\sigma_{ik} = c_{iklm} \epsilon_{lm}$$

to give

$$\rho \ddot{u}_i = c_{iklm} \partial_k \epsilon_{lm} .$$

From the definition of strain (1.9) we have

$$\epsilon_{lm} = \tfrac{1}{2}(\partial_m u_l + \partial_l u_m) ,$$

whence

$$\rho \ddot{u}_i = \tfrac{1}{2} c_{iklm}(\partial_k \partial_m u_l + \partial_k \partial_l u_m) .$$

Since c_{iklm} is symmetrical with respect to l and m, we may interchange l and m in the first term in brackets, which gives

$$\rho \ddot{u}_i = c_{iklm} \partial_k \partial_l u_m . \tag{2.15}$$

We now assume plane harmonic travelling waves of the form

$$u_i = A_i \exp[i(k_j x_j - \omega t)] , \tag{2.16}$$

where A_i are the amplitudes of the displacement components and k_j are the components of the wave vector. A_i may also be written as $A\alpha_i$, where

A is the displacement amplitude and the α_i are the direction cosines of the particle displacements. Substituting equation (2.16) in (2.15) gives

$$\rho\omega^2 u_i = c_{iklm}k_k k_l u_m . \tag{2.17}$$

This equation may be written in a more homogeneous form by putting $u_i = u_m \delta_{im}$, where δ_{im} is the unit tensor:

$$(\rho\omega^2 \delta_{im} - c_{iklm}k_k k_l)u_m = 0 . \tag{2.18}$$

Equation (2.18) was developed by Christoffel (1877) and is now often referred to as **Christoffel's equation**. It represents a set of three homogeneous equations of the first degree with u_1, u_2, and u_3 as unknowns. They have non-zero solutions only if the determinant of the coefficients is zero, that is

$$|c_{iklm}k_k k_l - \rho\omega^2 \delta_{im}| = 0 . \tag{2.19}$$

Evaluation of equation (2.19) leads to a cubic equation in ω^2 (or alternatively in terms of c^2). The three roots of this equation are in general different, leading to three different velocities of propagation.

It is convenient to rewrite equation (2.18) in the form

$$(\lambda_{im} - \rho c^2 \delta_{im})u_m = 0 , \tag{2.20}$$

where the tensor λ_{im} is defined as

$$\lambda_{im} = c_{iklm}n_k n_l , \tag{2.21}$$

c is the phase velocity of the waves ($= \omega/k$) and n_i denotes the direction cosines of the normal to the wavefront. The following determinant corresponding to equation (2.19) must now be evaluated

$$|\lambda_{im} - \rho c^2 \delta_{im}| = 0 . \tag{2.22}$$

The factors λ_{im} depend both on the crystal symmetry and on the orientation of the waves. We can write this equation out in full:

$$\begin{vmatrix} (\lambda_{11} - \rho c^2) & \lambda_{12} & \lambda_{13} \\ \lambda_{12} & (\lambda_{22} - \rho c^2) & \lambda_{23} \\ \lambda_{13} & \lambda_{23} & (\lambda_{33} - \rho c^2) \end{vmatrix} = 0 , \tag{2.23}$$

where

$$\lambda_{11} = c_{11}n_1^2 + c_{66}n_2^2 + c_{55}n_3^2 + 2c_{56}n_2 n_3 + 2c_{15}n_1 n_3 + 2c_{16}n_1 n_2 ,$$

$$\lambda_{12} = c_{16}n_1^2 + c_{26}n_2^2 + c_{45}n_3^2 + (c_{46} + c_{25})n_2 n_3 + (c_{14} + c_{56})n_1 n_3$$

$$+ (c_{12} + c_{66})n_1 n_2 , \tag{2.24}$$

$$\cdots \quad \cdots \quad \cdots \quad \cdots \quad \cdots \quad \cdots \quad \cdots \quad \cdots \quad \cdots \quad \cdots \quad \cdots \quad \cdots \quad \cdots .$$

The matrix notation for the elastic constants used here is discussed in section 1.7.

Examination of equation (2.23) shows that the displacement vectors associated with each eigenvalue, ρc^2, are mutually perpendicular. For a given direction of propagation, defined by the wave vector k, three waves are possible with mutually perpendicular displacement vectors but with different velocities. In general these waves will not be pure longitudinal or pure transverse. However, for certain directions of propagation in a given crystal lattice (for which k is an eigenvector of λ_{ik}), one wave is pure longitudinal and the other two are pure transverse.

Christoffel showed that the direction cosines, α_i, of the particle displacements on the wavefront are related to the corresponding wave velocities by

$$\alpha_1\lambda_{11} + \alpha_2\lambda_{12} + \alpha_3\lambda_{13} = \alpha_1\rho c^2 ,$$
$$\alpha_1\lambda_{12} + \alpha_2\lambda_{22} + \alpha_3\lambda_{23} = \alpha_2\rho c^2 , \qquad\qquad (2.25)$$
$$\alpha_1\lambda_{13} + \alpha_2\lambda_{23} + \alpha_3\lambda_{33} = \alpha_3\rho c^2 ,$$

where each value of c is substituted in turn.

2.3.1 Cubic crystals

Solutions of Christoffel's equation for a cubic crystal will now be considered. For a cubic crystal there are three independent elastic constants:

$$c_{11} = c_{22} = c_{33} ,$$
$$c_{12} = c_{21} = c_{13} = c_{31} = c_{23} = c_{32} ,$$
$$c_{44} = c_{55} = c_{66} ,$$

all others being zero.

As shown by Borgnis (1955), a pure longitudinal wave and two pure transverse waves may be transmitted in the [100], [110], and [111] directions. These directions for a cubic lattice are shown in figure 2.1.

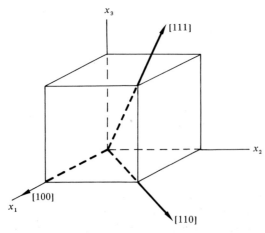

Figure 2.1. Major crystal directions in a cubic lattice.

For a pure longitudinal wave, the displacement vector should be parallel to the wavefront normal, that is $u \times n = 0$, where u is the displacement vector and n is the unit vector normal to the wavefront. For a pure transverse wave the displacement vector should be perpendicular to the wavefront normal, that is $u.n = 0$.

2.3.1.1 Plane waves in the [100] direction

For this orientation, $n_1 = 1$; $n_2 = n_3 = 0$, if we assume that the direction of propagation is in the x_1 direction, as shown in figure 2.1. Equation (2.23) then simplifies to

$$\begin{vmatrix} c_{11} - \rho c^2 & 0 & 0 \\ 0 & c_{44} - \rho c^2 & 0 \\ 0 & 0 & c_{44} - \rho c^2 \end{vmatrix} = 0 \,,$$

whence $(c_{11} - \rho c^2)(c_{44} - \rho c^2)(c_{44} - \rho c^2) = 0$.

The three roots of this equation are

$$c_1 = \left(\frac{c_{11}}{\rho}\right)^{\frac{1}{2}}, \quad c_2 = c_3 = \left(\frac{c_{44}}{\rho}\right)^{\frac{1}{2}}. \tag{2.26}$$

On solving equation (2.25) we find $\alpha_1 = 1$, so that c_1 is the velocity of a longitudinal wave. For c_2, $\alpha_2 = 1$, and for c_3, $\alpha_3 = 1$, so that $c_2 = c_3$ is the velocity of a transverse wave. It follows that for the transverse wave the particle displacement can be in any direction in the (100) plane.

2.3.1.2 Plane waves in the [110] direction

For this direction $n_1 = n_2 = 2^{-\frac{1}{2}}$; $n_3 = 0$. Equation (2.23) then yields

$$\begin{vmatrix} \frac{1}{2}(c_{11} + c_{44}) - \rho c^2 & \frac{1}{2}(c_{12} + c_{44}) & 0 \\ \frac{1}{2}(c_{12} + c_{44}) & \frac{1}{2}(c_{11} + c_{44}) - \rho c^2 & 0 \\ 0 & 0 & c_{44} - \rho c^2 \end{vmatrix} = 0 \,,$$

whence

$$[\tfrac{1}{2}(c_{11} + c_{44}) - \rho c^2][\tfrac{1}{2}(c_{11} + c_{44}) - \rho c^2][c_{44} - \rho c^2]$$
$$- [\tfrac{1}{2}(c_{12} + c_{44})][\tfrac{1}{2}(c_{12} + c_{44})][c_{44} - \rho c^2] = 0 \,.$$

The solutions of this equation are

$$c_1 = \left(\frac{c_{11} + c_{12} + 2c_{44}}{2\rho}\right)^{\frac{1}{2}}; \quad c_2 = \left(\frac{c_{44}}{\rho}\right)^{\frac{1}{2}}; \quad c_3 = \left(\frac{c_{11} - c_{12}}{2\rho}\right)^{\frac{1}{2}}. \tag{2.27}$$

Solution of equation (2.25) shows that:

for c_1: $\alpha_1 = \alpha_2 = 2^{-\frac{1}{2}}$, so that the particle displacement is in the [110] direction; c_1 is therefore the velocity of a longitudinal wave;

for c_2: $\alpha_3 = 1$, so that the particle displacement is in the [001] direction; c_2 is then the velocity of a transverse wave;

for c_3: $\alpha_1 = 2^{-\frac{1}{2}}$; $\alpha_2 = -2^{-\frac{1}{2}}$, so that the particle displacement is in the [1$\bar{1}$0] direction; c_3 also represents a transverse wave.

Since all three velocities in equation (2.27) are different, all three independent elastic constants can be obtained from measurement of the longitudinal and the two transverse velocities in the [110] direction of propagation.

2.3.1.3 Plane waves in the [111] direction

For the [111] direction $n_1 = n_2 = n_3 = 3^{-\frac{1}{2}}$. Solving equation (2.23) as above yields the following wave velocities:

$$c_1 = \left(\frac{c_{11} + 2c_{12} + 4c_{44}}{3\rho}\right)^{\frac{1}{2}} ; \qquad c_2 = c_3 = \left(\frac{c_{11} - c_{12} + c_{44}}{3\rho}\right)^{\frac{1}{2}} . \qquad (2.28)$$

The wave types may be verified from equation (2.25):

for c_1: $\alpha_1 = \alpha_2 = \alpha_3 = 3^{-\frac{1}{2}}$, so that the particle displacement is in the [111] direction; c_1 therefore represents a longitudinal wave;

for c_2 and c_3: $\alpha_1 = \alpha_2 = \alpha_3 = 0$, and the particle displacement can have any direction in the (111) plane.

Experiment E2.1 Measurement of elastic constants of germanium single crystals by an ultrasonic method

(McSkimin, H. J., 1953, J. Appl. Phys., 24, 988.)

Ultrasonic longitudinal and shear waves in the frequency range 10-30 MHz were transmitted down a fused silica rod through a polystyrene or silicone one-quarter wavelength seal into the solid specimen, as shown in figure E2.1. The assembly is mounted inside a copper chamber to prevent moisture from condensing on the specimen; a vacuum may be maintained in the chamber or it can be filled with dry nitrogen. The unit may be cooled with liquid nitrogen to approximately $-195°C$ and then allowed to warm gradually while measurements are taken. Only the specimen end need be cooled to the desired temperature so that at all times the quartz crystal transducer is maintained near room temperature. This technique avoids problems with bond failure between transducer and rod.

Wave packets, with pulse durations 1-50 μs, derived from a signal generator, enter the specimen through the seal and are reflected back and forth giving rise to a series of transmitted pulses, T, decaying exponentially with time. At certain discrete frequencies the emerging wavetrains will be in phase so that an observed pattern of envelopes may be displayed, as shown in figure E2.2. The number of wavelengths, n_0, in the complete path (twice the specimen thickness) is given by

$$n_0 = \frac{\nu_0}{\Delta\nu} , \qquad (E2.1)$$

where ν_0 is the critical frequency of interest and $\Delta\nu$ is the frequency separation between two consecutive critical frequencies. The velocity of propagation is then given by

$$c = \frac{2t\nu_0}{n_0 - (\varphi_0/2\pi)} , \qquad (E2.2)$$

where t is the specimen thickness and φ_0 is the phase shift per reflection. In general, this phase shift term may be neglected at frequencies near the seal $\frac{1}{4}\lambda$ frequency, so that the velocity of propagation reduces to

$$c = \frac{2t\nu_0}{n_0}. \qquad (\text{E2.3})$$

The longitudinal wave velocities were measured with the use of a 25 MHz X-cut quartz crystal and shear wave velocities with the use of a 13 MHz Y-cut crystal. In figure E2.3 is shown the variation of the velocity of longitudinal waves in a germanium single crystal, with propagation parallel to the [110] direction, as a function of temperature. Similar curves were obtained for shear waves. The dimensions of the specimen were $1 \cdot 9 \times 1 \cdot 3 \times 0 \cdot 6293$ cm. The edge dimensions were left rough in order to scatter energy reaching the outer boundaries.

The three independent elastic constants for germanium, c_{11}, c_{12}, c_{44}, were calculated from the velocity measurements with the aid of equations (2.27). In figure E2.4 is shown a plot of one of the constants, c_{11}, as a function of temperature. Similar measurements were also made on silicon single crystals and in specimens of fused silica.

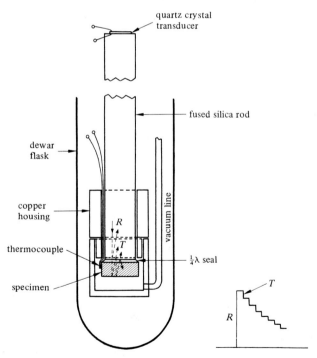

Figure E2.1. Apparatus for measuring elastic constants.

Figure E2.2. Pattern of envelopes, in-phase condition.

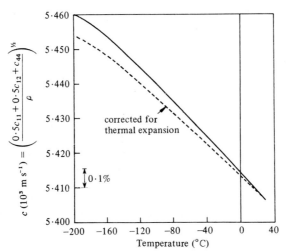

Figure E2.3. Velocity of longitudinal waves in germanium single crystal—propagation parallel to [110] direction.

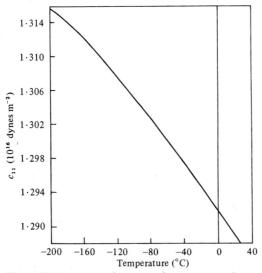

Figure E2.4. c_{11} as a function of temperature for germanium.

2.3.2 Hexagonal crystals

In the hexagonal system there are five independent elastic constants:
$c_{11} = c_{22}$; c_{12}; $c_{13} = c_{23}$; c_{33}; $c_{44} = c_{55}$; $c_{66} = \frac{1}{2}(c_{11} - c_{12})$. Application of the method described earlier shows that in the [001] direction (unique axis) the three wave velocities are given by

$$c_1 = \left(\frac{c_{33}}{\rho}\right)^{\frac{1}{2}} ; \qquad c_2 = c_3 = \left(\frac{c_{44}}{\rho}\right)^{\frac{1}{2}} . \tag{2.29}$$

c_1 represents a longitudinal wave, and $c_2 = c_3$ represents a transverse wave with particle displacement in any direction in the (001) plane.

For plane waves in the [100] direction, or any other direction perpendicular to the [001] axis, we have

$$c_1 = \left(\frac{c_{11}}{\rho}\right)^{1/2}; \qquad c_2 = \left(\frac{c_{44}}{\rho}\right)^{1/2}; \qquad c_3 = \left(\frac{c_{11}-c_{12}}{2\rho}\right)^{1/2}. \qquad (2.30)$$

c_1 is the velocity of a longitudinal wave, c_2 is the velocity of a transverse wave with particle displacement in the [001] direction, and c_3 is a transverse wave with particle displacement in the [010] direction.

By making measurements in these three directions four of the five constants may be determined. In order to measure the fifth constant a wave may be transmitted at an angle of $45°$ between the [100] and [001] directions. Thus, $n_1 = n_3 = 2^{-1/2}$; $n_2 = 0$ and the following two velocities are found:

$$c_1 = \left(\frac{c_{11}-c_{12}+2c_{44}}{4\rho}\right)^{1/2};$$

$$c_2 = c_3 = \left\{\frac{\frac{1}{2}(c_{11}+c_{33}+2c_{44}) \pm [\frac{1}{4}(c_{11}-c_{33})^2 + (c_{13}+c_{44})^2]^{1/2}}{2\rho}\right\}^{1/2}. \qquad (2.31)$$

Examination of the direction cosines of the particle displacements shows that, unless c_{11} is nearly equal to c_{33}, any attempt to launch either a longitudinal or a transverse wave will generate both types of wave. Under these circumstances the phenomenon of acoustic birefringence occurs.

2.3.3 Trigonal crystals

A number of investigations have been conducted into the determination of the elastic constants of trigonal crystals using wave propagation methods. Farnell (1961) considers the propagation of sound waves in α-quartz and sapphire and develops formulae for the components of the displacement and energy-flow vectors in relation to the orientation of the wave normal. Truell et al. (1969, appendix A), give details for the determination of the elastic constants of aluminium oxide. Pace and Saunders (1971) present plots of the elastic wave velocity surfaces as a function of orientation in arsenic, antimony, and bismuth, together with equations for particle displacements and energy flux vectors. The pure mode axes in each crystal are also given. By propagating waves in three different directions the six independent constants can be evaluated.

2.3.4 Velocity, wave and slowness surfaces

When a mechanical disturbance occurs at a point in an infinite isotropic solid, two concentric spherical wavefronts spread out. The wavefront with the greater velocity is associated with longitudinal motion of the medium and the slower wavefront with transverse motion. For spherical wavefronts the direction of energy flow is in the direction of the normal to the wavefronts. For nonspherical wavefronts, the direction of the energy flow (sometimes referred to as the ray) is oblique to the wavefront normal, as shown in figure 2.2. The wavefronts are also known as elastic **wave surfaces**. Other methods of plotting this information involve the **velocity surface**, which is the locus of the phase velocity vectors, and the **inverse** or **slowness surface**, which is the velocity surface inverted with respect to a sphere.

In general, a wave surface may be defined as the envelope of the plane wave elements characterised by $(n_j x_j - ct)$, where n_j are the direction cosines of the wavefront normals and c is the phase velocity measured in the direction n_j. In defining the slowness surface, a **slowness vector**, s_j $(= n_j/c)$, is introduced. In terms of this vector the equation for the displacement of a plane wave may be written:

$$u_i = A\alpha_i \exp[i\omega(s_j x_j - t)] .$$

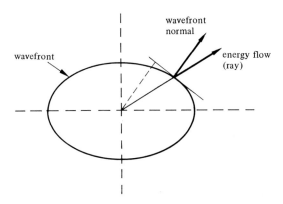

Figure 2.2. For a nonspherical wavefront the direction of the energy flow is oblique to the wavefront normal.

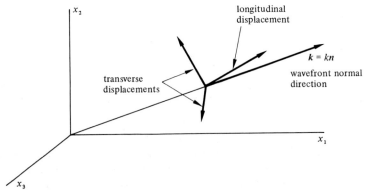

Figure 2.3. Relationships between the wavefront normal direction and the displacement vectors. The three particle displacements are mutually orthogonal but represent pure wave types only in certain directions of propagation.

The **energy flow vector**, as defined by Love (1927), is

$$E_i = \sigma_{ij} \dot{u}_j .$$ (2.32)

From a point source in an anisotropic solid, three wave surfaces spread out, the faster surface corresponding to longitudinal motion followed by two transverse wave surfaces with different velocities in general. These surfaces will mostly be non-spherical as will be the corresponding slowness and velocity surfaces. As before, **phase velocity** is defined as the velocity of propagation of planes of constant phase. Hence, the three possible

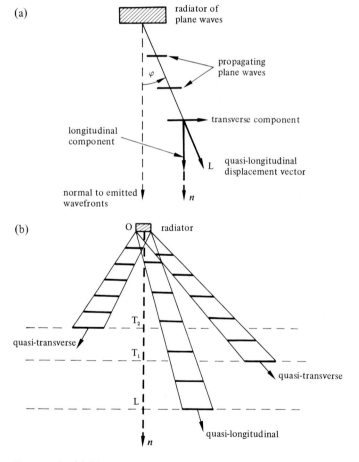

Figure 2.4. (a) The propagation of a quasi-longitudinal wave in an anisotropic solid. (b) The three possible energy beams that may radiate into an anisotropic solid. The dashed lines parallel to the radiator represent the planes which each type of vibration has just reached after time t. The distances OL, OT_1, and OT_2 are directly proportional to the three phase velocities for direction n. In general the three beams are not coplanar. (After Musgrave, 1959.)

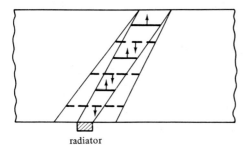

radiator

Figure 2.5. Oblique reflection of energy beam in an anisotropic solid with parallel sides. (After Musgrave, 1959.)

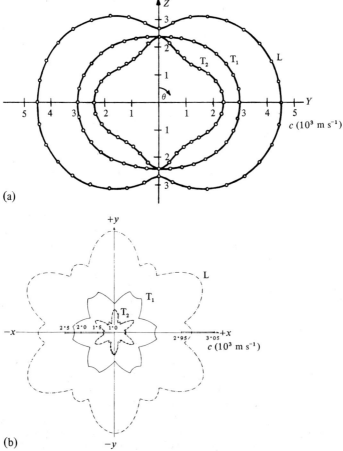

(a)

(b)

Figure 2.6. Velocity surface sections for (a) zinc (after Musgrave, 1961), and (b) $Bi_{1.60}Sb_{0.40}Te_3$ solid solution (after Akgoz *et al.*, 1972). L represents a quasi-longitudinal wave; T_1 and T_2 quasi-transverse waves.

phase velocities in an anisotropic solid are related to the velocities with which the corresponding wavefront normals propagate. The relationships between the particle displacements and the velocities are given by equations (2.25). The three particle displacements at any given point in the solid are mutually perpendicular, that is, the eigenvectors of equation (2.20) are orthogonal.

The relationships between the displacement vectors and the wave normals are shown in figure 2.3. The deviation φ of any of the displacement vectors from the wave normal is given by an expression of the form $\cos\varphi = n_i \alpha_i$ for each wave type. Figure 2.4 illustrates the result of propagating one type of plane wave from a plane radiator into an anisotropic solid, showing also the three possible beams that may radiate into the solid. It follows from these properties that energy may travel back and forth across a specimen with parallel sides along a path making an oblique angle with the normals to the specimen surfaces, as shown in figure 2.5. Further details concerning wave propagation in anisotropic solids may be found in Musgrave (1959; 1961). Examples of velocity surfaces are given in figure 2.6.

Experiment E2.2 Wave propagation effects in single crystals
(Markham, M. F., 1957, *Br. J. Appl. Phys. Suppl.,* 6, S.56, and Musgrave, M. J. P., 1959, *Rep. Prog. Phys.,* 22, 74.)

In some early attempts to measure the elastic constants of single crystals, waves were propagated in arbitrary directions. Markham gives an account of experiments with single crystals of copper and silver which were machined in octagonal form. When trying to measure the three wave velocities along normals to each of the faces, he found it in some cases impossible to receive all three modes of propagation. This effect has been explained by Musgrave (1959) who has shown that the energy of an elastic wave in an anisotropic medium will not, in general, travel along the same path as the normal to the plane wavefront but can deviate as much as $45°$ from this.

In order to determine the direction of maximum energy propagated from a source of finite size it is necessary to construct the wave surface, as distinct from the velocity surface. The wave surface is the surface on which all points will vibrate in the same phase at $t = 1$ when a point source of disturbance is discharged at the origin at $t = 0$. If Huygen's construction is made from a wave surface, the position of maximum energy in the plane of the wave may be found.

Figure E2.5 shows a typical longitudinal wave surface for a zinc crystal (hexagonal) projected on the yz plane. The wave surface has a circular section in the xy plane, but not in the yz plane. If a quasi-longitudinal wave is propagated along a direction ON which is inclined at $15°$ to the z axis, the position of maximum energy, which is given by Huygen's construction, is at L which is the point of contact with the plane normal to ON and the wave surface. It will be seen that the deviation of the energy path OL from ON is considerable. It should be noted that the wave travels with a phase velocity, c_ϱ, associated with the ON direction and not OL in spite of the fact that the energy has travelled along an oblique path.

Figure E2.6 shows a zinc crystal through which longitudinal waves are propagated. Quartz crystals T_x and R_x are attached to opposite faces of the zinc and it is found that no energy is received at **A** from a pulse transmitted from C (i.e. when the quartz crystals are directly opposite each other), but maximum energy is received when the upper crystal is moved to **B**.

Musgrave (1959) makes some additional comments and explanation. In figure E2.7a are shown the three possible plane waves which may be launched from the plane face of a zinc crystal whose normal is at $15°$ to the zonal axis. To excite the T_1 wave, a truly transverse displacement is necessary; the L and T_2 waves, being quasi-longitudinal and quasi-transverse respectively, are both generated by the application of a truly longitudinal displacement and either wave may be detected by means of the component of its displacement vector normal to the specimen surface, although the sensitivity to

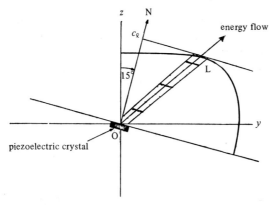

Figure E2.5. Partial section of longitudinal sheet of wave surface for zinc showing deviation of rays from wave normal direction ON. (The diagram is drawn approximately to scale.)

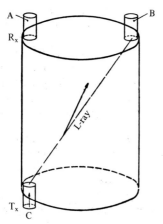

Figure E2.6. Oblique propagation of quasi-longitudinal waves in zinc crystal with the wave normal inclined $15°$ to z axis. **A**, no energy received here; **B**, maximum energy received here.

the T_2 wave is, of course, smaller than to the L wave. The necessary acoustic contact may be preserved by an oil film and the receiver can then be readily slid from one position to another on the side of the specimen remote from the transmitter. By this means the direction of propagation of the L wave in the zinc crystal specimen has been convincingly demonstrated as shown in figure E2.7b. A wave of much smaller amplitude and slower velocity was detected in another position of the receiver and this was almost certainly the T_2 wave which was excited and whose energy travelled along the beam reflected from the side of the specimen, the phase velocity along the axis of the specimen remaining constant.

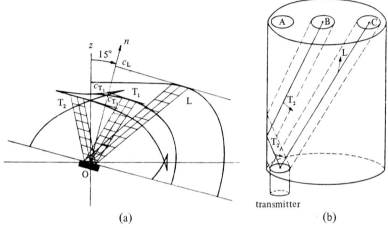

(a)   transmitter   (b)

Figure E2.7. (a) Theoretical. The possible waves of normal (l, m, n) where $n = \cos 15°$. (b) Experimental. Longitudinal displacement from the transmitter resolves into L and T_2 components. Receiver detects no energy at A, T_2 energy at B, L energy at C.

Experiment E2.3 Elastograms
(Schaefer, C., Bergmann, L., 1934; 1935, *Naturwissenschaften,* **22,** 685; **23,** 799.)

Debye and Sears (1932) and, independently, Lucas and Biquard (1932) first observed that when a beam of monochromatic light passed through a liquid perpendicular to travelling sound waves, the successive compressions and rarefactions of the sound beam act as a phase diffraction grating giving rise to an optical diffraction pattern. They found that the loss in intensity of the light beam at a given point was proportional to the sound intensity. This effect is shown schematically in figure E2.8.

Schaefer and Bergmann (1934) developed a technique for producing such diffraction patterns when ultrasonic waves pass through a solid. Initially their method was used to study glass. In figure E2.9 is shown their experimental arrangement for producing the diffraction patterns. A quartz transducer is cemented to the side of the specimen so that either a travelling or a standing ultrasonic beam crosses the path of the monochromatic beam of light. The transducer can excite a great number of standing waves of various polarisations and directions of propagation. In the case of a cube of isotropic material the diffraction pattern consists of two concentric circles as in figure E2.10. From the radius of the innermost circle the wavelength of the longitudinal waves may be found, while from the radius of the outer circle the wavelength of the transverse waves is found.

Usually only the first order diffraction pattern is observed. Each spot on the elastogram is due to the diffraction of light at the phase grating set up by an elastic wave with wave normal parallel to the line joining that spot with the centre of the elastogram. It is therefore desirable to excite as many elastic modes as possible at each frequency in order to obtain a full pattern.

The angle of diffraction, θ_1, of any one spot measured from the central spot is related to the wavelength of the corresponding sound beam, λ^*, and the wavelength of the light, λ, by

$$\lambda^* \sin\theta_1 = \lambda .\qquad\qquad\qquad\qquad\qquad\text{(E2.4)}$$

The spacing of the grating therefore corresponds to the length of the sound wave in the specimen.

Schaefer and Bergmann (1935) have extended the method to the measurement of the elastic constants of single crystals. For anisotropic single crystals the pattern of the diffracted light consists of three closed curves which display the symmetry of the

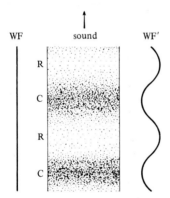

Figure E2.8. Plane light wave front WF incident on an ultrasonic beam and the emerging light wave front WF'. R denotes rarefaction and C denotes compression. Change in the wave front is greatly exaggerated in the diagram. (After Hargrove and Achyuthan, 1965.)

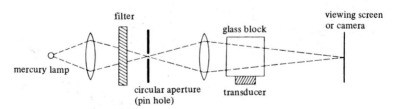

Figure E2.9. Schematic diagram of the experimental arrangement for Schaefer–Bergmann patterns. (After Hargrove and Achyuthan, 1965.)

lattice as viewed in the direction of the light. In figure E2.11 are shown a number of such patterns. From the radii of these figures the variation in velocity can be obtained, and hence the variation in elastic constants, as a function of orientation in the crystal. The method has the advantage that the different types of waves in the crystal may all be observed at the same time. Thus a single specimen may be used for the determination of all the elastic constants. The accuracy, however, is limited to ±1% under optimum conditions.

As pointed out by Joel (1961), an elastogram corresponds to a section of the slowness surface (Musgrave, 1959). In defining the slowness surface for each direction of propagation, vectors of length $1/c$ are drawn from a fixed origin, where c represents the values of the associated velocities (in general three). The section of this surface under consideration is the one determined by the plane of incidence. On the elastograms the distance of a spot from the centre is inversely proportional to the wavelength of the corresponding elastic wave travelling in that same direction. Hence this distance is inversely proportional to the velocity of the wave since the frequency is constant for any one experiment. Therefore the elastogram and the section of the slowness surface determined by the plane of propagation coincide except for a scale factor. The scale factor is equal to $f\lambda\nu$, where f is the focal length of the objective lens, λ is the wavelength of the light, and ν is the frequency of the elastic waves.

Variations of the above method that may be employed with opaque specimens have been described by Willard (1951).

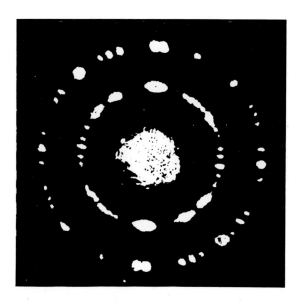

Figure E2.10. Elastogram for glass. The inner circle is a section of the slowness surface for longitudinal waves, the outer circle corresponds to transverse waves.

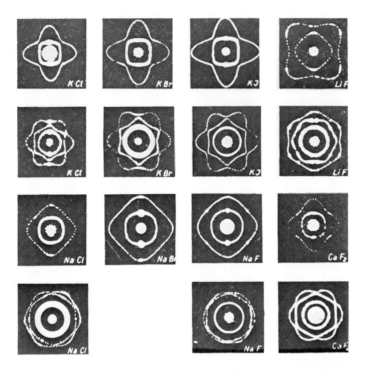

Figure E2.11. Elastograms for a number of alkali halide (cubic) crystals. According to the direction of the sound wave, 2 or 3 slowness surfaces are observed corresponding to the number of independent velocities. The first and third rows refer to transmission down the [100] axis, the second and fourth rows to transmission down the [111] axis.

2.4 Higher-order elastic constants

2.4.1 Third-order elastic constants

Third-order elastic constants are of importance in discussing properties of solids that depend on anharmonic terms in the interatomic potential, for example in the theory of thermal expansion and in dealing with the interaction of acoustic and thermal phonons in a crystal lattice. In the linear theory of elasticity it is assumed that the components of stress are linearly related to the components of strain. The elastic free energy is then a quadratic function of the strain tensor. When a monochromatic wave propagates under these conditions, it does so without any distortion of shape. The principle of superposition holds for a set of such waves.

If the wave amplitude is large, or if nonlinear interactions take place, there will be distortion of the wave so that higher order harmonics of the fundamental will appear. When nonlinear effects are present, the strain tensor contains higher order terms. Equation (1.9) must now be replaced by

$$\epsilon_{ik} = \frac{1}{2}\left(\frac{\partial u_i}{\partial x_k} + \frac{\partial u_k}{\partial x_i} + \frac{\partial u_l}{\partial x_l}\frac{\partial u_l}{\partial x_k} + ...\right). \tag{2.33}$$

For many problems one additional term is sufficient. The stress tensor will then include quadratic terms and the equations of motion will become nonlinear. The elastic free energy may be written

$$F = F_0 + F_1 + F_2 + F_3 + ... ,$$

where F_0 is the initial strain energy (usually assumed zero); F_1 is the potential energy due to a static load, if present; F_2 is given by equation (1.31), namely,

$$F_2 = \tfrac{1}{2}c_{iklm}\epsilon_{ik}\epsilon_{lm} ;$$

and F_3 is given by (Brugger, 1964)

$$F_3 = \tfrac{1}{6}c_{iklmno}\epsilon_{ik}\epsilon_{lm}\epsilon_{no} , \tag{2.34}$$

where c_{iklmno} is the third order elastic constant tensor. This is a sixth-order tensor having a possible 729 components which reduces to 56 independent constants for a triclinic material, 6 for a cubic material, and to 3 for an isotropic material. In matrix notation

$$F_3 = \tfrac{1}{6}c_{pqr}\epsilon_p\epsilon_q\epsilon_r , \tag{2.35}$$

where $p, q, r = 1, 2, ..., 6$.

The above equations have been written in terms of spatial (Eulerian) coordinates. Frequently they are written in material (Lagrangian) coordinates with the material strain components, η_{ik}, replacing ϵ_{ik}. Material coordinates are more appropriate in dealing with moving fluids or deformed solids, since the coordinate system is then attached to the moving medium.

2.4.2 Higher-order constants

A more general thermodynamic definition of higher-order elastic constants has been given by Brugger (1964). The main thermodynamic potentials may be written

$$dU = TdS + \frac{1}{\rho_0}t_{ik}d\eta_{ik} ,$$

$$dF = -SdT + \frac{1}{\rho_0}t_{ik}d\eta_{ik} ,$$

$$dH = TdS - \frac{1}{\rho_0}\eta_{ik}dt_{ik} ,$$

$$dG = -SdT - \frac{1}{\rho_0}\eta_{ik}dt_{ik} . \tag{2.36}$$

Here U is the internal energy, F the elastic free energy, H the enthalpy, and G Gibbs function, defined in terms of the conjugate variables S and T, and t_{ik} and η_{ik}/ρ_0, where t_{ik} are the thermodynamic tensions (see Thurston, 1964) and ρ_0 is the unstrained density.

The adiabatic and isothermal stiffnesses c and compliances s of the nth order, for $n \geqslant 2$, are

$$c_{ikpq...}^{S} = \rho_0 \left(\frac{\partial^n U}{\partial \eta_{ik} \partial \eta_{pq} \cdots} \right)_S ,$$

$$c_{ikpq...}^{T} = \rho_0 \left(\frac{\partial^n F}{\partial \eta_{ik} \partial \eta_{pq} \cdots} \right)_T ,$$

$$s_{ikpq...}^{S} = -\rho_0 \left(\frac{\partial^n H}{\partial t_{ik} \partial t_{pq} \cdots} \right)_S ,$$ (2.37)

$$s_{ikpq...}^{T} = -\rho_0 \left(\frac{\partial^n G}{\partial t_{ik} \partial t_{pq} \cdots} \right)_T .$$

As shown in Thurston and Brugger (1964), there are six independent third-order constants for cubic crystals as follows:

$$c_{111} = c_{222} = c_{333} ,$$

$$c_{144} = c_{255} = c_{366} ,$$

$$c_{112} = c_{223} = c_{133} = c_{113} = c_{122} = c_{233} ,$$

$$c_{155} = c_{244} = c_{344} = c_{166} = c_{266} = c_{355} ,$$ (2.38)

$$c_{123} ,$$

$$c_{456} ;$$

all others are zero. In comparison, the second order constants are:

$$c_{11} = c_{22} = c_{33} ,$$

$$c_{12} = c_{23} = c_{13} ,$$

$$c_{44} = c_{55} = c_{66} .$$

For isotropic media, there are three independent third-order constants which may be expressed in terms of the Lamé second order (λ, G) and third order (ν_1, ν_2, ν_3) coefficients as follows:

$$c_{123} = \nu_1 , \qquad c_{144} = \nu_2 , \qquad c_{456} = \nu_3 ,$$

$$c_{112} = \nu_1 + 2\nu_2 , \qquad c_{155} = \nu_2 + 2\nu_3 ,$$ (2.39)

$$c_{111} = \nu_1 + 6\nu_2 + 8\nu_3 .$$

The second-order constants are:

$$c_{11} = \lambda + 2G , \qquad c_{12} = \lambda , \qquad c_{44} = G .$$

Constants for other crystal systems have been given by Hearmon (1953).

A convenient method for determining the third-order constants experimentally is to measure the velocity of ultrasonic waves in a specimen that is subjected to a static strain, either uniaxial or hydrostatic. Expressions for the constants in terms of measured velocities in different crystalline directions are given in the paper of Thurston and Brugger (1964).

Experiment E2.4 Third-order elastic constants of polycrystalline metals
(Smith, R. T., Stern, R., Stephens, R. W. B., 1966, *J. Acoust. Soc. Am.*, **40**, 1002.)

An ultrasonic pulse method has been developed to determine the third-order elastic constants of polycrystalline metals, including steels, aluminium alloys, magnesium, tungsten, and molybdenum, when a uniaxial stress is applied to the specimens. The constants have been computed with the use of the theory of Thurston and Brugger (1964) for the propagation of elastic waves in homogeneously stressed solids. The practical range of uniaxial stress is limited since the theoretical analysis is not valid for deformations involving slip processes.

Since changes in the velocity of propagation of the ultrasonic waves must be measured to a few parts in 10^6, a pulse–echo interference method has been developed in which a comparison is made between a pulse which traverses the specimen and one derived directly from the pulsed oscillator. The two signals are added algebraically and the delay time between them is adjustable so that the display can be set in an antiphase condition. If the external load applied to the specimen is altered and the delay line adjusted to restore the antiphase condition, transit time changes of about 2×10^{-10} s can be detected. The signals displayed on the oscilloscope are coherent not only within one sweep of the time base but also from one sweep to the next, resulting in a stationary display.

Following Thurston and Brugger, a 'natural' velocity of propagation is introduced, defined as the unstressed path length divided by the transit time of the ultrasonic waves. This concept makes it possible to relate the raw experimental data on transit time changes versus applied stress directly to the third-order constants without having to correct for changes in path length and density.

If e_1 and e_2 represent the instantaneous voltages derived from the mth received pulse–echo and the long transmitter pulse respectively, then at time t

$$e_1 = a \sin[\omega t - k(2m-1)l_0 + 2(m-1)\gamma + \varphi] , \tag{E2.5}$$

$$e_2 = a \sin \omega t , \tag{E2.6}$$

where a is the amplitude, adjusted to be equal for both signals; $k = 2\pi\nu/W$ is the propagation constant expressed in terms of W, Thurston and Brugger's 'natural' velocity; l_0 is the unstrained specimen thickness; γ is the phase shift produced on reflection at either transducer; φ is some arbitrary phase angle. The parameter φ includes not only the phase shift introduced by the delay line but also any constant parasitic phase shifts between transmitter and receiver. The interference conditions are shown schematically in figure E2.12.

The algebraic sum of e_1 and e_2 is identically zero independently of time, provided that

$$\varphi = \frac{2\pi v l_0 (2m-1)}{W} + (2n+1)\pi - 2(m-1)\gamma .\tag{E2.7}$$

Equation (E2.7) may be written in terms of the delay line calibrated scale divisions

$$D = \frac{S\Delta W}{W_0} + C(v) ,\tag{E2.8}$$

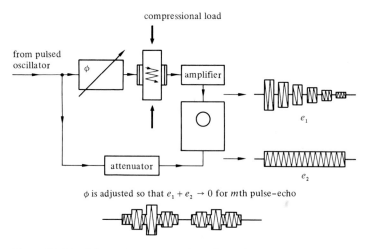

Figure E2.12. Schematic representation of interference conditions.

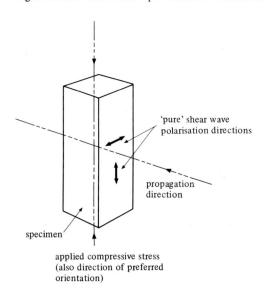

Figure E2.13. Polarisation scheme for shear waves.

where

$$S = -\frac{(2m-1)l_0}{W_0 d} \qquad (E2.9)$$

and d is the delay time per scale division of the delay line. $C(v)$ is constant for a given frequency. The ratio $\Delta W/W_0$ is related to the stress derivatives appearing in Thurston and Brugger's theory by the first-order approximation

$$\frac{\Delta W}{W_0} = \frac{\Delta\sigma}{2\rho_0 W_0^2}(\rho_0 W^2)_0' \qquad (E2.10)$$

where σ is the applied stress and the prime denotes the derivative with respect to σ at constant temperature. The derivatives $(\rho_0 W^2)_0'$ have been evaluated by Thurston and

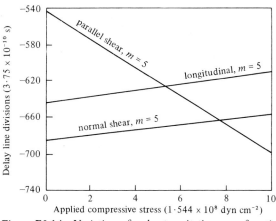

Figure E2.14. Variation of pulse transit time as a function of applied stress.

Table E2.1. Second- and third-order elastic constants for steel, molybdenum and tungsten.

	Specimen			
	Steel, Hecla 37[a]	Hecla ATV[b]	Molybdenum[c]	Tungsten[c]
Density (10^3 kg m^{-3})	7·823	8·065	9·71	14·16
2nd-order constants (10^5 bar)				
λ	11·1 ± 0·1	8·7 ± 0·2	15·7 ± 0·2	7·50 ± 0·05
G	8·21 ± 0·05	7·16 ± 0·3	11·0 ± 0·1	7·30 ± 0·1
3rd-order constants (10^5 bar)				
v_1	−35·8 ± 7·0	3·4 ± 2·0	−5·1 ± 5·0	−21·5 ± 3·0
v_2	−28·2 ± 3·0	−55·2 ± 8·0	−28·3 ± 3·0	−14·3 ± 1·5
v_3	−17·7 ± 0·8	−10·0 ± 1·0	−19·3 ± 0·4	−12·4 ± 0·2

[a] 0·4% C, 0·3% Si, 0·8% Mn; [b] austenitic, 36% Ni, 10% Cr, 1% Mn; [c] 99·99%, sintered.

Brugger for several orientations of propagation direction and stress direction. The third-order constants are then evaluated from the initial slope of the plot of D versus σ.

Specimens were cut in the form of rectangular parallelepiped bars, 1 inch square section, with the main axis along the direction of residual preferred orientation. The ultrasonic waves were propagated normal to the main axis of the bar and the uniaxial load was applied along the bar axis by a simple hydraulic system. Measurements were made with both longitudinal and shear waves, the latter being polarised either parallel or normal to the uniaxial load, as shown in figure E2.13.

In figure E2.14 are shown typical plots of the transit time changes of the ultrasonic waves as a function of uniaxial load and in table E2.1 are listed some of the second- and third-order constants.

Experiment E2.5 Fourth-order elastic constants
(Graham, R. A., 1972, *J. Acoust. Soc. Am.*, **51**, 1576.)

So far the only method that has been used to obtain fourth order elastic constants is the shock compression technique. When subjected to shock-wave compression a number of solids have been observed to exhibit unusually large elastic limits. A technique utilising a planar impact of samples with flat-faced projectiles allows the determination of fourth order constants over a continuous stress range from 1 kbar to several hundreds of kilobars.

When a solid is rapidly compressed (in less than 10^{-8} s) over a large planar area, the inertial response of the sample produces a well defined state of one-dimensional strain in the direction of shock propagation. The shock-wave propagation velocity through the sample may be detected by optical techniques, quartz gauges, charged pins or electrical response measurements. An accuracy of about $\pm 1\%$ can be achieved in measurements of the shock velocity.

Using the principle of conservation of mass, we can show that

$$\theta = \frac{\Delta V}{V_0} = \frac{v}{U} \, , \tag{E2.11}$$

where θ is the linear compression (dilatation), ΔV is the change in specific volume imparted by the shock front, V_0 is the original specific volume, v is the particle velocity imparted by the shock front, and U is the shock wave velocity. Fowles (1967) has shown that the expansion of the strain energy in a power series may be written

$$\sigma_x = \frac{V}{V_0}(c_{xx}P_x + \tfrac{1}{2}c_{xxx}P_x^2 + \tfrac{1}{6}c_{xxxx}P_x^3 + \ldots) \tag{E2.12}$$

Table E2.2. High-order elastic constants (10^4 bar) measured at $25°C$ for sapphire (4% maximum strain) and fused quartz (6% maximum strain).

Elastic constants	Sapphire	Fused quartz
c_{111}	$-3 \cdot 3 \pm 0 \cdot 3$	$0 \cdot 55 \pm 0 \cdot 01$
c_{333}	$-3 \cdot 3 \pm 0 \cdot 3$	
c_{1111}	50 ± 15	11 ± 1
c_{3333}	50 ± 15	

where σ_x is the longitudinal component of stress in the shock propagation direction (assumed to be the x axis), c_{xxxx} is the longitudinal fourth order elastic constant, and $P_x \equiv \theta(\frac{1}{2}\theta - 1)$.

In analysing the data it was found that the second-order longitudinal constants described the experimental data satisfactorily for compressions up to about 1%, the second- and third-order elastic constants give a good description for stresses up to about 2%, whereas the fourth-order contribution must be utilised to give a good fit to the observed data at compressions from 2% to 4%. The values found for the observed third- and fourth-order constants for sapphire (single crystal Al_2O_3) and for fused quartz are shown in table E2.2. This was the first occasion the fourth-order constants were reported for sapphire and fused quartz.

In a further paper (*Phys. Rev.*, **B6**, 4779), Graham (1972) used this method to obtain similar data for crystalline α-quartz together with values for linear and non-linear piezoelectric constants.

References

Akgoz, Y. C., Saunders, G. A., Sumengen, Z., 1972, *J. Mater. Sci.,* **7**, 279.

Borgnis, F. E., 1955, *Phys. Rev.,* **98**, 1000.

Brugger, K., 1964, *Phys. Rev.,* **A133**, 1611.

Christoffel, E. B., 1877, *Ann. di Mat.,* (Ser.2), 8, 193.

Debye, P., Sears, F., 1932, *Proc. Nat. Acad. Sci.,* **18**, 409.

Farnell, G. W., 1961, *Canad. J. Phys.,* **39**, 65.

Fowles, R., 1967, *J. Geophys. Res.,* **72**, 5729.

Hargrove, L. E., Achyuthan, K., 1965, in *Physical Acoustics,* Volume 2B, Ed. W. P. Mason (Academic Press, New York).

Hearman, R. F. S., 1953, *Acta Cryst.,* **6**, 331.

Joel, N., 1961, *Proc. Phys. Soc.,* **78**, 38.

Love, A. E. H., 1927, *Mathematical Theory of Elasticity,* 4th edition (Cambridge University Press, Cambridge).

Lucas, R., Biquard, P., 1932, *Compt. Rend.,* **194**, 2132.

Markham, M. F., 1957, *Br. J. Appl. Phys.,* Suppl. No.6, S.56.

Musgrave, M. J. P., 1959, *Rep. Prog. Phys.,* **22**, 74.

Musgrave, M. J. P., 1961, *Prog. Solid Mechanics,* 2, 63.

Pace, N. G., Saunders, G. A., 1971, *J. Phys. Chem. Solids,* **32**, 1585.

Schaefer, C., Bergmann, L., 1934, *Naturwissenschaften,* **22**, 685.

Schaefer, C., Bergmann, L., 1935, *Naturwissenschaften,* **23**, 799.

Thurston, R. N., 1964, in *Physical Acoustics,* Ed. W. P. Mason, Volume 1A (Academic Press, New York).

Thurston, R. N., Brugger, K., 1964, *Phys. Rev.,* **A133**, 1604.

Truell, R., Elbaum, C., Chick, B. B., 1969, *Ultrasonic Methods in Solid State Physics* (Academic Press, New York).

Willard, G. W., 1951, *J. Acoust. Soc. Am.,* **23**, 83.

REVIEW QUESTIONS
2.1   Show how to find the wave equations for (a) dilatational (longitudinal) waves, (b) rotational (transverse) waves, in an infinite elastic solid. (§2.2)
2.2   Discuss a method for setting up Christoffel's equation governing the propagation of plane waves in an infinite elastic anisotropic solid. (§2.3)
2.3   Find the equations for the velocities of longitudinal and transverse waves travelling in the [100] direction in a cubic crystal. (§2.3.1)
      How may the wave types be identified (mathematically)?
2.4   What is meant by a velocity surface, a wave surface, a slowness surface? (§2.3.4)
2.5   Show how to define the third-order elastic constant tensor. (§2.4.1)

PROBLEMS
2.6   Derive expressions (2.28) for the velocities of longitudinal and transverse waves travelling in the [111] direction in a cubic crystal.
2.7   Verify equations (2.30) for longitudinal and transverse waves travelling in the [100] direction in a hexagonal crystal.

Waves in bounded solids

3.1 Introduction

When mechanical wave energy strikes a boundary between two media, in general there will be some energy reflected back into the first medium and some energy transmitted into the second medium. The simplest physical system involves the reflection of a longitudinal wave travelling in a perfect fluid and incident on a plane boundary separating the fluid from a vacuum. Since longitudinal waves only can exist in a perfect fluid, and no mechanical waves in a vacuum, there can only be a reflected longitudinal wave which obeys the ordinary law of reflection. If the second medium is also a perfect fluid there will be some transmitted energy which will obey the law of refraction.

If one medium is a solid, two wave types are now possible and the observed effects become much more complicated. When a plane wave (longitudinal or transverse) is incident on a plane boundary separating two isotropic solids, there may be two reflected and two transmitted waves even when only one wave type is incident on the boundary. If the two solids are anisotropic then three wave types may exist in each solid. In the latter case the relations between the wavefront normals and the particle displacements become very complex.

In addition to these bulk wave effects there are waves which exist only near the surface of a solid. In the study of earthquake waves the received waves at a given point on the Earth's surface are found to consist of four different types: a primary wave (longitudinal) which has the greatest velocity, a secondary wave (transverse), and two types of surface wave. Rayleigh surface waves are two-dimensional having a longitudinal component in the direction of propagation coupled to a transverse component perpendicular to the surface. The amplitude of the transverse component decays rapidly with depth below the surface. A second type of surface wave, known as a Love wave, has two transverse displacement components both of which are perpendicular to the direction of propagation.

3.2 Fluid–vacuum interface

Although the main interest in this chapter is in the behaviour of solids with boundaries, it is instructive to consider firstly reflection and transmission phenomena in fluids for which the analysis procedures may be set up under simpler physical conditions. We shall therefore consider a plane harmonic wave travelling in a perfect isotropic fluid with the wave vector in the $x_1 x_3$ plane, as shown in figure 3.1. The wave is incident on an infinite plane surface in the $x_2 x_3$ plane which separates the fluid from a vacuum.

It is convenient to write the displacement vector u_i in terms of a **scalar potential** ϕ and a **vector potential** ψ_i, so that

$$u_i = \text{grad}\,\phi + \text{curl}\,\psi_i \ . \tag{3.1}$$

This procedure has two advantages. It allows the separation of the longitudinal and the transverse components of the displacement, if both are present, and it presents the solutions in a form that can readily be related to other variables such as particle velocity, fluid pressure, stress, etc. In the present case, since in a perfect fluid only a longitudinal wave is possible, $\text{curl}\,\psi_i = 0$ and

$$u_i = \text{grad}\,\phi \ . \tag{3.2}$$

Solutions are therefore required of the wave equation

$$\nabla^2\phi = \frac{1}{c_{\varrho}^2}\frac{\partial^2\phi}{\partial t^2} \tag{3.3}$$

subject to suitable boundary conditions.

Plane harmonic wave solutions will be assumed which may be written for the incident and reflected waves as follows:

$$\phi_{\text{inc}} = \hat{A}\exp[i(k_j x_j - \omega t)] = \hat{A}\exp[ik(n_1 x_1 + n_3 x_3) - i\omega t] \tag{3.4}$$

$$\phi_{\text{refl}} = \hat{R}\hat{A}\exp[i(k_j x_j - \omega t)] = \hat{R}\hat{A}\exp[ik(n_1' x_1 + n_3' x_3) - i\omega t] \tag{3.5}$$

where $\hat{A}$ is the complex potential amplitude of the incident wave, $\hat{R}$ is the **reflection coefficient** (in general complex, since there may be a change in phase on reflection as well as a change in amplitude), $k = \omega/c_{\varrho}$ is the magnitude of the wave vector in the fluid, and n_i and n_i' are the direction cosines of the incident and reflected wavefront normals respectively.

In terms of the angle of incidence, α, and the angle of reflection, β, as shown in figure 3.1, equations (3.4) and (3.5) may be written [neglecting the time factor $\exp(-i\omega t)$ which appears in all equations]

$$\phi_{\text{inc}} = \hat{A}\exp[ik(x_1\cos\alpha + x_3\sin\alpha)] \tag{3.6}$$

$$\phi_{\text{refl}} = \hat{R}\hat{A}\exp[ik(-x_1\cos\beta + x_3\sin\beta)] \ . \tag{3.7}$$

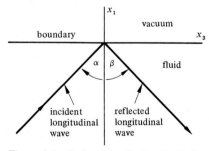

Figure 3.1. Reflection of a longitudinal wave at a fluid–vacuum interface.

We now introduce the **boundary conditions** at the surface. In general both the force and the particle displacement, normal and tangential to the surface, must be continuous. In place of force it is often convenient to use pressure or stress. In this particular case, since there is no medium above the surface, the resultant pressure arising from the incident and reflected waves must be zero at the surface. The two boundary conditions may therefore be written

$$u_{inc} + u_{refl} = 0 , \tag{3.8}$$

$$p_{inc} + p_{refl} = 0 . \tag{3.9}$$

Equation (3.9) is sufficient for the solution of the problem provided the pressure can be related to the potential ϕ. From equation (1.38) p is related to the dilatation by $p = -K\theta$, where K is the bulk modulus. But $\theta = \text{div}u_i = \nabla^2\phi$. Thus

$$p = -K\nabla^2\phi = -\rho\frac{\partial^2\phi}{\partial t^2}$$

using equation (3.3). Finally, if the time dependence is $\exp(-i\omega t)$,

$$p = -\rho\omega^2\phi . \tag{3.10}$$

Equation (3.9) may therefore be written

$$p_{res} = \rho\omega^2(\phi_{inc} + \phi_{refl}) = 0 , \tag{3.11}$$

where p_{res} is the resultant pressure at the surface.

If the point of intersection of the waves on the surface is taken as $x_1 = 0$, then combining equations (3.6), (3.7), and (3.11) gives

$$p_{res} = \rho\omega^2[\hat{A}\exp(ikx_3\sin\alpha) + \hat{R}\hat{A}\exp(ikx_3\sin\beta)] = 0 .$$

If this equation is to be zero for all values of x_3, then

$$\hat{A}\exp(ikx_3\sin\alpha) = -\hat{R}\hat{A}\exp(ikx_3\sin\beta) .$$

Hence

$$\sin\alpha = \sin\beta ;$$

that is

$$\alpha = \beta \tag{3.12}$$

Also

$$\hat{R} = -1 .$$

Equation (3.12) is a statement of the law of reflection.

An interpretation of the equation $\hat{R} = -1$ can be made as follows. Since $\hat{R}$ is a complex number, we may write $\hat{R} = |R|\,e^{i\delta}$, where $|R|$ is the magnitude of $\hat{R}$ and δ is an associated phase angle giving a measure of the difference in phase between the incident and reflected waves. In this case $|R| = 1$ and $\delta = \pi$ (since $e^{i\pi} = -1$). Thus, there is no loss of amplitude on reflection but there is a change in phase of π radians.

In the region near the boundary where both waves coexist the resultant field pattern is described by the total potential, ϕ:

$$\phi = \phi_{inc} + \phi_{refl} = \hat{A}[\exp(ikx_1\cos\alpha) - \exp(-ikx_1\cos\alpha)]\exp(ikx_3\sin\alpha)] . \tag{3.13}$$

Equation (3.13) may be interpreted as an interference or standing wave pattern in the x_1 direction due to the combination of incident and reflected plane travelling waves.

As may be deduced from figure 3.1, there is also a disturbance propagated along the boundary in the $+x_3$ direction with wave vector $k_3 = k\sin\alpha$. The phase velocity, c_3, of this boundary wave is

$$c_3 = \frac{\omega}{k_3} = \frac{\omega}{k\sin\alpha} = \frac{c_\varrho}{\sin\alpha} . \tag{3.14}$$

We see from this expression that c_3 is never less than c_ϱ and $\to\infty$ as $\alpha \to 0$. A similar boundary wave may be observed when sea waves strike a wall at an angle. In the latter case the wall is rigid and solution of the wave equation with appropriate boundary conditions shows that waves are reflected according to the law of reflection but with no phase change.

3.3 Fluid–fluid interface

Consider a similar situation to that of section 3.2 but with the boundary separating two different fluids. As well as the reflected wave there will now be a transmitted wave in the second fluid with angle of refraction β'. The wave equation is again equation (3.3) which must be solved subject to the boundary conditions:

$$u_{inc} + u_{refl} = u_{trans} , \tag{3.15}$$

$$p_{inc} + p_{refl} = p_{trans} . \tag{3.16}$$

Plane wave solutions will be assumed for the incident, reflected, and transmitted waves as follows:

$$\phi_{inc} = \hat{A}\exp[ik(x_1\cos\alpha + x_3\sin\alpha)] ,$$

$$\phi_{refl} = \hat{R}\hat{A}\exp[ik(-x_1\cos\beta + x_3\sin\beta)] , \tag{3.17}$$

$$\phi_{trans} = \hat{T}\hat{A}\exp[ik'(x_1\cos\beta' + x_3\sin\beta')] ,$$

where $\hat{T}$ is the **transmission coefficient** and $k' = \omega/c_\varrho'$ is the magnitude of the wave vector in the second fluid.

Combining equations (3.2) and (3.15) gives for the normal component of u (with the abbreviated notation introduced in section 1.4)

$$\partial_1\phi_{inc} + \partial_1\phi_{refl} = \partial_1\phi_{trans} , \tag{3.18}$$

and combining equations (3.16) and (3.10) gives the pressure equation

$$\rho\omega^2(\phi_{inc} + \phi_{refl}) = \rho'\omega^2\phi_{trans} . \tag{3.19}$$

Equations (3.17) may now be substituted into equation (3.18) giving

$$ik\cos\alpha\phi_{inc} - ik\cos\beta\phi_{refl} = ik'\cos\beta'\phi_{trans} .$$

That is,

$$k\cos\alpha\exp(ikx_3\sin\alpha) - k\cos\beta\hat{R}\exp(ikx_3\sin\beta) = k'\cos\beta'\hat{T}\exp(ik'x_3\sin\beta'), \tag{3.20}$$

where the point of intersection of rays on the surface has been taken as the origin so that $x_1 = 0$ at the surface.

If equations (3.17) are now substituted into equation (3.19) we obtain

$$\frac{\rho}{\rho'}[\exp(ikx_3\sin\alpha) + \hat{R}\exp(ikx_3\sin\beta)] = \hat{T}\exp(ik'x_3\sin\beta') . \tag{3.21}$$

Dividing equation (3.20) by $k'\cos\beta'$ and equating to equation (3.21) yields

$$\left(\frac{k\cos\alpha}{k'\cos\beta'} - \frac{\rho}{\rho'}\right)\exp(ikx_3\sin\alpha) = \hat{R}\left(\frac{k\cos\beta}{k'\cos\beta'} + \frac{\rho}{\rho'}\right)\exp(ikx_3\sin\beta) . \tag{3.22}$$

If equation (3.22) is to hold for all values of x_3, then

$$\sin\alpha = \sin\beta . \tag{3.23}$$

This result confirms the **law of reflection**.

A **law of refraction** may be found by substituting equation (3.23) in equation (3.21):

$$k\sin\alpha = k'\sin\beta' . \tag{3.24}$$

An expression for $\hat{R}$ may be found from equation (3.22):

$$\hat{R} = \frac{k\rho'\cos\alpha - k'\rho\cos\beta'}{k\rho'\cos\alpha + k'\rho\cos\beta'} = \frac{Z'-Z}{Z'+Z} \tag{3.25}$$

where $Z \equiv \rho c/\cos\alpha$ is defined as the **wave impedance**. At normal incidence, the impedance is ρc and is then called the characteristic impedance of the medium.

The transmission coefficient may now be found from equations (3.21) and (3.25):

$$\hat{T} = \frac{\rho}{\rho'}(1+\hat{R}) = \frac{\rho}{\rho'}\frac{2Z'}{Z'+Z} . \tag{3.26}$$

Equations (3.25) and (3.26) are identical to the corresponding equations for reflection and transmission of electromagnetic waves.

The results of section 3.2 may easily be confirmed since, when $Z' = 0$, $\hat{R} = -1$ and $\hat{T} = 0$.

3.3.1 Condition for no reflection

If $Z = Z'$, then equation (3.25) yields $\hat{R} = 0$. Thus, if the densities and wave velocities satisfy this condition, there will be a transmitted wave with no reflection. This phenomenon is analogous to the Brewster angle for electromagnetic waves.

An expression may be found for the corresponding angle of incidence as follows: in equation (3.25) put $m = \rho'/\rho$ and $n = c_\varrho/c_\varrho'$, then

$$\hat{R} = \frac{m \cos\alpha - (n^2 - \sin^2\alpha)^{\frac{1}{2}}}{m \cos\alpha + (n^2 - \sin^2\alpha)^{\frac{1}{2}}} \; .$$

Hence, when $Z = Z'$,

$$\sin^2\alpha = \frac{m^2 - n^2}{m^2 - 1} \; . \tag{3.27}$$

The condition for α to be a real angle is

$$0 < \frac{m^2 - n^2}{m^2 - 1} < 1 \; .$$

Hence, when $m > 1$, the condition for n is $1 < n < m$; and when $m < 1$, n must satisfy $1 > n > m$.

3.3.2 Grazing incidence

If $c_\varrho > c_\varrho'$ and $\alpha \to 90°$, then $\hat{R} = -1$. Thus, at grazing incidence there is complete reflection independent of the values of the two impedances.

3.3.3 Total internal reflection

From equation (3.24) we see that

$$\sin\beta' = \frac{k}{k'}\sin\alpha = \frac{c_\varrho'}{c_\varrho}\sin\alpha \; .$$

β' is therefore real only when $\sin\beta' \leqslant 1$. When $\sin\beta' = 1$, there is a **critical angle** of incidence, α_{cr}, given by

$$\sin\alpha_{cr} = \frac{c_\varrho}{c_\varrho'} = n$$

provided $c_\varrho < c_\varrho'$. Under these conditions, total internal reflection occurs when $\sin\alpha > n$.

For instance, if a longitudinal wave is incident from air ($\rho = 1 \cdot 3$ kg m^{-3}, $c_\varrho = 333$ m s^{-1}) onto water ($\rho' = 10^3$ kg m^{-3}, $c_\varrho' = 1500$ m s^{-1}), the above condition is satisfied and a critical angle occurs at $\alpha_{cr} = 13°42'$. Total internal reflection will occur for angles of incidence greater than this value.

As shown by Brekhovskikh (1960), total internal reflection does not occur at any angle of incidence if absorption in the second medium is taken into account. Under these circumstances a wave penetrates into the second medium as an inhomogeneous wave whose amplitude decreases exponentially with distance from the boundary separating the two media.

3.3.4 Energy relations

The energy flowing per second through unit area normal to the direction of propagation is called the **intensity** of the wave. For a periodic wave the intensity is usually evaluated as the mean value over one cycle. As shown in elementary texts, the mean intensity is related to the pressure amplitude, $|p|$, by the equation

$$I = \frac{|p|^2}{2\rho c} \;.$$

Equations (3.10) and (3.17) yield then the following values for the incident, reflected, and transmitted intensities:

$$I_{\text{inc}} = \frac{(\rho \omega^2 A)^2}{2\rho c} \;, \tag{3.28}$$

$$I_{\text{refl}} = \frac{(\rho \omega^2 RA)^2}{2\rho c} = R^2 I_{\text{inc}} \;, \tag{3.29}$$

$$I_{\text{trans}} = \frac{(\rho' \omega^2 TA)^2}{2\rho' c'} = \frac{\rho c}{\rho' c'}(1+R)^2 I_{\text{inc}} \;. \tag{3.30}$$

The **energy reflection coefficient** is

$$\frac{I_{\text{refl}}}{I_{\text{inc}}} = R^2 = \left(\frac{Z'-Z}{Z'+Z}\right)^2 \cdot \frac{\cos\alpha}{\cos\alpha} \tag{3.31}$$

The **energy transmission coefficient** is

$$\frac{I_{\text{trans}}}{I_{\text{inc}}} = \frac{\rho c}{\rho' c'}(1+R)^2 = \frac{\rho c}{\rho' c'}\left(\frac{2Z'}{Z'+Z}\right)^2 \cdot \frac{\cos\gamma}{\cos\alpha} \tag{3.32}$$

At normal incidence

$$\left.\frac{I_{\text{trans}}}{I_{\text{inc}}}\right|_{\alpha=0} = \frac{4ZZ'}{(Z'+Z)^2} \;. \tag{3.33}$$

Equation (3.33) shows that, at normal incidence, the same value of the transmission coefficient is obtained irrespective of the direction of the energy flow. This result does not necessarily hold for oblique incidence. It is assumed that the energy transmitted is always equal to the sum of the incident and reflected energies.

In terms of pressure, the transmitted pressure depends on the direction of the wave motion. For instance, for a sound wave passing from air into water, $R \approx 1$ and $p_{\text{water}} \approx 2 p_{\text{air}}$. When the wave is reversed and passes from water into air, $R \approx -1$ and $p_{\text{air}} \ll p_{\text{water}}$.

3.4 Isotropic solid–vacuum interface

Consider a plane boundary (of infinite extent) in the $x_2 x_3$ plane which separates an isotropic perfectly elastic solid from a vacuum. In practice, in comparison with a solid, air is almost indistinguishable from a vacuum.

A plane harmonic wave with wave propagation vector k in the $x_1 x_3$ plane is incident on the surface with an angle of incidence α, as shown in figure 3.2.

The displacement vector, u_i, will again be written as

$$u_i = \operatorname{grad} \phi + \operatorname{curl} \psi_i , \tag{3.34}$$

so that both longitudinal and transverse waves may be accommodated. In general, then, two wave equations must be satisfied

$$\nabla^2 \phi = \frac{1}{c_\varrho^2} \frac{\partial^2 \phi}{\partial t^2} , \tag{3.35}$$

$$\nabla^2 \psi_i = \frac{1}{c_t^2} \frac{\partial^2 \psi_i}{\partial t^2} . \tag{3.36}$$

Boundary conditions

The boundary conditions at the surface are (a) the normal and tangential components of the displacement should be continuous, and (b) the normal and tangential components of stress should be continuous.

In considering condition (b), since there can be no resultant stress in a vacuum, we have a free surface and equation (1.49) is applicable, i.e. $\sigma_{ik} n_k = 0$. Now n_k is a vector in the direction of the normal to the boundary plane, so that $n_1 = 1$; $n_2 = n_3 = 0$. Hence the only stress components that are possible in this case are σ_{11}, σ_{21}, and σ_{31}. The boundary conditions for stress therefore lead to

$$\sigma_{11} = \sigma_{21} = \sigma_{31} = 0 . \tag{3.37}$$

Equation (3.37) may be written out in full with the aid of Hooke's law:

$$\sigma_{ik} = \lambda \theta \delta_{ik} + 2G\epsilon_{ik} ,$$

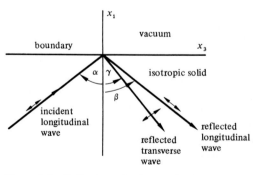

Figure 3.2. Reflection of a longitudinal wave at an isotropic solid–vacuum interface.

where $\epsilon_{ik} = \frac{1}{2}(\partial_k u_i + \partial_i u_k)$ and u_i is given by equation (3.34). Equation (3.37) then becomes

$$\sigma_{11} = \lambda\nabla^2\phi + 2G\left[\frac{\partial^2\phi}{\partial x_1^2} + \frac{\partial^2\psi_3}{\partial x_1\partial x_2} - \frac{\partial^2\psi_2}{\partial x_1\partial x_3}\right] = 0 ,\tag{3.38}$$

$$\sigma_{21} = \sigma_{12} = G\left[2\frac{\partial^2\phi}{\partial x_1\partial x_2} + \frac{\partial^2\psi_3}{\partial x_2^2} - \frac{\partial^2\psi_3}{\partial x_1^2} + \frac{\partial^2\psi_1}{\partial x_1\partial x_3} - \frac{\partial^2\psi_2}{\partial x_2\partial x_3}\right] = 0 ,\tag{3.39}$$

$$\sigma_{31} = \sigma_{13} = G\left[2\frac{\partial^2\phi}{\partial x_1\partial x_3} + \frac{\partial^2\psi_2}{\partial x_1^2} - \frac{\partial^2\psi_2}{\partial x_3^2} + \frac{\partial^2\psi_3}{\partial x_2\partial x_3} - \frac{\partial^2\psi_1}{\partial x_1\partial x_2}\right] = 0 .\tag{3.40}$$

Examination of equations (3.38) to (3.40) shows that, since each of them involves terms in both ϕ and ψ_i, the boundary conditions for stress cannot be satisfied in general by the reflection of only one type of wave. Even if the incident wave is purely longitudinal in character, there will be a reflected transverse wave as well as a reflected longitudinal wave. This phenomenon is referred to as **mode conversion** and is often observed experimentally.

3.4.1 Incident wave longitudinal
If the incident wave is longitudinal and after reflection there is both a longitudinal and transverse reflected wave, the required solutions of the wave equations (3.35) and (3.36) have the form

$$\phi_{inc} = \hat{A}\,\exp[ik_\varrho(x_1\cos\alpha + x_3\sin\alpha)] ,$$

$$\phi_{refl} = \hat{R}_l\hat{A}\,\exp[ik_\varrho(-x_1\cos\beta + x_3\sin\beta)] ,\tag{3.41}$$

$$\psi_{refl} = \hat{R}_t\hat{A}\,\exp[ik_t(-x_1\cos\gamma + x_3\sin\gamma)] ,$$

where β is the angle of reflection of the longitudinal wave, and γ is the angle of reflection of the transverse wave.

Since there is no motion postulated in the x_2 direction, the boundary conditions (3.38) to (3.40) reduce to

$$\sigma_{11} = \lambda\nabla^2\phi + 2G\left(\frac{\partial^2\phi}{\partial x_1^2} - \frac{\partial^2\psi_2}{\partial x_1\partial x_3}\right) = 0 ,$$

$$\sigma_{31} = G\left[2\frac{\partial^2\phi}{\partial x_1\partial x_3} + \frac{\partial^2\psi_2}{\partial x_1^2} - \frac{\partial^2\psi_2}{\partial x_3^2}\right] = 0 .\tag{3.42}$$

In addition, the condition that the normal component of the displacement be continuous yields

$$\partial_1\phi_{inc} + \partial_1\phi_{refl} - \partial_3\psi_{refl} = 0 .\tag{3.43}$$

Substituting equations (3.41) in equation (3.43) we obtain

$$k_\varrho\cos\alpha\exp(ik_\varrho x_3\sin\alpha) - \hat{R}_\varrho k_\varrho\cos\beta\exp(ik_\varrho x_3\sin\beta)$$

$$-\hat{R}_t k_t\sin\gamma\exp(ik_t x_3\sin\gamma) = 0 .\tag{3.44}$$

Equation (3.44) holds for all values of x_3 provided

$$k_\varrho \sin\alpha = k_\varrho \sin\beta = k_t \sin\gamma .$$ (3.45)

We see that, for the reflected longitudinal wave, the law of reflection holds, i.e. $\alpha = \beta$. For the reflected transverse wave

$$k_\varrho \sin\alpha = k_t \sin\gamma ;$$

that is

$$\frac{\sin\alpha}{\sin\gamma} = \frac{c_\varrho}{c_t} .$$ (3.46)

Since $c_\varrho/c_t > 1$, $\sin\gamma < \sin\alpha$.

Equation (3.44) shows that if the longitudinal wave is incident normally on the surface, $\alpha = \beta = \gamma = 0$ and there is no reflected transverse wave. Under these conditions, $\hat{R}_\varrho = -1$ and the reflected longitudinal wave is out of phase by π with the incident wave.

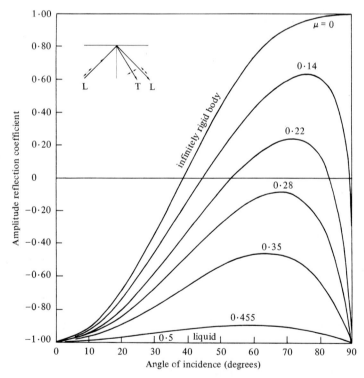

Figure 3.3. Curves for the amplitude reflection coefficients for longitudinal waves reflected from an isotropic solid-vacuum interface as a function of angle of incidence and Poisson's ratio, μ (after Arenberg, 1948).

General expressions for $\hat{R}_\varrho$ and $\hat{R}_t$ may be found by combining equations (3.41) with equations (3.42):

$$\hat{R}_\varrho = \frac{(c_t/c_\varrho)^2 \sin 2\beta \sin 2\gamma - \cos^2 2\gamma}{(c_t/c_\varrho)^2 \sin 2\beta \sin 2\gamma + \cos^2 2\gamma} \tag{3.47}$$

$$\hat{R}_t = -\frac{4(c_t/c_\varrho)\sin\gamma \cos\beta \cos 2\gamma}{(c_t/c_\varrho)^2 \sin 2\beta \sin 2\gamma + \cos^2 2\gamma} \cdot \tag{3.48}$$

Note that these are the reflection coefficients for the displacement potentials. In terms of displacement or particle velocity, equation (3.47) remains unaltered but equation (3.48) must be multiplied by c_ϱ/c_t.

In figure 3.3 are shown plots of the amplitude reflection coefficient for longitudinal waves as a function of angle of incidence and Poisson's ratio μ. The latter is a convenient parameter, since

$$\frac{c_\varrho}{c_t} = \left[\frac{2 - 2\mu}{1 - 2\mu}\right]^{\frac{1}{2}} \cdot \tag{3.49}$$

It may be observed that, for $\mu < 0 \cdot 26$, there are two angles of incidence for which $\hat{R}_\varrho = 0$. At these angles the incident longitudinal wave is completely converted into a reflected transverse wave.

3.4.2 Incident wave transverse
There are two possible cases depending on the plane of polarisation of the incident wave.

Case A. Particle displacement in the x_2 direction
The incident wave may be represented by

$$\psi_{\text{inc}} = \hat{A} \exp[ik_t(x_1 \cos\alpha + x_3 \sin\alpha)]$$

and the boundary conditions (3.38) to (3.40) reduce to

$$\sigma_{11} = \sigma_{13} = 0 ,$$

$$\sigma_{12} = G\left[\frac{\partial^2 \psi_1}{\partial x_1 \partial x_3} - \frac{\partial^2 \psi_3}{\partial x_1^2}\right] = 0 .$$

Thus, the boundary conditions may be satisfied by a reflected transverse wave only of the form

$$\psi_{\text{refl}} = \hat{R}_t \hat{A} \exp[ik_t(-x_1 \cos\gamma + x_3 \sin\gamma)] .$$

Combining the boundary conditions with the assumed wave solutions shows that

$$\alpha = \gamma , \qquad \hat{R}_t = 1 .$$

We conclude that (i) the angle of incidence is equal to the angle of reflection, (ii) there is no loss of amplitude on reflection, and (iii) there is no phase change on reflection.

Case B.   Transverse wave with particle displacement in the x_1x_3 plane
The boundary conditions for the stress in this case are equations (3.38) to
(3.40). Continuity of the normal displacement requires that

$$-\partial_3\psi_{\text{inc}} - \partial_3\psi_{\text{refl}} + \partial_1\phi_{\text{refl}} = 0 \ .$$

It follows that these boundary conditions can only be satisfied by the
appearance of two reflected waves, one longitudinal and one transverse.
Solutions of the wave equations for this case are then assumed to be

$$\psi_{\text{inc}} = \hat{A}\exp[ik_t(x_1\cos\alpha + x_3\sin\alpha)]\ ,$$

$$\phi_{\text{refl}} = \hat{R}_\varrho\hat{A}\exp[ik_\varrho(-x_1\cos\beta + x_3\sin\beta)]\ ,$$

$$\psi_{\text{refl}} = \hat{R}_t\hat{A}\exp[ik_t(-x_1\cos\gamma + x_3\sin\gamma)]\ .$$

If we now combine the boundary conditions with the assumed solutions
we find that
(i) for the reflected transverse wave the law of reflection holds, that is
$\alpha = \gamma$;
(ii) for the reflected longitudinal wave

$$\sin\beta = \frac{c_\varrho}{c_t}\sin\alpha\ ;$$

since $\sin\beta$ cannot exceed unity and $c_\varrho/c_t > 1$, a critical angle exists for
α such that $\sin\alpha_{\text{cr}} = c_t/c_\varrho$. Beyond this angle the transverse wave is
reflected without loss of amplitude; figure 3.4 illustrates the conditions
at the critical angle;
(iii) the reflection coefficients for the transverse and longitudinal waves are

$$\hat{R}_t = \frac{(c_t/c_\varrho)^2\sin 2\gamma\sin 2\beta - \cos^2 2\gamma}{(c_t/c_\varrho)^2\sin 2\gamma\sin 2\beta + \cos^2 2\gamma}\ , \qquad (3.50)$$

$$\hat{R}_\varrho = \frac{\sin 4\alpha}{(c_t/c_\varrho)^2\sin 2\gamma\sin 2\beta + \cos^2 2\gamma}\ ; \qquad (3.51)$$

graphs corresponding to equation (3.50) are shown in figure 3.5.

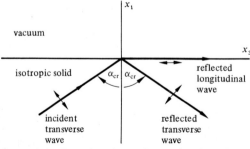

Figure 3.4. Critical angle conditions for the reflection of a transverse wave (polarised
in the plane of the diagram) at an isotropic solid–vacuum interface.

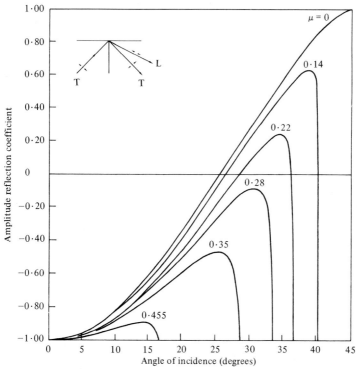

Figure 3.5. Curves for the amplitude reflection coefficients for transverse waves reflected from an isotropic solid-vacuum interface as a function of angle of incidence and Poisson's ratio, μ (after Arenberg, 1948).

Experiment E3.1 Mode conversion in a LiF crystal
(Joel, N., 1961, *Proc. Phys. Soc.*, **78**, 38.)

In this experiment an 8 MHz ultrasonic wave incident on a LiF-air boundary was converted almost completely into a transverse wave. The method used involved a Schaefer-Bergmann elastogram.

The LiF crystal was cut into the shape of a triangular prism, as shown in figure E3.1a, and an X-cut quartz plate was attached to the base so that a longitudinal wave was incident on the (010) and (001) faces at 45°. The complete elastogram for this orientation should show 8 spots; only 6 spots were observed. On an elastogram the distance of a spot from the centre is inversely proportional to the wavelength of the corresponding elastic wave, travelling in the same direction.

As shown in chapter 2, an elastogram is a section of a slowness surface. In figure E3.1b is shown the (100) section of the slowness surface of LiF. The two spots on the inner curve, above and below the central point O, correspond to the longitudinal waves incident on the two faces inclined at 45°. The reflected slowness vectors, and hence also the corresponding points on the elastogram, are determined by drawing through point I lines perpendicular to the axes X_2 and X_3. This is shown in figure

E3.1b for the waves reflected on the face parallel to X_2 giving rise to the intersection points L and T. Thus, OL is the direction of the reflected longitudinal slowness vector (and wave normal) and OT is the direction of the transverse vector. The inclined face parallel to X_3 gives rise to reflected waves symmetrical with the waves reflected from the face parallel to X_2, relative to the line OI. Also, for each elastic wave two spots appear on the elastogram, symmetrical relative to the central point O. Hence, the expected diffraction pattern consists of 8 spots as marked.

The experimentally observed elastogram is shown in figure E3.1c. It is noted that the two spots left and right of centre, corresponding to the reflected longitudinal waves, are absent. Only the spots corresponding to the incident waves and to the reflected transverse waves are present.

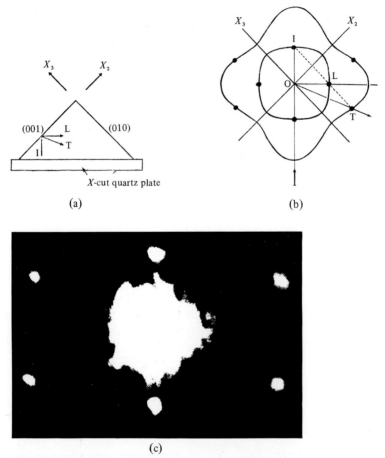

(a)   (b)

(c)

Figure E3.1. (a) Shape and orientation of the LiF crystal. (b) (100) section of the slowness surface of LiF; the dots correspond to the incident wave and to the expected reflected waves. (c) Experimental elastogram; the spots should correspond geometrically with the dots in (b); the central patch corresponds to the incident beam of light (5780 Å) used in producing the elastogram.

A theoretical computation of the corresponding energy fluxes shows that only 0·1% of the incident energy is to be expected as reflected longitudinal energy with 99·9% of the energy being carried by the reflected transverse wave. Further computation shows that complete mode conversion would occur for an angle of incidence of 47°.

Experiment E3.2 Ultrasonic goniometer
(Bradfield, G., 1968a, *Non-destructive Testing,* 1, 165; 1968b, 1, 370.)

Bradfield has described devices using the goniometer principle for the measurement of the elasticity of solid surfaces. In one form of the instrument, the specimen rests on a table immersed in a tank of water. A beam of ultrasonic waves is incident on the specimen and the reflected beam is picked up by a scanning transducer which is geared to rotate at twice the rate at which the specimen table rotates. The reflected beam is then received by the transducer whatever the angle of incidence. The system may be modified to use a common transmitting and receiving transducer. A movable reflector then replaces the scanning transducer, as shown in figure E3.2.

When ultrasonic waves are transmitted into the tank, the incident wave strikes the surface of the specimen and is split into a reflected and a refracted wave. The reflected wave travels away from the surface at an angle equal to the angle of incidence, α. The refracted wave enters the specimen at an angle of refraction β' given by equation (3.52), namely, $c'_\varrho \sin\alpha = c_\varrho \sin\beta'$.

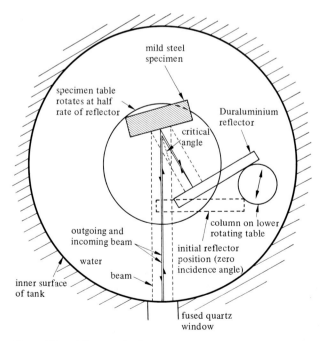

Figure E3.2. Water-tank ultrasonic goniometer with common transmitting/receiving transducer and movable reflector.

In general both longitudinal and transverse waves will be generated in the specimen, as shown in figure E3.3. When the angle β' becomes equal to $90°$, a surface wave is generated in the specimen and there is a sudden drop in intensity of the reflected wave. The ultrasonic goniometer can therefore be used to measure the velocities of longitudinal, shear, and surface waves as well as the variation of elasticity with depth below the surface. Angular measurements may be made with a precision of $\pm 1°$.

A similar type of instrument in which it is unnecessary to immerse the specimen in a water tank has been developed by the author. A Perspex semi-cylinder rests with its flat surface on the specimen with a thin film of oil between specimen and Perspex. The transducers (barium titanate) are mounted on two Perspex shoes of the same width as the semi-cylinder. In figure E3.4 the device is shown, with the cylinder divided for measurement of longitudinal wave velocity.

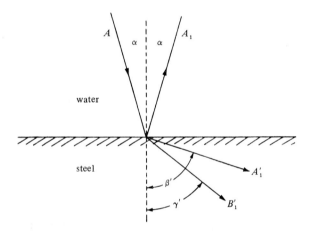

Figure E3.3. Beams at water/steel interface. A is amplitude of incident wave in water; A_1 is amplitude of reflected wave in water; A'_1 and B'_1, are amplitudes of refracted longitudinal and shear waves, respectively, in steel; α is angle of incidence and reflection in water; β' is angle of refraction of longitudinal wave in steel; γ' is angle of refraction of shear wave in steel.

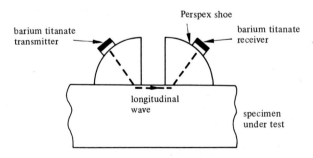

Figure E3.4. Perspex type of goniometer as used to measure longitudinal wave velocity.

Bradfield (1968b) used the water tank goniometer with reflector to measure the reflected wave amplitude as a function of angle of incidence for steel. Curve (a) in figure E3.5, shows the values expected from plane wave theory; curve (b) shows the experimentally determined results. The discrepancy occurs because of the finite distance between transmitting and receiving transducers. The experiment consequently operates in the Fresnel-Fraunhofer border region. The author has developed the necessary modified theory.

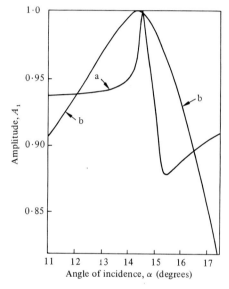

Figure E3.5. Amplitude of reflected wave in water plotted against angle of incidence in water: (a) predicted from plane wave theory; (b) observed after two reflections from the water/steel interface.

3.5 Isotropic solid–solid interface

When either a longitudinal or a transverse wave falls on the boundary between two isotropic solids, there will in general be both a longitudinal and a transverse reflected wave, together with a longitudinal and a transverse transmitted wave. In figure 3.6 is shown an incident longitudinal wave with associated reflected and transmitted waves. When the equations for the possible waves are combined with the appropriate boundary conditions, by a similar procedure to that used in the previous sections, the following condition is found

$$\frac{\sin\alpha}{c_\varrho} = \frac{\sin\beta}{c_\varrho} = \frac{\sin\gamma}{c_t} = \frac{\sin\beta'}{c_\varrho'} = \frac{\sin\gamma'}{c_t'} \ , \tag{3.52}$$

where the primes refer to the second medium, as before.

Reflection and transmission coefficients may be evaluated and are of the form

$$\hat{R} = \frac{Z_{\text{tot}} - Z}{Z_{\text{tot}} + Z} , \qquad \hat{T} = \frac{\rho}{\rho'}(1 + \hat{R}) . \qquad (3.53)$$

where Z is the incident wave impedance and Z_{tot} is the total impedance at the boundary due to the reflected and transmitted waves. Full expressions for the various wave forms involved are given by Redwood (1960). For a normally incident longitudinal wave there is no reflected or transmitted transverse wave.

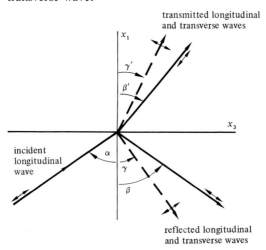

Figure 3.6. Longitudinal wave incident on an isotropic solid–solid interface with reflected and transmitted longitudinal and transverse waves.

3.6 Anisotropic solid–solid interface

When a plane elastic wave is incident on a plane boundary separating two anisotropic media, there will in general be three reflected waves and three transmitted waves. These waves may be represented as follows: for the displacement of the incident wave, we may write

$$u = \hat{A} \exp[i\omega(r \cdot s - t)] ,$$

where s is the slowness vector; the reflected waves may be written in the form

$$R_i \hat{A}_i \exp[i\omega(r \cdot s_i - t)] ,$$

where i $(= 1, 2, 3)$ represents an L, T_1, or T_2 wave; the transmitted waves then have the form

$$T_i \hat{A}'_i \exp[i\omega(r' \cdot s'_i - t)] .$$

Application of the boundary conditions for continuity of displacement and stress leads to a set of six equations and hence to the condition that the rate of change of phase for each wave along the boundary must be the same.

The various amplitudes may be found by the previous method leading again to six equations which are in general complex. The coefficients in the amplitude equations involve the elastic constants of the two media and the orientation of the waves with respect to the boundary.

For further details, including a discussion of critical angle conditions for each type of wave and an example involving two crystals of copper of different orientation, see Musgrave (1959).

3.7 Rayleigh surface waves

Rayleigh (1885) showed theoretically that it was possible for a two-dimensional wave to propagate on the free surface of a solid. The particle motion at the surface is elliptical, there being a longitudinal wave component parallel to the surface and in the direction of propagation, and a transverse component normal to the surface. Both the longitudinal and the transverse components attenuate rapidly with depth so that most of the acoustical energy is confined to a layer about one wavelength thick.

There are a number of other types of surface wave including Love waves (two-dimensional transverse waves involving a transverse component along the surface with particle displacement perpendicular to the direction of propagation), Stoneley waves (waves at the interface between two solids, the wave motion on each side of the interface being of the Rayleigh type), and Lamb waves which propagate in a plate.

3.7.1 Velocity of Rayleigh waves—isotropic solid

Consider a free infinite plane surface separating an isotropic solid from a vacuum. A plane monochromatic wave travelling along the surface in the x_1 direction must satisfy the wave equation for the displacement

$$\nabla^2 u_i = \frac{1}{c^2} \frac{\partial^2 u_i}{\partial t^2} . \tag{3.54}$$

The form of solution that represents a Rayleigh wave (with displacement vector in the $x_1 x_3$ plane) is

$$u_i = f(x_3) \exp[i(kx_1 - \omega t)] , \tag{3.55}$$

where $f(x_3)$ must include damping for x_3 negative (increasing depth). As shown in Landau and Lifshitz (1970), $f(x_3)$ has the form

$$f(x_3) = \mathscr{C} \exp(Kx_3) \tag{3.56}$$

where $\mathscr{C}$ is a constant and

$$K = \left(k^2 - \frac{\omega^2}{c^2} \right)^{1/2} .$$

Further, u_i may be written in terms of the longitudinal displacement $u_{\varrho,i}$ and the transverse displacement $u_{t,i}$.

Then the *transverse* part of the displacement must satisfy the condition

$\operatorname{div} u_{t,i} = 0$;

that is

$$\frac{\partial u_{t,1}}{\partial x_1} + \frac{\partial u_{t,3}}{\partial x_3} = 0 \ . \tag{3.57}$$

Equation (3.55) may be written for this displacement

$$u_{t,i} = \mathscr{C} \exp(K_t x_3) \exp[\mathrm{i}(kx_1 - \omega t)] \ , \tag{3.58}$$

where

$$K_t = \left(k^2 - \frac{\omega^2}{c_t^2}\right)^{\frac{1}{2}} \ .$$

On combining equations (3.57) and (3.58) we find

$$\frac{u_{t,1}}{u_{t,3}} = -\frac{K_t}{\mathrm{i}k} \ .$$

Hence

$$\begin{aligned} u_{t,1} &= K_t a \exp[K_t x_3 + \mathrm{i}(kx_1 - \omega t)] \ , \\ u_{t,3} &= -\mathrm{i}ka \exp[K_t x_3 + \mathrm{i}(kx_1 - \omega t)] \ , \end{aligned} \tag{3.59}$$

where a is a constant.

The *longitudinal* part of the displacement must satisfy the condition

$\operatorname{curl} u_{\varrho,i} = 0$;

that is

$$\frac{\partial u_{\varrho,1}}{\partial x_3} - \frac{\partial u_{\varrho,3}}{\partial x_1} = 0 \ . \tag{3.60}$$

On combining equations (3.57) and (3.60), we obtain

$$\mathrm{i}k u_{\varrho,3} - K_\varrho u_{\varrho,1} = 0 \ .$$

Hence

$$\begin{aligned} u_{\varrho,1} &= kb \exp[K_\varrho x_3 + \mathrm{i}(kx_1 - \omega t)] \ , \\ u_{\varrho,3} &= -\mathrm{i}K_\varrho b \exp[K_\varrho x_3 + \mathrm{i}(kx_1 - \omega t)] \ , \end{aligned} \tag{3.61}$$

where b is a constant and

$$K_\varrho = \left(k^2 - \frac{\omega^2}{c_\varrho^2}\right)^{\frac{1}{2}} \ .$$

The *boundary conditions* to be satisfied are equations (1.49) for a free surface: $\sigma_{ik} n_k = 0$. Since n_k is the unit vector in the direction of the normal to the surface and is parallel to the x_3 axis, then $\sigma_{13} = \sigma_{23} = \sigma_{33} = 0$. Hence, from Hooke's law, equation (1.36),

$$\epsilon_{13} = 0 ,$$
$$\epsilon_{23} = 0 , \tag{3.62}$$
$$\lambda(\epsilon_{11} + \epsilon_{22}) + (\lambda + 2G)\epsilon_{33} = 0 .$$

If we make use of the definition of strain, equation (1.9), the first and third of equations (3.62) become

$$\frac{\partial u_1}{\partial x_3} + \frac{\partial u_3}{\partial x_1} = 0$$

$$c_\ell^2 \frac{\partial u_3}{\partial x_3} + (c_\ell^2 - 2c_t^2) \frac{\partial u_1}{\partial x_1} = 0 \tag{3.63}$$

where

$$u_1 = u_{\ell,1} + u_{t,1} ; \qquad u_3 = u_{\ell,3} + u_{t,3} .$$

If we now combine all the above conditions we obtain the **Rayleigh wave velocity equation**

$$r^6 - 8r^4 + 8r^2(3 - 2s^2) - 16(1 - s^2) = 0 , \tag{3.64}$$

where

$$r = \frac{c_R}{c_t} ,$$

$$s = \frac{c_t}{c_\ell} = \left(\frac{1 - 2\mu}{2(1 - \mu)} \right)^{\frac{1}{2}} ,$$

c_R is Rayleigh wave velocity, and μ is Poisson's ratio. Since the range of possible values of μ is 0 to $0 \cdot 5$, the corresponding range of r is $0 \cdot 87$ to $0 \cdot 96$. As r does not depend on frequency, there is no dispersion of Rayleigh waves on an isotropic solid. In practice Rayleigh waves propagate with relatively little attenuation and, as shown above, have a velocity somewhat less than that of bulk transverse waves. They are a major component of the waves arising from earthquakes and account for the large disturbances recorded after the initial longitudinal and transverse waves have been received.

Computation shows that the amplitudes of the longitudinal and transverse components of a surface wave decrease exponentially with distance below the surface, as shown in figure 3.7. The longitudinal displacement decays to zero in a depth of $\sim 0 \cdot 2\lambda$, where λ is the wavelength. At this depth there is a plane in which there is no motion parallel to the surface. At greater depths the displacement once again becomes finite but is of opposite phase. For the transverse component, the displacement first of all increases with depth, reaching a maximum at a depth of $0·08\lambda$ and then decreases monotonically. At a depth of one wavelength the displacement amplitude has fallen to $0·2$ of its value at the surface.

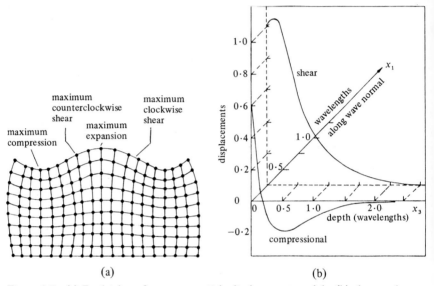

(a)   (b)

Figure 3.7. (a) Rayleigh surface wave particle displacement model; (b) shear and compressional displacements in a Rayleigh wave as a function of depth and distance along wave normal (after de Klerk, 1971).

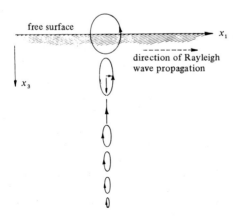

Figure 3.8. Combined particle displacements for a Rayleigh surface wave. The sense of the rotation reverses below a depth of approximately $0·2\lambda$ when the longitudinal component reverses phase (after Sabine and Cole, 1971).

The particle trajectories at the surface are ellipses with the major axis normal to the surface. The displacement in the x_1 direction is retarded by $\frac{1}{2}\pi$ compared with the displacement in the x_3 direction. It is found that at high frequencies the propagation of surface waves becomes frequency dependent. The attenuation with depth is then greater than at low frequencies. This effect may be compared with the skin effect which occurs with high-frequency electromagnetic waves. In figure 3.8 are shown combined particle displacements as a function of depth.

3.7.2 Velocity of Rayleigh waves—anisotropic media

On the surface of an anisotropic solid two different transverse modes are possible, in addition to the longitudinal mode. Thus, a more general surface wave is observed whose displacements are three-dimensional. Such a wave may be regarded as a combination of Rayleigh and Love types. A pure Rayleigh wave (or a pure Love wave) can then only be observed when the crystal directions and surface orientation are such that both component waves propagate as pure modes. Stoneley (1955) investigated these conditions in a cubic crystal and showed that, for Rayleigh and Love waves to propagate over a {100} plane, propagation must take place along the [100] or [110] axes. For instance, the velocity equation for a Rayleigh wave propagating over the (001) plane along the [100] direction is

$$c_{11}(\rho c^2 - c_{44})\left[\rho c^2 - \left\{c_{11} - \frac{c_{12}^2}{c_{11}}\right\}\right]^2 = c_{44}(\rho c^2 - c_{11})(\rho c^2)^2 \ . \tag{3.65}$$

This equation reduces to the equation for an isotropic solid if $c_{12} = c_{11} - 2c_{44}$, the condition discussed in section 1.8.

Further details concerning surface waves on cubic crystals are given by Farnell (1970).

Experiment E3.3   An instrument for making surface waves visible

(Adler, R., Korpel, A., Desmares, P., 1968, *IEEE Trans. Sonics and Ultrasonics,* **SU-15,** 157.)

Here an instrument is described which renders Rayleigh waves visible on a television screen. In it, a well-focused unmodulated monochromatic light beam is directed at the substrate and the lateral motion of the reflected beam is picked up by a photocell equipped with a sharply defined aperture. The photocell output then includes an AC component due to the sound wave. The method gives a high signal-to-noise ratio since the acoustic part of the output is filtered, amplified and detected, all non-acoustic signals being discarded. The general optical arrangement and an improved arrangement, involving a correcting lens which allows for the scanning of a larger area, are shown in figures E3.6 and E3.7. A laser flying-spot scanner is used as light source which involves a He–Ne laser whose beam is swept horizontally and vertically at standard television rates while being focused to an essentially diffraction-limited spot. The light flux into the illuminated rectangle is between 10 and 20 mW. Substrates of steel or ceramic were used and were polished to a mirror-like finish. On ceramic surfaces a thin layer of gold was deposited to enhance the reflectivity.

During the scanning process the light beam sweeps across the substrate at high speed. This produces a Doppler shift in the signal frequency appearing at the photocell. The horizontal scanning velocity corresponds to about 240 m s^{-1} or 8% of the Rayleigh wave velocity on steel surfaces (3000 m s^{-1}). Rayleigh waves were produced on the surface by applying an 8 MHz signal to a wedge-type transducer attached to the surface. The corresponding Doppler shift is +640 kHz if the sound wave travels against the direction of the horizontal scan, and −640 kHz if it travels in the same direction as the scan. The Doppler shift can assume any value between these two extremes for a wave travelling at an arbitrary angle. In figure E3.8 is shown the

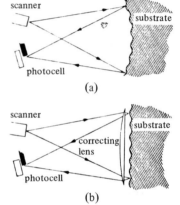

(a)

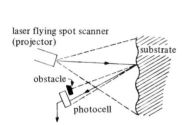

(b)

Figure E3.6. Arrangement for optical scanning of a pattern of surface waves.

Figure E3.7. Improved arrangements to allow scanning of larger areas: (a) spherical substrate; (b) correcting lens and flat substrate.

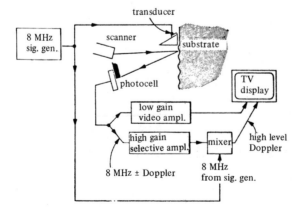

Figure E3.8. Circuit used in experiments. Lower branch selects, amplifies, and processes surface wave signals. Upper branch transmits conventional image of substrate.

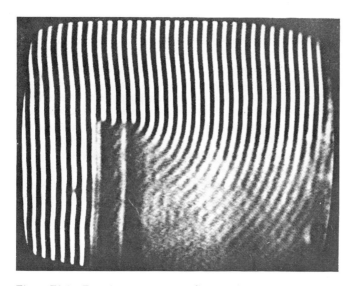

Figure E3.9. Two deep grooves in a flat steel plate obstruct the propagation of the straight parallel wave fronts. Surface waves are diffracted around the corner of the second groove, forming concentric circles.

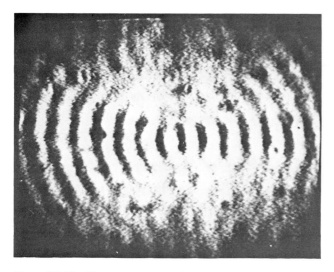

Figure E3.10. Plane wave coming from the right is reflected from curved edge (beyond left boundary of picture) of a round steel disk. Reflected wave, which goes through focus, is enhanced by choice of Doppler frequency while incident plane wave is rejected. This picture was taken with 7 mm horizontal scan.

circuit used in the experiments. After filtering and amplifying, the acoustic signal is processed so as to provide on the television screen a stationary pattern corresponding to a stroboscopic view of the travelling surface waves. This function is carried out by heterodyning the Doppler-shifted 8 MHz signal with the original unshifted signal taken directly from the sound signal generator. Thus the system becomes a synchronous detection process in which the acoustic phase at each spot is compared with the reference phase from the signal generator. As the scanning beam sweeps over a sequence of spots, the output from the synchronous detector becomes positive or negative according to the local acoustic phase at each spot. Positive and negative outputs are then displayed as black and white on the television screen. The upper branch shown in the circuit permits a display of the conventional video signals produced by grooves, edges, and other normally visible features of the substrate. This serves as background on which the surface wave patterns are superimposed. The lowest intensity surface waves which can be made visible with this arrangement correspond to amplitudes of the order of $0 \cdot 1$ Å. Among the illustrations given in the paper is the diffraction of surface waves around an edge (figure E3.9) and the effect of reflection from a curved surface (figure E3.10).

References
Arenberg, D. L., 1948, *J. Acoust. Soc. Am.,* **20**, 1.
Brekhovskikh, L. M., 1960, *Waves in Layered Media* (Academic Press, New York).
de Klerk, J., 1971, *Ultrasonics,* **9**, 35.
Farnell, G. W., 1970, in *Physical Acoustics,* Eds. W. P. Mason, R. N. Thurston, volume 6 (Academic Press, New York), p.109.
Landau, L. D., Lifshitz, E. M., 1970, *Theory of Elasticity,* 2nd edition (Pergamon Press, Oxford).
Mason, W. P., 1958, *Physical Acoustics and the Properties of Solids* (Van Nostrand, New York).
Musgrave, M. J. P., 1959, *Rep. Prog. Phys.,* **22**, 74.
Rayleigh, Lord, 1885, *Proc. London Math. Soc.,* **17**, 4.
Redwood, M., 1960, *Mechanical Waveguides* (Pergamon Press, Oxford).
Sabine, H., Cole, P. H., 1971, *Ultrasonics,* **9**, 103.
Stoneley, R., 1955, *Proc. R. Soc. A,* **232**, 447.

REVIEW QUESTIONS

3.1 Outline a method for finding the reflection coefficient for plane waves incident obliquely on a fluid–fluid interface. (§3.3)

3.2 Under what conditions can total internal reflection occur when plane waves are incident on a fluid–fluid interface? (§3.3.3)

3.3 Write down the set of plane wave equations and boundary conditions required to investigate the reflection of a longitudinal wave from an isotropic solid–vacuum interface. How do you explain the appearance of a reflected transverse wave? (§3.4)

3.4 Discuss the wave types that are observed at angles (a) less than, (b) greater than, the critical angle when a transverse wave (particle displacements in the $x_1 x_3$ plane) is reflected from an isotropic solid–vacuum interface. (§3.4.2)

3.5 Draw a diagram to illustrate the possible wave types that arise when a longitudinal wave is incident on an isotropic solid–solid interface. (Figure 3.6)

3.6 What are the assumptions made in deriving the Rayleigh wave velocity equation? (§3.7)

3.7 Discuss the displacement functions involved when a Rayleigh wave travels along the surface of an isotropic solid. (§3.7.1)

PROBLEMS

3.8 Obtain equation (3.13) directly by solving the wave equation by the method of separation of variables and taking into account the boundary conditions.

3.9 Show that the definition of the wave impedance, $Z = \rho c / \cos\alpha$, is in agreement with the usual definition in terms of the ratio of the acoustic pressure to the particle velocity for a plane pressure wave.

3.10 Verify equation (3.27).

3.11 Verify equations (3.28) to (3.30).

3.12 Find the condition under which an incident longitudinal wave will be reflected from an isotropic solid–vacuum interface as a transverse wave only.

4

Acoustic waveguides

4.1 Introduction
In wave propagation experiments on materials the specimen is usually in the form of a bar, cylinder, or plate. One object of such experiments is to determine the relevant material parameters; hence any factors dependent on shape must be eliminated. Before discussing experimental methods of measurement it is therefore necessary to examine the effects of specimen geometry on the wave propagation.

Travelling waves and standing waves in bars and plates are subject to guided wave conditions rather than free-field conditions. Longitudinal waves are possible under some circumstances; other wave types include torsional, flexural, as well as surface. Higher-order modes become possible which involve zigzag propagation paths. When only one propagation mode is in operation, transmission line theory may simplify the analysis.

While the main interest in this text is in solid waveguides, the simplest system to study initially is the fluid waveguide. This system does have some practical importance and allows all the main features of more complex systems to be studied without excessive mathematical complication.

4.2 Fluid plate with free boundaries
A fluid plate with free boundaries appears to be rather an academic concept, but in practice this condition can be approximated by means of fluid enclosed within pressure-release boundaries, such as can be provided by sheets of special rubber. We shall consider a waveguide formed by a plate of ideal fluid of thickness $2a$ in the x_1 direction and infinite in the x_2 and x_3 directions. The possible modes of propagation may be found either by considering the reflections of waves at the boundaries or by finding directly the solutions of the wave equation subject to the boundary conditions.

4.2.1 Wave reflection method
Consider a plane dilatational wave incident on the upper boundary of the waveguide at an angle α to the normal, as shown in figure 4.1. There will be a reflected dilatational wave at an angle α with no motion in the x_2 direction. The reflected wave will then travel obliquely across the plate before being reflected from the lower boundary. A standing wave pattern will therefore be set up in the x_1 direction (the expression for this pattern is found by adding the equations for the incident and reflected waves) together with a boundary effect which propagates in the x_3 direction. At the boundaries the resultant pressure is zero, and a combination of this condition with the standing-wave expression yields the allowed values of the wave vector k for the different modes of propagation. Details of this method are given in Redwood (1960).

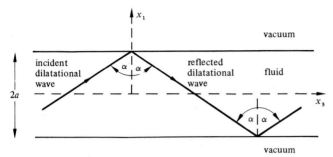

Figure 4.1. Mode propagation in a fluid plate with free boundaries.

4.2.2 Direct method
Solutions are required for the wave equation

$$\nabla^2 \phi = \frac{1}{c_\ell^2}\frac{\partial^2 \phi}{\partial t^2} \tag{4.1}$$

subject to the boundary condition

$$p_{res}|_{x_1 = \pm a} = 0 , \tag{4.2}$$

where ϕ is the displacement potential and p_{res} is the resultant pressure at the boundary.

Plane wave solutions may be found by the method of separation of variables. Thus, put

$$\phi = f_1(x_1)f_3(x_3)g(t) . \tag{4.3}$$

Substitution in the wave equation yields the three ordinary differential equations

$$\frac{d^2 f_1}{dx_1^2} + k_1^2 f_1 = 0 , \tag{4.4}$$

$$\frac{d^2 f_3}{dx_3^2} + k_3^2 f_3 = 0 , \tag{4.5}$$

$$\frac{d^2 g}{dt^2} + \omega^2 g = 0 . \tag{4.6}$$

It is convenient to write the solution of equation (4.4) in terms of either a sine or cosine function. The solution of equation (4.5) must represent a travelling wave in the $+x_3$ direction and will be written in terms of $\exp(ik_3 x_3)$. Equation (4.6) yields the usual harmonic time variation in the form $\exp(-i\omega t)$. Hence equation (4.3) becomes

$$\phi = A\begin{Bmatrix}\sin\\\cos\end{Bmatrix}k_1 x_1 \exp[i(k_3 x_3 - \omega t)] . \tag{4.7}$$

D

The boundary condition requires that

$$p_{res}|_{x_1 = \pm a} = \left(-\rho \frac{\partial^2 \phi}{\partial t^2}\right)_{x_1 = \pm a} = \rho \omega^2 \phi|_{x_1 = \pm a} = 0 \ .$$

If we consider first the sine terms of equation (4.7), we find that the boundary condition is satisfied if

$$\sin k_1 a = 0 \ ,$$

that is, if

$$k_1 a = \tfrac{1}{2} m \pi \ , \qquad m = 2, 4, 6, \dots \ . \tag{4.8}$$

If we now take the cosine terms of equation (4.7), we find that the boundary condition is satisfied if

$$\cos k_1 a = 0 \ ,$$

that is, if

$$k_1 a = \tfrac{1}{2} m \pi \ , \qquad m = 1, 3, 5, \dots \ , \tag{4.9}$$

m determines the mode of propagation. Condition (4.8) governs the **asymmetric modes**; condition (4.9) governs the **symmetric modes**. These names derive from the symmetry or otherwise of ϕ and p about the x_3 axis. For $m = 0$, $\phi = 0$ everywhere and therefore no wave can propagate in this mode.

The solutions (4.7) may now be written in full:

asymmetric modes, m even

$$\phi = A \sin\frac{m\pi}{2a} x_1 \exp[i(k_{3,m} x_3 - \omega t)] \ , \tag{4.10}$$

symmetric modes, m odd

$$\phi = A \cos\frac{m\pi}{2a} x_1 \exp[i(k_{3,m} x_3 - \omega t)] \ . \tag{4.11}$$

A subscript has been added to k_3, since k_3 will now depend on m. This follows since $k_1^2 + k_2^2 + k_3^2 = k^2 = (\omega/c_\ell)^2$ and hence

$$k_3^2 = \left(\frac{\omega}{c_\ell}\right)^2 - k_1^2 = \left(\frac{\omega}{c_\ell}\right)^2 - \left(\frac{m\pi}{2a}\right)^2 \ . \tag{4.12}$$

$k_2 = 0$ since there is no propagation in the x_2 direction. In figure 4.2 are shown the pressure distributions ($p = \rho \omega^2 \phi$) for the first two symmetric and the first two asymmetric modes. There is no plane-wave mode corresponding to $m = 0$.

More generally, the modes are designated as (m, n), where m determines the variation in the x_1 direction and n the variation in the x_2 direction. For the fluid plate with free boundaries, only $(m, 0)$ modes are possible. In figure 4.2 it may be observed that there are m loops in the x_1 direction.

If the waveguide is excited by a source having a pressure distribution identical with one of the modes, then only that mode will be excited. If the pressure distribution of the source differs from that of any one of the modes, then several modes may be excited simultaneously. When a source produces a pressure pulse of short duration, then all modes will be excited for which a Fourier component of the pulse exists.

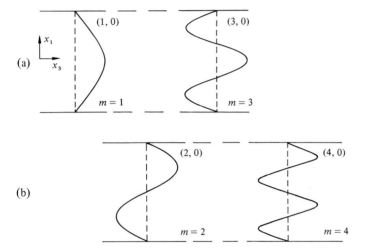

Figure 4.2. Pressure distributions for (a) the first two symmetric modes, (b) the first two asymmetric modes, for a fluid plate with free boundaries.

4.2.3 Phase velocity

The phase velocity, c_p, of the waves in the guide is the velocity with which a point of constant phase moves along the boundary. The dependence of c_p on the mode of propagation may be found with the use of equation (4.12):

$$c_p \equiv c_3 = \frac{\omega}{k_3} = \frac{c_\varrho}{[1 - (k_1 c_\varrho / \omega)^2]^{1/2}} = \frac{c_\varrho}{[1 - (m\pi/2a)^2 (c_\varrho / \omega)^2]^{1/2}} . \qquad (4.13)$$

It may be noted that
(i) c_p is always greater than c_ϱ.
(ii) $c_p \to \infty$ as $k_1 c_\varrho / \omega \to 1$.
  When $k_1 c_\varrho / \omega = 1$, $c_p = \infty$ and the particular mode is said to be 'cut-off'; the **cut-off frequency**, ν_0, is then given by

$$\nu_0 = \frac{k_1 c_\varrho}{2\pi} = \frac{m c_\varrho}{4a} . \qquad (4.14)$$

Reference to figure 4.1 makes it clear that

$$c_3 = \frac{\omega}{k_3} = \frac{\omega}{k \sin \alpha} = \frac{c_\varrho}{\sin \alpha} .$$

At cut-off, when $c_3 \to \infty$, $\sin\alpha \to 0$. The angle of incidence of the constituent plane waves on the boundary planes is therefore zero at cut-off. (iii) c_p is a function of frequency, that is there is dispersion.

4.2.4 Group velocity

Consider a wave packet or pulse containing more than one component monochromatic wave. Then, since c_p is a function of frequency, each component wave will travel with its own phase velocity. The envelope of the pulse will change shape as the pulse progresses along the waveguide. Sommerfeld (1950) shows that the energy carried by a pulse travels with the group velocity and not the phase velocity.

The **phase velocity** of a monochromatic wave is defined as

$$c_p = \frac{\omega}{k} \ .$$

The **group velocity**, c_g, of a wave packet is defined as

$$c_g = \frac{d\omega}{dk} \ . \tag{4.15}$$

In the absence of dispersion c_p is independent of k; therefore, because $\omega = c_p k$, we can write $d\omega = c_p dk$, whence it follows that $c_g = c_p$.

In general, c_p is a function of k, and therefore

$$d\omega = c_p dk + k dc_p \ .$$

Thus

$$c_g \equiv \frac{d\omega}{dk} = c_p + k \frac{dc_p}{dk} = c_p - \lambda \frac{dc_p}{d\lambda} \ . \tag{4.16}$$

Hence, for the fluid plate with free boundaries, it follows from equations (4.16) and (4.13) that

$$c_g = c_\varrho \left[1 - \left(\frac{k_1 c_\varrho}{\omega} \right)^2 \right]^{\frac{1}{2}} \ . \tag{4.17}$$

It may be noted that
(i) $c_g c_p = c_\varrho^2$.
(ii) $c_g < c_\varrho$ unless $\nu = \infty$.
(iii) At the cut-off frequency, $c_g = 0$ and no energy is propagated.
The phase and group velocity dispersion curves for $m = 1$ and $m = 3$ are shown in figure 4.3.

For a given mode to be excited, the pressure pulse must not only have the right spatial component of pressure, but the frequency must correspond to finite values of the phase velocity for that mode. At high frequencies many modes are possible, but as the frequency is reduced the higher modes are cut off, that is c_p goes to infinity for a particular mode and no energy is propagated. Eventually, at low enough frequencies, all modes are cut off in the fluid plate with free boundaries.

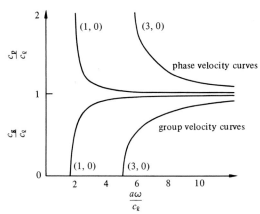

Figure 4.3. Phase and group velocity curves for the first two symmetric modes for a fluid plate with free boundaries.

4.3 Fluid plate with rigid boundaries

As before, solutions are sought for the wave equation

$$\nabla^2 \phi = \frac{1}{c_\ell^2} \frac{\partial^2 \phi}{\partial t^2} \ ,$$

but for rigid boundaries, subject to the boundary condition that the particle velocity normal to the boundary is zero, we have

$$\dot{u}_1 \bigg|_{x_1 = \pm a} = \frac{\partial^2 \phi}{\partial x_1 \partial t} \bigg|_{x_1 = \pm a} = 0 \ . \tag{4.18}$$

Assume solutions of the form

$$\phi = A \begin{Bmatrix} \sin \\ \cos \end{Bmatrix} k_1 x_1 \exp[i(k_3 x_3 - \omega t)] \ .$$

Then the boundary condition becomes

$$\dot{u}_1 = i\omega k_1 A \begin{Bmatrix} \cos \\ -\sin \end{Bmatrix} k_1 x_1 \exp[i(k_3 x_3 - \omega t)] = 0 \ .$$

If we consider first the sine part of the solution, we find that the boundary condition is satisfied if $\cos k_1 a = 0$, that is, if

$$k_1 a = \tfrac{1}{2} m\pi \ , \qquad m = 1, 3, 5, \dots \ .$$

If we now consider the cosine part of the solution, we find the boundary condition is satisfied if $\sin k_1 a = 0$, that is, if

$$k_1 a = \tfrac{1}{2} m\pi \ , \qquad m = 0, 2, 4, \dots \ .$$

The solutions may thus be written as follows:

asymmetric modes, m odd

$$\phi = A \sin\frac{m\pi}{2a} x_1 \exp[i(k_{3,m} x_3 - \omega t)] \ ; \tag{4.19}$$

symmetric modes, m even

$$\phi = A \cos\frac{m\pi}{2a} x_1 \exp[i(k_{3,m}x_3 - \omega t)] . \tag{4.20}$$

It can be seen that a solution exists for $m = 0$, namely

$$\phi = A \exp[i(k_3 x_3 - \omega t)] .$$

This represents a plane wave for which the phase velocity, from equation (4.13), is

$$c_p = \frac{\omega}{k_3} = c_\varrho .$$

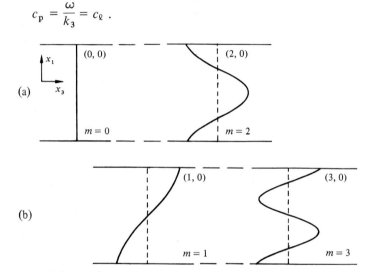

Figure 4.4. Pressure distributions for (a) the first two symmetric modes, (b) the first two asymmetric modes, for a fluid plate with rigid boundaries.

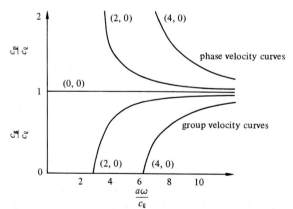

Figure 4.5. Phase and group velocity curves for the first three symmetric modes for a fluid plate with rigid boundaries.

The plane-wave mode, designated (0, 0), has no cut-off since the phase velocity does not change with frequency. A given wave pulse will therefore propagate without change of shape. In figure 4.4 are shown the pressure distributions for the first two symmetric modes and the first two asymmetric modes of a fluid plate with rigid boundaries. Phase and group velocity curves for three symmetric modes are plotted in figure 4.5. It may be observed that at sufficiently low frequencies only the plane wave mode can propagate.

Experiment E4.1   Propagation in a fluid waveguide
(Proud, J. M., Tamarkin, P., Kornhauser, E. T., 1956, *J. Acoust. Soc. Am.*, 28, 80.)

This paper deals with a scaled-down laboratory experiment, the purpose of which is to investigate the propagation characteristics of simple sound pulses in a medium of accurately known and regulated dispersive properties. The dispersive medium is provided by the water in a 3·6 m long channel of rectangular cross-section with pressure-release walls that serves as a waveguide. The sides and bottom of the channel are lined with 1·2 cm thick Celltite rubber which approximates an air-water interface for the water. The top of the guide is left open thus permitting adjustment of the cut-off frequency in the guide by varying the water depth. The available inside width of the guide is 10·2 cm. The source is a barium titanate plate, 5·0 cm on a side, located so as to provide optimal coupling for the particular mode under investigation and driven by a pulsed oscillator. The detector is a small barium titanate probe that can be accurately positioned in any part of the waveguide by a carriage with fine and coarse screw adjustments. The pulse is in the form of a rectangular envelope containing a carrier frequency which is a shape that allows ready comparison between experiment and theory. Sound pulses which contain many cycles of a carrier frequency provide an approximation to a simple wave group since the Fourier spectrum is then sharply peaked about the carrier frequency.

From a theory for the transient motion of sound in tubes developed by Pearson (1953), the phase velocity in a rectangular waveguide with free boundaries is given by

$$c_\mathrm{p} = c_\varrho \left[1 - \left(\frac{\pi c_\varrho}{\omega} \right)^2 \left(\frac{m^2}{a^2} + \frac{n^2}{b^2} \right) \right]^{-\frac{1}{2}}, \qquad (E4.1)$$

where c_p is the phase velocity; c_ϱ is the velocity of a longitudinal wave in the unbounded medium; m, n are integers denoting the (m, n) mode in the guide; a and b are the lengths of the sides of the guide. It may be noted that this equation is analogous to equation (4.13) for a fluid plate with free boundaries. Experimental values are compared in figure E4.1 with the theoretical curves for two different modes. Standing waves were set up in making these measurements.

The second part of the paper deals with the measurement of pulse shapes at different distances along the waveguide. Again, comparison of the measured shapes with those computed from Pearson's theory shows good agreement. Fluctuations in the pulse envelope are observed since the waveguide is dispersive. The initial transient portion of the pulse is observed to beat with the main signal. This effect decreases with time in amplitude and period. A representative pulse shape is shown in figure E4.2.

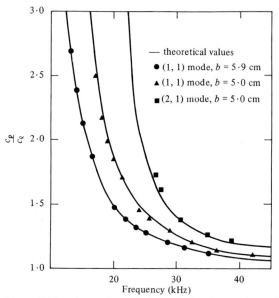

Figure E4.1. Comparison of theoretical and experimental values of the phase velocity in a rectangular waveguide with free boundaries; $a = 10 \cdot 2$ cm.

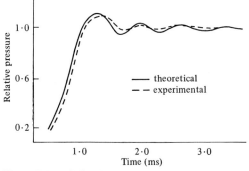

Figure E4.2. Pulse shape for a given group velocity ($0 \cdot 48\ c_\varrho$) at a distance of $1 \cdot 0$ m from the source. Frequency of measurement is 17 kHz. The origin corresponds to the arrival time of the initial transient which is found to travel at velocity c_ϱ.

4.4. Solid cylinder

Attention will now be given to solid waveguides, the simplest geometry being that of the cylinder. There is a long history of experimental and theoretical investigations into the behaviour of elastic waves in cylinders. Because of the difficulty in satisfying the boundary conditions, it is not possible to obtain exact solutions of the wave equation for finite cylinders (and plates). A number of satisfactory approximation methods have been devised, some of which have been discussed by Green (1960).

In the case of elastic waves in an infinitely long isotropic elastic cylinder exact solutions of the wave equation have been obtained independently by Pochhammer (1876) and Chree (1889). Three types of wave motion are possible in a cylinder: longitudinal, torsional, and flexural. For each wave type a series of waveguide modes is obtained for which the phase velocity is a function of the parameter a/λ, where a is the radius of the cylinder and λ is the wavelength. Before discussing the Pochhammer–Chree theory let us briefly consider earlier theories.

In the classical one-dimensional theory a plane longitudinal wave is considered to be travelling along an infinitely long cylinder of an isotropic solid with free lateral boundaries. The cylinder is considered to be very thin so that lateral motion may be excluded. If a simple extension or compression takes place along the axis of the cylinder, which will be taken in the x_1 direction, then application of the boundary condition (1.49) at the free surfaces confirms that the only non-zero stress component is σ_{11}. Then, from the definitions of Young's modulus and strain, equations (1.39) and (1.9), we find

$$\sigma_{11} = E\epsilon_{11} = E\frac{\partial u_1}{\partial x_1} \, ,$$

where E is the adiabatic Young's modulus. Substituting in the general equations of motion (1.15) gives

$$\rho\ddot{u}_1 = \frac{\partial \sigma_{11}}{\partial x_1} = E\frac{\partial^2 u_1}{\partial x_1^2}$$

or,

$$\frac{\partial^2 u_1}{\partial x_1^2} = \frac{1}{c_L^2}\frac{\partial^2 u_1}{\partial t^2} \, , \tag{4.21}$$

where

$$c_L = \left(\frac{E}{\rho}\right)^{\frac{1}{2}} \tag{4.22}$$

is the velocity of propagation of longitudinal waves in the cylindrical rod. This velocity is less than the corresponding velocity c_ϱ for a solid infinite in all directions.

The simple treatment presented above ignores any lateral motion of the rod due to the Poisson effect. Rayleigh (1894) developed a theory taking into account 'lateral inertia' and found for the phase velocity

$$c_L' = \frac{\omega}{k_1} = c_L\left[1 - \mu^2\pi^2\left(\frac{a}{\lambda}\right)^2\right] \tag{4.23}$$

where c_L' is the measured phase velocity, μ is Poisson's ratio and a is the radius of the rod. The measured phase velocity is now a function of the parameter a/λ.

Experiment E4.2 Dynamic elastic moduli of cylinders
(Pollard, H. F., 1964, *Aust. J. Phys.*, 17, 8.)

The dynamic Young's modulus may be determined in principle by measuring the frequency, ν_1, of the fundamental longitudinal mode of a long cylinder. According to classical theory,

$$E = \rho c_L^2 = \rho(\lambda_1 \nu_1)^2 \; ; \tag{E4.2}$$

ρ and λ_1 are found from the specimen dimensions. However, equation (E4.2) ignores lateral motion. The measured phase velocity, c_L', of longitudinal waves in a rod of radius a is found to depend on the ratio a/λ. A typical dispersion curve for a brass rod is shown in figure E4.3; the measured fundamental frequency was 1994 Hz, the highest harmonic included being the 60th for which $\nu_{60} = 113370$ Hz. Measurements were made using a mechanical impulse method described in this paper (see also Experiment E5.1). In order to find the theoretical phase velocity, c_L, for an infinitely thin rod, and hence to determine Young's modulus, it is first necessary to measure Poisson's ratio, μ.

Poisson's ratio
From Rayleigh's relationship, equation (4.23), the theoretical phase velocity is

$$c_L = c_L' \left[1 + \mu^2 \pi^2 \left(\frac{a}{\lambda} \right)^2 \right] . \tag{E4.3}$$

Since the wavelength for a given mode of vibration is fixed by the length of the specimen, equation (E4.3) may also be written

$$\nu_n = \nu_n' \left[1 + \mu^2 \pi^2 \left(\frac{a}{\lambda_n} \right)^2 \right] , \tag{E4.4}$$

where ν_n' is the measured frequency of the nth normal mode in a rod of radius a and ν_n is the frequency of the nth normal mode in an infinitely thin rod of the same length. Thus, in terms of a finite cylindrical rod, equation (E4.2) may be written

$$E = \rho c_L^2 = \rho(\lambda_n \nu_n)^2 = \frac{4\rho l^2 (\nu_n')^2}{n^2} \left[1 + \left(\frac{n\pi\mu a}{2l} \right)^2 \right]^2 , \tag{E4.5}$$

where l is the length of the rod. This expression for E holds for small values of a/λ. When a/λ approaches unity, a more precise calculation of c_L is required which involves the application of the Pochhammer–Chree theory, as discussed by Bancroft (1941).

It is seen from equation (E4.4) that Poisson's ratio may be found if ν_n' is measured for a number of different values of a/λ. It is then convenient to determine μ from the slope of the graph of $(1 - \nu_n'/n\nu_1)$ versus $(a/\lambda)^2$. In figure E4.4 is shown such a graph for a brass rod, the plotted points corresponding to the 15th, 25th, 35th, and 45th harmonics. Some uncertainty arises as to the correct value of ν_1. If an incorrect value is assumed, the straight line does not pass through the origin. A new line drawn parallel to the original one applies the necessary correction to the value of ν_1 without altering the value of μ, which depends only on the slope. Although measurements at only two harmonics are sufficient, additional check points are useful with increasing a/λ, in order to determine whether Rayleigh's relationship is still valid. The value of μ calculated from figure E4.4 is 0·377. The corresponding values of c_L and E are 3020 m s^{-1} and 7·72 × 10^{10} N m^{-2}, respectively.

Shear modulus

The dynamic shear modulus G may be found by measuring the frequency of the fundamental torsional mode. Since there is no dispersion for this mode, G may be found simply from

$$G = \rho c_T^2 = \frac{4\rho l^2 \nu_n^2}{n^2} \tag{E4.6}$$

where c_T is the phase velocity of torsional waves in a rod, l is the length of the specimen, and n is the mode number. For the specimen considered here, $\nu_1 = 1516$ Hz, $c_T = 2297$ m s^{-1}, $G = 4 \cdot 47 \times 10^{10}$ N m^{-2}.

If the specimen were isotropic, Poisson's ratio could be found from E and G by means of the standard relationship

$$\mu = \frac{E}{2G} - 1 . \tag{E4.7}$$

However, in practice most metal rods exhibit some degree of anisotropy introduced by the manufacturing processes so that values of μ calculated by means of equation (E4.7) may be in serious error.

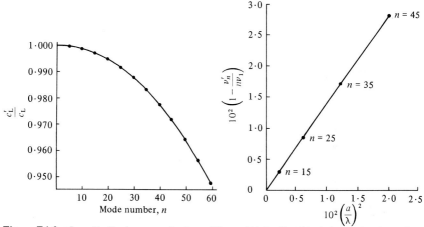

Figure E4.3. Longitudinal wave velocity dispersion curve for a brass rod (61% Cu, 36% Zn, 3% Pb; density $8 \cdot 48 \times 10^3$ kg m^{-3}; length $0 \cdot 7576$ m; radius $4 \cdot 77$ mm). The abscissae may alternatively be stated in terms of a/λ since $a/\lambda = 0 \cdot 00315 n$ for this specimen.

Figure E4.4. Graphical determination of Poisson's ratio from a plot of $(1 - \nu_n'/n\nu_1)$ versus $(a/\lambda)^2$ for the 15th, 25th, 35th, and 45th harmonics of the brass rod quoted in figure E4.3.

4.4.1 Pochhammer–Chree theory

Wave motion in an infinite isotropic elastic solid must satisfy the wave equations

$$\nabla^2 \phi = \frac{1}{c_\ell^2} \frac{\partial^2 \phi}{\partial t^2} , \qquad \nabla^2 \psi_i = \frac{1}{c_t^2} \frac{\partial^2 \psi_i}{\partial t^2}$$

where the displacement u_i has been expressed in terms of a scalar potential ϕ

and a vector potential ψ_i: $u_i = \mathrm{grad}\,\phi + \mathrm{curl}\,\psi_i$. Solutions of the wave
equations are sought in terms of the cylindrical coordinates r, θ, and x_1
(the axis of the cylinder being taken in the x_1 direction) together with the
appropriate boundary conditions for free lateral surfaces. The wave
equations in cylindrical coordinates are

$$\frac{\partial^2 \phi}{\partial r^2} + \frac{1}{r}\frac{\partial \phi}{\partial r} + \frac{\partial^2 \phi}{\partial x_1^2} = \frac{1}{c_\ell^2}\frac{\partial^2 \phi}{\partial t^2} \ ,$$

$$\frac{\partial^2 \psi}{\partial r^2} + \frac{1}{r}\frac{\partial \psi}{\partial r} + \frac{\partial^2 \psi}{\partial x_1^2} = \frac{1}{c_t^2}\frac{\partial^2 \psi}{\partial t^2} \ . \tag{4.24}$$

When the particle displacements are independent of θ, the solutions
correspond to longitudinal waves. If they are independent of r and x_1,
the solutions correspond to torsional waves. If the displacements are a
function of r, θ, and x_1, the solutions correspond to flexural waves.

4.4.1.1 *Solutions for longitudinal waves*
Since the particle displacements are in this case independent of θ, harmonic
solutions are assumed of the form

$$\phi = \phi_0(r)\exp[\mathrm{i}(k_1 x_1 - \omega t)] \ , \qquad \psi = \psi_0(r)\exp[\mathrm{i}(k_1 x_1 - \omega t)] \ . \tag{4.25}$$

Their substitution into equations (4.24) yields the two differential
equations

$$\frac{\partial^2 \phi_0}{\partial r^2} + \frac{1}{r}\frac{\partial \phi_0}{\partial r} + \left[\left(\frac{\omega}{c_\ell}\right)^2 - k_1^2\right]\phi_0 = 0 \ ,$$

$$\frac{\partial^2 \psi_0}{\partial r^2} + \frac{1}{r}\frac{\partial \psi_0}{\partial r} + \left[\left(\frac{\omega}{c_t}\right)^2 - k_1^2\right]\psi_0 = 0 \ . \tag{4.26}$$

These equations may now be solved by the method of separation of
variables, and the solutions written in the form

$$\phi = A J_0(k_\varrho r)\exp[\mathrm{i}(k_1 x_1 - \omega t)] \ ,$$

$$\psi = B J_0(k_t r)\exp[\mathrm{i}(k_1 x_1 - \omega t)] \ , \tag{4.27}$$

where

$$k_\varrho^2 = \left(\frac{\omega}{c_\varrho}\right)^2 - k_1^2 \ , \qquad k_t^2 = \left(\frac{\omega}{c_t}\right)^2 - k_1^2$$

and $J_0(kr)$ is a zero-order Bessel function.

The boundary conditions for stress at the free lateral surfaces of the
cylinder are that the normal and tangential stresses must be zero, that is

$$\sigma_{rr} = \lambda\left(\frac{u_r}{r} + \frac{\partial u_r}{\partial r} + \frac{\partial u_{x_1}}{\partial x_1}\right) + 2G\frac{\partial u_r}{\partial r} = 0 \ ,$$

$$\sigma_{x_1 r} = G\left(\frac{\partial u_r}{\partial x_1} + \frac{\partial u_{x_1}}{\partial r}\right) = 0 \ , \tag{4.28}$$

when $r = a$, where a is the radius of the cylinder. The radial and axial displacements are

$$u_r = \frac{\partial \phi}{\partial r} + \frac{\partial^2 \psi}{\partial r \partial x_1} \, ,$$

$$u_{x_1} = \frac{\partial \phi}{\partial x_1} - \frac{\partial^2 \psi}{\partial r^2} - \frac{1}{r} \frac{\partial \psi}{\partial r} \, .$$

As shown by Love (1927), and Redwood (1960), application of the boundary conditions leads to the frequency equation

$$k_1^2 \frac{k_t J_0(k_t a)}{J_1(k_1 a)} - \frac{1}{2}\left(\frac{\omega}{c_t}\right)^2 \frac{1}{a} + \left[\frac{1}{2}\left(\frac{\omega}{c_t}\right)^2 - k_1^2\right]^2 \frac{J_0(k_\varrho a)}{k_\varrho J_1(k_\varrho a)} = 0 \, . \tag{4.29}$$

This equation may be solved for any frequency to give the phase velocity corresponding to a particular mode, as shown by Bancroft (1941) and by Davies (1948). In a similar way expressions may be found for the particle displacements u_r and u_{x_1}. In figure 4.6a are shown phase velocity curves for the first three longitudinal modes of a solid cylinder having a Poisson's ratio of $0 \cdot 29$ together with the results of the classical theory and a curve corresponding to Rayleigh's formula (4.23). Group velocity curves for the first two modes, calculated from the slopes of the curves in figure 4.6a are shown in figure 4.6b.

For the first longitudinal mode, also described as the (1, 1) mode or Young's modulus mode, the phase velocity is constant at low frequencies and is given by the classical formula (4.22). A reasonable upper limit of validity of this simple relationship occurs when $a/\lambda \geqslant \frac{1}{5}$. In the (1, 1) mode the wave fronts are planes whose normals are in the direction of the axis of the cylinder. At high frequencies, that is for increased values of a/λ, the phase velocity of the (1, 1) mode approaches the velocity of Rayleigh surface waves.

The **higher order longitudinal modes** exhibit a cut-off phenomenon, similar to the cut-off effect in fluid and electromagnetic waveguides. For these higher modes the wavefronts are incident at an angle α to the lateral surfaces of the cylinder with the phase velocity along the axis given by $c = c_\varrho/\sin\alpha$. Cut-off occurs when $\alpha \to 0$ and hence $c \to \infty$. With increasing frequency the phase velocity of the higher modes approaches that for transverse waves.

It is of interest to note that the Bessel functions in equation (4.29) may be expanded to give approximate solutions. Since

$$J_0(ka) = 1 - (\tfrac{1}{2}ka)^2 + \frac{(\tfrac{1}{2}ka)^4}{1^2 \times 2^2} - \frac{(\tfrac{1}{2}ka)^6}{1^2 \times 2^2 \times 3^2} + \cdots$$

and

$$J_1(ka) = \tfrac{1}{2}ka - \frac{(\tfrac{1}{2}ka)^3}{1^2 \times 2} + \frac{(\tfrac{1}{2}ka)^5}{1^2 \times 2^2 \times 3} - \cdots \, ,$$

by taking first-order terms only we can write

$$\frac{\omega}{k_1} \equiv c_L = \left(\frac{E}{\rho}\right)^{\frac{1}{2}} ,$$

where c_L is the low-frequency velocity of plane longitudinal waves in a thin rod. This is the simple classical equation.

When second-order terms in a are included, we obtain

$$\frac{\omega}{k_1} = c_L \left[1 - \mu^2 \pi^2 \left(\frac{a}{\lambda}\right)^2 \right] ,$$

which is Rayleigh's formula.

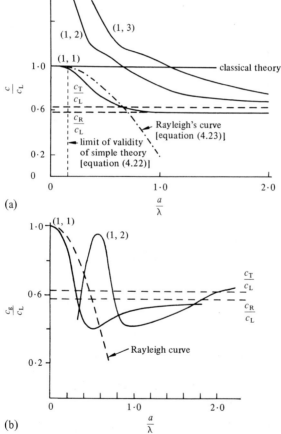

Figure 4.6. (a) Phase velocity distributions for the first three longitudinal modes of a solid cylinder having Poisson's ratio 0·29. c_L, c_T, and c_R represent the velocities of longitudinal, torsional, and Rayleigh waves respectively. (After Davies, 1948.)
(b) Group velocity distributions for the first two longitudinal modes of a solid cylinder as in (a) above.

4.4.1.2 *Solutions for torsional waves*

If $u_r = 0$, $u_{x_1} = 0$, and u_θ is finite and independent of θ, then the appropriate solutions of the wave equations, expressed in terms of the displacements, that satisfy the boundary conditions are

$$u_\theta = A J_1(k_t r) \exp[i(k_1 x_1 - \omega t)] \qquad \text{when } k_t \neq 0 \qquad (4.30a)$$

$$u_\theta = B r \exp[i(\omega x_1/c_t - \omega t)] \qquad \text{when } k_t = 0 . \qquad (4.30b)$$

Equation (4.30a) gives a series of modes with cut-off frequencies similar to the higher modes for longitudinal waves. Equation (4.30b) represents the fundamental torsional mode for which the phase velocity is independent of frequency; a pulse of arbitrary shape will therefore be transmitted without distortion in this mode. The fundamental torsional mode has frequently been used in the design of delay lines. In figure 4.7 are shown the phase velocity distributions for the first three torsional modes in a cylinder having a Poisson's ratio of $0 \cdot 29$.

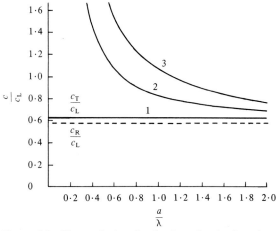

Figure 4.7. Phase velocity distributions for the first three torsional modes of a solid cylinder having Poisson's ratio $0 \cdot 29$. (After Davies, 1956.)

4.4.1.3 *Solutions for flexural waves*

The solutions of the wave equations for flexural waves may be found by a similar process, although they will be more complicated since the displacements depend on r, θ, and x_1. Sittig (1957) has derived the frequency equation in the form of the determinant

$$\begin{vmatrix} n^2 - 1 - a^2 k_z^2(x-1) & n^2 - 1 - a^2 k_z^2(2x-1) & 2(n^2-1)[\gamma_n(k_t a) - n] - a^2 k_z^2(2x-1) \\ \gamma_n(k_t a) - n - 1 & \gamma_n(k_t a - n - 1) & 2n^2 - 2[\gamma_n(k_t a) - n] - a^2 k_z^2(2x-1) \\ \gamma_n(k_t a) - n & -(x-1)[\gamma_n(k_t a) - n] & n^2 \end{vmatrix} = 0 ,$$

$$(4.31)$$

where

$$\gamma_n(ka) = \frac{ka\,J_{n-1}(ka)}{J_n(ka)} \quad , \qquad x = \frac{c^2}{2c_t^2} \quad , \qquad n = 0, 1, 2, \dots \,.$$

The longitudinal and torsional modes are found as special cases of equation (4.31) by putting $n = 0$. The phase velocity distributions for the first three flexural modes for a cylinder with Poisson's ratio $0 \cdot 29$ are shown in figure 4.8. As a/λ increases the phase velocity of the first flexural mode approaches the velocity of Rayleigh surface waves, while for the higher modes the phase velocity approaches the velocity of transverse waves.

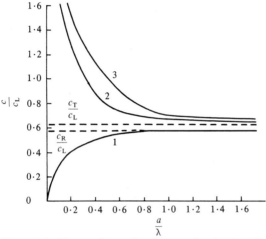

Figure 4.8. Phase velocity distributions for the first three flexural modes of a solid cylinder having Poisson's ratio $0 \cdot 29$. (After Abramson, 1957.)

4.4.1.4 *Pulse response*

Although the Pochhammer–Chree theory applies strictly only for continuous waves it may be used to describe the propagation of pulses provided $a/\lambda < 1$. At high frequencies, when $a/\lambda > 1$, 'trailing pulses' appear owing to mode conversion at the boundaries. The modified theory proposed by Redwood (1959) may then be employed. For instance, for an initial longitudinal pulse the solutions must provide for the fact that transverse waves, generated by mode conversion at the boundaries, cannot appear until after the initial longitudinal waves have reached the boundary. The Bessel function solutions assume both types of waves present from the start. Redwood (1959) has overcome this problem by the use of a Hankel function in place of the Bessel function in the second of equations (4.27). The modified solutions are

$$\phi = A J_0(k_\varrho r) \exp[i(k_1 x_1 - \omega t)] \,,$$

$$\psi = B H_0^{(1)}(k_t r) \exp[i(k_1 x_1 - \omega t)] \,.$$

$$(4.32)$$

On applying the boundary conditions we obtain the modified frequency equation

$$k_1^2 \frac{k_t H_0^{(1)}(k_t a)}{H_1^{(1)}(k_t a)} - \frac{1}{2}\left(\frac{\omega}{c_t}\right)^2 \frac{1}{a} + \left[\frac{1}{2}\left(\frac{\omega}{c_t}\right)^2 - k_1^2\right]^2 \frac{J_0(k_\varrho a)}{k_\varrho J_1(k_\varrho a)} = 0 \,. \qquad (4.33)$$

Further discussion concerning the propagation of pulses is given in Redwood (1960), where various approximate theories for the cylinder, and the theory of solid waveguides of other cross sections such as rectangular, elliptical, and cylindrical shells are also discussed. In the case of rods of square cross section the dispersion curves are similar to those for a cylinder provided the cross-sectional areas of rod and cylinder are equal.

Experiment E4.3 Pulse transmission in a cylinder–mode conversion
(Kolsky, H., 1954, *Phil. Mag.*, **45**, 712.)

Kolsky investigated the transmission of a brief pulse through a short cylinder. The pulse was generated by a small explosive charge set off at the centre of one face of a steel cylinder and a capacitive detector was placed at the centre of the opposite face, as shown in figure E4.6. The cylinder was 10·4 cm long and 15·2 cm in diameter. The signal received by the detector is shown in figure E4.5 and the possible paths taken by the waves between source and detector are shown in figure E4.6.

Kolsky interpreted the results by assuming that both longitudinal and transverse waves were generated by the explosion, and these waves spread out spherically from the source. The direct waves (path 1) are the first to arrive and correspond to waves that travel with the dilatational velocity, that is the velocity of a longitudinal disturbance in an infinite medium. Path 2 corresponds to a longitudinal wave which arrives at the detector after one reflection from the side wall. Path 3 corresponds to a longitudinal wave which is reflected from the side wall as a transverse wave before reaching the detector. Path 5 corresponds to a transverse wave reflected from the side wall as a transverse wave. The measured arrival times of the individual signals were found to agree with the values calculated from the velocity of longitudinal waves: $c_\varrho = [(\lambda + G)/\rho]^{1/2}$ and the velocity of transverse waves: $c_t = [G/\rho]^{1/2}$.

Even with the shape of cylinder employed, only the first few received pulses are clearly separated. For a specimen in the shape of a rod, or with longer pulses, the received signal is a complex mixture of direct and reflected waves. Under these circumstances elastic waveguide theory must be applied and the velocity of propagation will lie between c_ϱ and c_t.

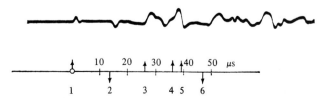

Figure E4.5. Received signal arising from the transmission of a pulse through a short cylinder. The numbers refer to the paths shown in figure E4.6.

This experiment clearly shows that, in the transient response of an elastic waveguide, some part of the energy will travel with the velocity of the dilatational waves. Classical theory, which deals with steady-state conditions, does not predict this effect nor does it predict the observed loss of amplitude of the main signal, nor the presence of any of the secondary signals. A modified theory, referred to in section 4.4.1.4, to account for these effects has been proposed by Redwood (1959).

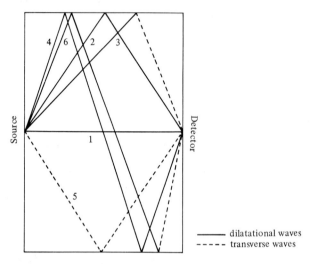

— dilatational waves
- - - - transverse waves

Figure E4.6. Possible wave paths involved in the transmission of a pulse through a short cylinder.

4.5 Acoustic transmission line theory

When dealing with solid cylinders of finite length, the system usually possesses a number of discrete resonant modes. Instead of attempting to solve the basic wave equations for each problem, it is often more convenient to use a method based on normal mode analysis or transmission line theory. The latter method is particularly useful when dealing with only one mode, such as the Young's modulus mode. In developing the theory, use is made of the mathematical analogies that exist between electrical and mechanical systems.

4.5.1 Mechanical-electrical analogies

Consider the mechanical system shown in figure 4.9a. Application of Newton's second law shows that

$$M\frac{\mathrm{d}v}{\mathrm{d}t} + R_{\mathrm{m}}v + \frac{1}{C_{\mathrm{m}}}\int v\,\mathrm{d}t = f(t) , \tag{4.34}$$

where M is mass, R_{m} is mechanical resistance, C_{m} is compliance (reciprocal of stiffness) and v is velocity.

An electrical circuit that obeys a similar equation is shown in figure 4.9b. Applying Kirchhoff's voltage law gives

$$L\frac{di}{dt} + Ri + \frac{1}{C}\int i\,dt = e(t)\ .\tag{4.35}$$

The various mechanical and electrical quantities may be considered as analogs according to the following scheme:

Mechanical system		*Electrical system* (force–voltage analogy)	
force	f	voltage	e
velocity	v	current	i
mass	M	inductance	L
mechanical resistance	R_m	electrical resistance	R
compliance	C_m	capacitance	C
mechanical impedance	$Z_m = f/v$	electrical impedance	$Z = e/i\ .$

Alternatively, the dual circuit may be considered for which the current equation is

$$C\frac{de}{dt} + \frac{e}{R} + \frac{1}{L}\int e\,dt = i(t)\ .\tag{4.36}$$

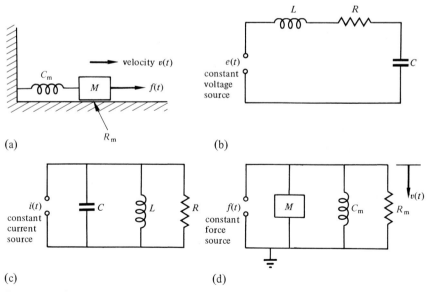

Figure 4.9. (a) Mechanical system; (b) analogous series electrical circuit (force-voltage analogy); (c) analogous parallel electrical circuit (force-current analogy); (d) mechanical circuit diagram.

In this case the following analogies may be drawn between equations (4.34) and (4.36):

Mechanical system		Electrical system (force–current analogy)	
force	f	current	i
velocity	v	voltage	e
mass	M	capacitance	C
mechanical resistance	R_m	conductance	$1/R$
compliance	C_m	inductance	L
mechanical impedance	Z_m	electrical admittance	Y .

The force–current analogy has some advantages when dealing with mechanical systems, particularly those involving lumped elements. A mechanical circuit may be drawn in which all points moving with the same velocity are connected to the same horizontal line, as in figure 4.9d. The circuits shown in figures 4.9c and 4.9d are then identical. Thus, in the force–current analogy each junction in the mechanical system corresponds to a junction (node) in the analogous electrical system, the velocity measured across a mechanical element is analogous to the voltage measured across an electrical element, the force acting through a mechanical element is analogous to the current through an electrical element. In addition, all points on a rigid mass are considered as the same junction and one terminal of the analogous capacitance is always connected to earth. It is relatively easy to draw an analogous circuit from inspection of a mechanical system if the force–current analogy is employed.

In the force–voltage analogy a junction in the mechanical system corresponds to a loop in the electrical system, the velocity across a mechanical element is analogous to the current through an electrical element, the force through a mechanical element is analogous to the voltage across an electrical element. If the circuit corresponding to the force–voltage analogy is required, it is usually easier to construct the circuit for the force–current analogy and then to draw the dual circuit by the 'dot method' as described by Beranek (1954). For problems involving wave motion the question of the ease of drawing circuits does not arise, and it is usual to employ the force–voltage analogy directly.

More generally, we may introduce a potential (or across) variable, $\alpha(t)$, and a flow (or through) variable, $\beta(t)$. A basic operator equation may then be formed:

$$\alpha(t) = \left[A\frac{\mathrm{d}}{\mathrm{d}t} + B + \frac{1}{D}\int \mathrm{d}t \right] \beta(t) .$$ (4.37)

For periodic signals with time variation $\exp(i\omega t)$, the differentiator $A(\mathrm{d}/\mathrm{d}t)$ becomes $i\omega A$, the multiplier remains equal to B, the integrator $(1/D)\int\mathrm{d}t$ becomes $1/i\omega D$; or we may write

$$\alpha(t) = Z\beta(t)$$ (4.38)

where

$$Z \equiv \frac{\alpha}{\beta} = B + i\omega A + \frac{1}{i\omega D}$$

is the impedance operator. All practical analogous systems have an equation identical to equation (4.37) with the appropriate substitutions for α, β, A, B, and D.

The basic dual equations for mechanical, electrical, and acoustical oscillations are:

mechanical

$$f = \left[M \frac{d}{dt} + R_m + \frac{1}{C_m} \int dt \right] v \; ,$$

$$v = \left[C_m \frac{d}{dt} + \frac{1}{R_m} + \frac{1}{M} \int dt \right] f \; ;$$

electrical

$$e = \left[L \frac{d}{dt} + R + \frac{1}{C} \int dt \right] i \; ,$$

$$i = \left[C \frac{d}{dt} + \frac{1}{R} + \frac{1}{L} \int dt \right] e \; ;$$

acoustical

$$p = \left[M_a \frac{d}{dt} + R_a + \frac{1}{C_a} \int dt \right] U \; ,$$

$$U = \left[C_a \frac{d}{dt} + \frac{1}{R_a} + \frac{1}{M_a} \int dt \right] p \; ;$$

where p is the acoustic pressure, M_a is the acoustic mass, R_a is the acoustic resistance, C_a is the acoustic compliance, and U is the volume flow or flux.

Table 4.1. Quantities required to form analogous equations.

Basic operator equation: $\alpha(t) = \left[A \frac{d}{dt} + B + \frac{1}{D} \int dt \right] \beta(t)$

System	α	A	B	D	β
Mechanical	f	M	R_m	C_m	v
	v	C_m	$1/R_m$	M	f
Electrical	e	L	R	C	i
	i	C	$1/R$	L	e
Acoustical	p	M_a	R_a	C_a	U
	U	C_a	$1/R_a$	M_a	p

In table 4.1 are listed the quantities that must be substituted into equation (4.37) in order to form the appropriate equation for either analogy. Additional notes and examples are given in Appendix 4.1. Further details may be found in Paul (1965).

4.5.2 Boundary conditions for a finite rod

At a free surface of a solid there must be no resultant normal or tangential stresses. Consider a stress pulse within a rod travelling in the axial direction, which will be taken as the positive x direction, and incident on a free end. The resultant stress at the free end is zero, that is

$$\sigma_{res} = \sigma_i + \sigma_r = 0 \ ,$$

where σ_i is the incident stress and σ_r is the reflected stress. Hence $\sigma_i = -\sigma_r$. The change of sign implies that, for instance, an incident compression pulse will be reflected as a tensile pulse.

The pulse does not change shape on reflection, as may be deduced from the following argument. The displacement at any point in the rod must satisfy the general wave equation whose solution may be written in d'Alembert's form as

$$u(x, t) = f(ct - x) + g(ct + x) \ . \tag{4.39}$$

The first term may be taken as representing an incident pulse (travelling in the $+x$ direction) and the second term a reflected pulse (travelling in the $-x$ direction), it being assumed that there is no loss of energy on reflection. From Hooke's law the axial stress is $\sigma_{xx} = E\epsilon_{xx} = E(\partial u_x/\partial x)$, where E is Young's modulus. Thus the resultant stress at the end is

$$\sigma_{res} = \sigma_i + \sigma_r = E[g'(ct + x) - f'(ct - x)] = 0 \ .$$

For convenience we shall put $x = 0$ at the end of the rod; then

$$g'(ct) = f'(ct) \ . \tag{4.40}$$

Integration of equation (4.40) shows that the functions f and g are identical, apart from a constant which represents a displacement of the whole system. Hence the shape of the stress pulse is unaltered on reflection from a free end, but there is a change of sign. This effect is illustrated in figure 4.10a.

The resultant displacement at the free end is

$$u_{res} = u_i + u_r = f(ct) + g(ct) = 2f(ct) \ .$$

Thus at a free end the displacement is doubled. A similar result holds for the particle velocity at a free end. In figure 4.10b is shown the reflection of a velocity pulse at a free end.

At a clamped end both the resultant displacement and velocity must be zero. Thus

$$u_{res} = u_i + u_r = f(ct) + g(ct) = 0 \ ,$$

where the origin for x is at the clamped end. Hence

$$f(ct) = -g(ct) \ ;$$

that is a displacement or velocity pulse is reflected without loss of amplitude, but with the particle motion reversed.

The resultant stress is

$$\sigma_{res} = \sigma_i + \sigma_r = E[g'(ct) - f'(ct)] \ .$$

But in our case $f(ct) = -g(ct)$. Hence $\sigma_{res} = 2Ef'(ct)$; that is the stress at a clamped end is doubled.

The behaviour of the mechanical system may now be compared with that of an electrical transmission line. A voltage or current pulse travelling along a line experiences only the characteristic impedance Z_0 of the line until it reaches an end termination. If the termination also has impedance Z_0 the pulse continues as if in an infinite line and there is no reflection.

For an electrical transmission line which is short-circuited at one end the end voltage is zero, and hence an incident voltage pulse will be reflected with phase reversal. A current pulse is reflected from a short-circuited end with phase unchanged and hence has double amplitude at the end.

For an electrical transmission line which is open-circuited at one end the end current is zero, and hence a current pulse will be reflected with phase reversal. On the other hand a voltage pulse is reflected with phase

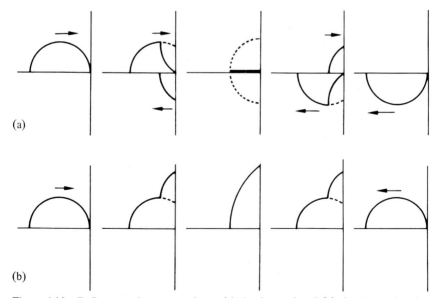

(a)

(b)

Figure 4.10. Reflection of a stress pulse at (a) the free end and (b) the clamped end of a rod. The resultant stress at the free end is always zero and at the clamped end is always twice the normal value. Diagrams (a) and (b) describe also the reflection of a velocity pulse at a clamped end and at a free end, respectively.

unchanged, and hence has double amplitude at the end. If force (or stress) is taken as analogous to voltage and particle velocity analogous to current, it may be concluded that the electrical analogue of a rod with free ends is a short-circuited transmission line. Similarly, the electrical analogue of a rod with clamped ends is an open-circuited transmission line. The theory of the acoustical transmission line will now be developed with the aid of mechanical–electrical analogies.

4.5.3 Transmission in an infinite rod

Consider plane longitudinal waves travelling along an infinite isotropic rod whose axis is in the x direction. Lateral motion due to the Poisson effect may be ignored at low frequencies when $\lambda \gg a$. The wave equation, in terms of displacement, that must be satisfied is equation (4.21), namely

$$\frac{\partial^2 u_x}{\partial x^2} = \frac{1}{c_L^2}\frac{\partial^2 u_x}{\partial t^2} \,, \qquad c_L^2 = \frac{E}{\rho} \,.$$

The general solution for harmonic plane waves may be written in the form

$$\hat{u}_x = \hat{A}\exp[i(\omega t - \hat{k}x)] + \hat{B}\exp[i(\omega t - \hat{k}x)] \,, \tag{4.41}$$

where $\hat{u}_x$ is the complex displacement, $\hat{A}$ the complex amplitude of a wave proceeding in the $+x$ direction, $\hat{B}$ the complex amplitude of a wave proceeding in the $-x$ direction, $\hat{k} = k - i\alpha$ is the complex wave vector, $k = \omega/c_L$, and α is the **attenuation coefficient**. In engineering texts a complex propagation constant γ is introduced, where $\gamma = i\hat{k} = \alpha + ik$.

[Note: In chapters 2 and 3 it was more convenient to write for a positive-going wave $A\exp[i(kx - \omega t)]$ so that the sign of the term in x gave the direction of the wave.]

The particle velocity, v_x, corresponding to equation (4.41) is

$$v_x \equiv \frac{\partial u_x}{\partial t} = i\omega\hat{A}\exp[i(\omega t - \hat{k}x)] + i\omega\hat{B}\exp[i(\omega t + \hat{k}x)] \,. \tag{4.42}$$

If we assume that Hooke's law holds, we can find the stress σ_{xx} from $\sigma_{xx} = E\epsilon_{xx} = E(\partial u_x/\partial x)$. Thus

$$\sigma_{xx} = -i\hat{k}E\hat{A}\exp[i(\omega t - \hat{k}x)] + i\hat{k}E\hat{B}\exp[i(\omega t + \hat{k}x)] \,. \tag{4.43}$$

If we introduce the **mechanical impedance**, $\hat{Z}_m$, defined as the ratio of force to velocity at any point x along the rod, then

$$\hat{Z}_m = -\frac{\sigma_{xx}S}{v_x} = \hat{Z}_0\frac{\hat{A}\exp(-i\hat{k}x) - \hat{B}\exp(i\hat{k}x)}{\hat{A}\exp(-i\hat{k}x) + \hat{B}\exp(i\hat{k}x)} \,, \tag{4.44}$$

where $\hat{Z}_0 = \hat{k}ES/\omega$ is the **characteristic impedance** of the medium and S is the area of cross section. (An impedance may be defined as the ratio of stress to velocity; it is then known as the specific acoustic impedance.) If the attenuation is small, Z_0 is real and is given by $Z_0 \approx ES/c_L = \rho c_L S$.

The negative sign appears in equation (4.44) since a positive particle velocity, v_x, would result from a compression applied to the rod for which σ_{xx} is customarily taken as negative. For a travelling wave in an infinite rod in the $+x$ direction, $\hat{B} = 0$ and hence $\hat{Z}_m = \hat{Z}_0$, that is, in the absence of any backward wave arising by virtue of reflection, the impedance experienced by the wave is the characteristic impedance of the medium.

4.5.4 The finite rod

Let us consider now a finite rod of length l with ends at $x = 0$ and $x = l$. Consider a generator connected at $x = 0$ and let the termination at $x = l$ be represented by a **terminating impedance** $\hat{Z}_T$. In the extreme cases, if the rod has a free end, $\hat{Z}_T = 0$; if the end is clamped, $Z_T = \infty$. The terminating impedance at $x = l$ is found from equation (4.44)

$$\hat{Z}_T = \hat{Z}_0 \frac{\hat{A}\exp(-i\hat{k}l) - \hat{B}\exp(i\hat{k}l)}{\hat{A}\exp(-i\hat{k}l) + \hat{B}\exp(i\hat{k}l)} . \tag{4.45}$$

Hence

$$\frac{\hat{B}}{\hat{A}}\exp(2i\hat{k}l) = \frac{\hat{Z}_0 - \hat{Z}_T}{\hat{Z}_0 + \hat{Z}_T} = \hat{R} , \tag{4.46}$$

where $\hat{R}$ is the complex reflection coefficient. From equations (4.43) and (4.46) it is observed that the **stress reflection coefficient** is $\hat{R}(\sigma) = -\hat{R}$. Also, from equations (4.42) and (4.46) the **velocity reflection coefficient** is $\hat{R}(v) = \hat{R}$.

It follows that for a free end $Z_T = 0$, $\hat{R}(\sigma) = -1$ and $\hat{R}(v) = 1$. It is thus confirmed that in terms of stress, an incident wave is reflected with phase reversed giving a resultant stress of zero at the free end. In terms of velocity an incident wave is reflected without phase change giving rise to a doubling in velocity amplitude at the free end.

At a fixed or clamped end $Z_T = \infty$, $\hat{R}(\sigma) = 1$, and $\hat{R}(v) = -1$. An incident stress wave is therefore reflected without phase change giving rise to a doubling in stress at a fixed end. An incident velocity wave is reflected with phase reversal, so that the resultant velocity is zero at a fixed end.

When $\hat{Z}_T = \hat{Z}_0$, $\hat{R} = 0$ and no reflection takes place when a rod is terminated by its characteristic impedance. The results developed above are identical to those applicable to electrical transmission lines (Slater, 1959) when force and velocity are replaced by voltage and current respectively. The comparison is further emphasised in the calculation of the input and transfer impedances for a finite rod.

4.5.4.1 *Input impedance*

The **input impedance**, $\hat{Z}_{11}$, at $x = 0$ is defined as

$$\hat{Z}_{11} = \frac{f_1}{v_1}$$

where f_1 is the input force and v_1 is the input velocity. From equation (4.45), we have

$$\hat{Z}_{11} = \hat{Z}_0 \frac{1 - \hat{B}/\hat{A}}{1 + \hat{B}/\hat{A}} .$$

On combining this with equation (4.46) we obtain

$$\hat{Z}_{11} = \hat{Z}_0 \frac{\hat{Z}_0 \sinh i\hat{k}l + \hat{Z}_T \cosh i\hat{k}l}{\hat{Z}_0 \cosh i\hat{k}l + \hat{Z}_T \sinh i\hat{k}l} . \tag{4.47}$$

Equation (4.47) is identical to the equation giving the input impedance of an electrical transmission line of length l with terminating impedance $\hat{Z}_T$. Important special cases of equation (4.47) arise for the extremes of $\hat{Z}_T = 0$ and $\hat{Z}_T = \infty$:

for $\hat{Z}_T = 0$,

$$\hat{Z}_{11} = \hat{Z}_0 \tanh i\hat{k}l ; \tag{4.48}$$

for $\hat{Z}_T = \infty$,

$$\hat{Z}_{11} = \hat{Z}_0 \coth i\hat{k}l . \tag{4.49}$$

Equation (4.48) gives the ratio of force to velocity at the input end of a rod with the other end free, which is analogous to a short-circuited transmission line. Equation (4.49) applies to a rod with the far end fixed or an open-circuited transmission line. Both equations represent a steady-state solution and describe the set of standing waves in a rod when the velocity is measured at the same end face to which the force is applied. It also follows that

$$\hat{Z}_{11}(\text{free}) \times \hat{Z}_{11}(\text{fixed}) = \hat{Z}_0^2 .$$

4.5.4.2 Transfer impedance

If the force is applied at one end of a rod and the velocity measured at the opposite end, the **transfer impedance**, $\hat{Z}_{12}$, is a more appropriate parameter. $\hat{Z}_{12}$ is defined as

$$\hat{Z}_{12} = \frac{f_1}{v_2} ,$$

where f_1 is the input force and v_2 is the output velocity. By analogy with the corresponding electrical equations (Cherry, 1949), the force and velocity at any point along a rod may be written

$$f_x = f_1 \cosh i\hat{k}x - v_1 \hat{Z}_0 \sinh i\hat{k}x ,$$

$$v_x = v_1 \cosh i\hat{k}x - \frac{f_1}{\hat{Z}_0} \sinh i\hat{k}x . \tag{4.50}$$

From equations (4.48) and (4.50) the transfer impedance of a rod of
length l with free ends is

$$\hat{Z}_{12} = \frac{f_1}{v_2} = \hat{Z}_0 \sinh i\hat{k}l .$$ (4.51)

In terms of the input impedance

$$\hat{Z}_{12} = \hat{Z}_{11} \cosh i\hat{k}l .$$ (4.52)

Similarly, for a rod with fixed ends

$$\hat{Z}_{12} = \hat{Z}_0 \operatorname{cosech} i\hat{k}l = \hat{Z}_{11} \operatorname{sech} i\hat{k}l .$$ (4.53)

Thus, depending on the experimental conditions, either $\hat{Z}_{11}$ or $\hat{Z}_{12}$ may be
used to compute the total response of the system.

4.5.5 Resonant transmission line
The equations developed in the preceding section apply to a line which is
excited at one end by an appropriate generator and is either short-circuited
or open-circuited at the other. It is of interest therefore to find the
conditions under which these equations may be applied to a resonant line
which may be short-circuited or open-circuited at both ends or may be
terminated at either end by an arbitrary impedance.

We may imagine the input end of a short-circuited line connected to a
generator for a short time and then short-circuited again before the signal
from the far end returns to the input end. The signal will then be
multiply reflected within the line and the variation in impedance of the
line as a function of either length or frequency will presumably be given
by equations (4.48) and (4.51).

Consider a short-circuited line for which the attenuation coefficient is
negligible. Then equation (4.48) may be written with k real:

$$Z_{11} = Z_0 \tanh ikl = iZ_0 \tan kl .$$

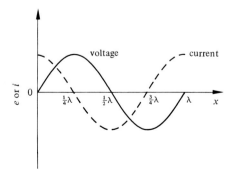

Figure 4.11. Current (velocity) and voltage (stress) wave patterns on a resonant
transmission line.

It is seen that $Z_{11} = 0$ whenever $kl = n\pi$, $n = 0, 1, 2, ...$ That is, a lossless short-circuited line acts as a resonator when $l = \frac{1}{2}n\lambda$. This is the condition for the occurrence of voltage (force) nodes, since when $Z_{11} = 0$, e_1 or $f_1 = 0$. The condition for current (velocity) nodes is that $Z_{11} = \infty$, that is, $kl = (n+\frac{1}{2})\pi$, $n = 0, 1, 2, ...$ or $l = \frac{1}{2}(n+\frac{1}{2})\lambda$. The variations of current and voltage on a resonant line are therefore out of step by a distance $\frac{1}{4}\lambda$, as shown in figure 4.11.

4.5.5.1 Resonant line with attenuation

When attenuation is present, instead of the ideal zero and infinite values of the impedance there will be minima and maxima at which the value of the impedance will depend on the value of the attenuation. It is often more convenient in practice to consider the admittance function rather than the impedance, since the velocity or current has a maximum at a resonance and hence the admittance will have a maximum. From equation (4.48) we can write for a short-circuited line

$$Y_{11} = \frac{1}{Z_{11}} = \frac{1}{Z_0} \coth(\alpha + ik)l ,\qquad(4.54)$$

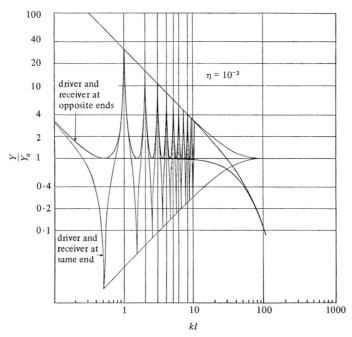

Figure 4.12. Computed response curves (relative admittance) for a rod vibrating longitudinally having a loss factor $\eta = Q^{-1} = 10^{-2}$. (After Biesterfeldt et al., 1960.)

where $(\alpha + i k)$ has been written for $i\hat{k} = i(k - i\alpha)$. On using the expansion formula for the hyperbolic function we obtain

$$|Y_{11}| = \frac{1}{Z_0}\left[\frac{\sinh^2\alpha l\cosh^2\alpha l + \sin^2 kl\cos^2 kl}{(\sinh^2\alpha l\cos^2 kl + \cosh^2\alpha l\sin^2 kl)^2}\right]^{\frac{1}{2}}. \qquad (4.55)$$

A graph of equation (4.55) is shown in figure 4.12 in relation to longitudinal waves in a rod with free ends. On the graph is also shown the result of carrying out a similar expansion for Y_{12} based on equation (4.51). Resonant peaks occur whenever the admittance has a maximum, that is whenever $kl = n\pi$ or $l = \frac{1}{2}n\lambda$, where n is an integer. When driver and receiver are at opposite ends of the rod, the minima correspond to an impedance equal to the characteristic impedance, Z_0, of the rod. When driver and receiver are at the same end, sharp antiresonances appear since the contributions of adjacent modes are out of phase.

4.5.5.2 Behaviour near resonance

In examining the behaviour of a resonant transmission line near a resonance it is useful to write the impedance in the form $\hat{Z} = R + iX$, where R is the resistance and X is the reactance. For instance, the input impedance of a short-circuited line may be written as $\hat{Z}_{11} = R_{11} + iX_{11}$. R_{11} and X_{11} may be found by expanding equation (4.48) and separating real and imaginary parts, assuming Z_0 to be real. Thus

$$R_{11} = Z_0\frac{\sinh\alpha l\cosh\alpha l}{\cosh^2\alpha l\cos^2 kl + \sinh^2\alpha l\sin^2 kl}, \qquad (4.56)$$

$$X_{11} = Z_0\frac{\sin kl\cos kl}{\cosh^2\alpha l\cos^2 kl + \sinh^2\alpha l\sin^2 kl}. \qquad (4.57)$$

These equations may be simplified if it is assumed that near resonance (i) the damping is small so that $\sinh\alpha l \to \alpha l$ and $\cosh\alpha l \to 1$; (ii) l can be taken as a multiple of $\frac{1}{4}\lambda$, that is $l = \frac{1}{4}m\lambda$ or $kl = \frac{1}{2}m\pi$, where m is an integer.

If m is odd, $\cos kl = 0$ and $\sin kl = 1$. Thus, for the $\frac{1}{4}\lambda$ line $R_{11} = Z_0/\alpha l$ and $X_{11} = 0$. If m is even, $\cos kl = 1$ and $\sin kl = 0$. For a $\frac{1}{2}\lambda$ line, $R_{11} = Z_0\alpha l$ and $X_{11} = 0$. As shown in elementary vibration texts, there is a **quality factor**, Q, associated with each resonance:

$$Q = \frac{\nu_0}{\Delta\nu} = \frac{k}{2\alpha}, \qquad (4.58)$$

where ν_0 is the resonant frequency and $\Delta\nu$ is the bandwidth of the resonance curve. Thus, for a line whose length is a multiple of $\frac{1}{4}\lambda$

$$\alpha_n l = \frac{m\pi}{4Q_n}, \qquad (4.59)$$

where n is the mode number ($m = 2n$). The resistance values computed earlier can therefore be written in the form

$$R_{11} = \frac{Z_0}{\alpha_n l} = \frac{4Q_n Z_0}{m\pi} , \qquad \text{if } m \text{ is odd,} \tag{4.60}$$

$$R_{11} = Z_0 \alpha_n l = \frac{m\pi Z_0}{4Q_n} , \qquad \text{if } m \text{ is even.} \tag{4.61}$$

The normal acoustical line is a half-wavelength resonator for which m is even and R_{11} varies as $1/Q_n$. For a highly resonant specimen R_{11} will be very low. Antiresonances occur at quarter wavelength intervals when m is odd. R_{11} now varies directly as Q_n and will generally be high.

A similar treatment to the above may be applied to the transfer impedance of a short-circuited line. Expanding equation (4.51) and applying the same approximations as before we obtain

$$R_{12} = 0 , \qquad X_{12} = Z_0 , \qquad \text{for } m \text{ odd} \tag{4.62}$$

$$R_{12} = Z_0 \alpha_n l = \frac{m\pi Z_0}{4Q_n} , \qquad X_{12} = 0 , \qquad \text{for } m \text{ even .} \tag{4.63}$$

For an open-circuited line similar expressions may be found by expanding equations (4.49) and (4.53). For this case resonances occur when m is odd and antiresonances when m is even.

4.5.6 Equivalence of transmission line and normal mode analysis near resonance

The classical method for describing the behaviour of mechanical and electrical systems is to find solutions of the appropriate wave equation subject to prescribed boundary conditions. For a continuous system which possesses normal modes of vibration this method is often unnecessarily tedious. Instead, the response of the system may be described in terms of normal coordinates appropriate to each mode and a set of mode parameters which may be chosen to represent physically measurable properties of the system, such as resonant frequencies, bandwidths and impedances. Once these parameters are known for each mode of vibration of the system, the response of the system may be computed for any given excitation function.

The essential feature of a normal mode is that it behaves like a simple oscillator and may therefore be described in terms of an equivalent mass–spring–damper model. The equation of motion for each mode will be

$$m_n \ddot{\xi}_n + r_n \dot{\xi}_n + \frac{\xi_n}{c_n} = F_n = \kappa_n F_0 , \tag{4.64}$$

where m_n, r_n, and c_n represent the mode mass, resistance, and compliance, respectively; ξ_n is a normal coordinate; F_n is the generalised force applicable to the nth mode; F_0 is the total driving force at a given instant of time; and κ_n is an **excitation constant**. κ_n is the Fourier component

of force that excites the nth mode divided by the total driving force. It
is defined by

$$\kappa_n F_0 = \int_V F(x, y, z) \frac{\xi_n(x, y, z)}{\xi_n(A)} \, dV \tag{4.65}$$

where

$$F_0 = \int_V F(x, y, z) \, dV \ .$$

$F(x, y, z)$ is the force distribution; $\xi_n(A)$ is the displacement function at
an arbitrary point of measurement specified by (x_A, y_A, z_A). κ_n therefore
gives a measure of the fraction of the total driving force that is available
for the excitation of a particular mode with reference to a particular
point of observation. In general, ξ_n and F are functions of x, y, z, and t.

As proposed by Skudrzyk (1958; 1968), the constant κ_n may be
absorbed in the mode parameters. Equation (4.64) may then be written

$$M_n \ddot{\xi}_n + R_n \dot{\xi}_n + \frac{\xi_n}{C_n} = F_0 \ . \tag{4.66}$$

The **modified mode parameters** are given by the following relationships

$$M_n = \frac{q_n}{\kappa_n} M \ , \qquad R_n = \frac{\omega_n^2}{\omega} \eta_n M_n \ , \qquad C_n = \frac{1}{\omega_n^2 M_n} \ . \tag{4.67}$$

q_n is a mode constant which represents the mean square of the displacement
at an arbitrary point A in the system and is defined by

$$q_n = \frac{\langle \xi_n^2 \rangle}{\xi_n^2(A)} \tag{4.68}$$

where $\langle \xi_n^2 \rangle$ is the space average of ξ_n^2 over the system. The form of the
mode constant q_n arises from the definition of M_n in terms of the fraction
of the total kinetic energy that is involved in the motion of the nth mode.
M is the total mass of the system and $\eta = 1/Q$ is the loss factor.

If ξ varies harmonically with time, equation (4.66) may be written

$$\left(-\omega^2 M_n + i\omega R_n + \frac{1}{C_n} \right) \xi_n = F_0 \ . \tag{4.69}$$

The solution of this equation in terms of displacement is

$$\xi_n = \frac{F_0}{i\omega \hat{Z}_n} \ ,$$

where

$$\hat{Z}_n = R_n + i\omega M_n + \frac{1}{i\omega C_n} \tag{4.70}$$

is the mode impedance. In terms of velocity the solution is

$$\hat{v}_n = \hat{\xi}_n = \frac{F_0}{\hat{Z}_n} = F_0 \hat{Y}_n \,,$$ (4.71)

where $\hat{Y}_n = 1/\hat{Z}_n$ is the mode admittance.

We can apply the principle of linear superposition to obtain the complete solution for all modes:

$$\hat{v} = \frac{F_0}{\hat{Z}} = F_0 \hat{Y} \,,$$ (4.72)

where

$$\hat{v} = \sum_n \hat{v}_n \,, \qquad \hat{Y} = \sum_n \hat{Y}_n = \sum_n \frac{1}{\hat{Z}_n} \,.$$

$\hat{v}$ and $\hat{Z}$ are usually evaluated for a particular point A in the system. If the system is freely suspended, additional impedances may be added to represent the motion of the centre of mass and any rotation around it. The form of the solution (4.72) suggests that the motion of a continuous system may be represented by a circuit containing an infinite number of parallel branches of series-resonant circuits, each of which represents a normal mode of the system. Such a circuit, based on the force–voltage analogy, is shown in figure 4.13.

As defined above, the mode parameters M_n, R_n, C_n, κ_n and q_n all depend on the nature of the driving force and its point of application, the shape of the normal mode functions of the system, and the coordinates of the point of observation. Likewise, Z_n is an impedance associated with the nth normal mode, the particular type of force, and the point of observation in the system relative to the point of application of the force. The mode impedance reduces to $Z_n = R_n$ at resonance, and for points just off resonance may be written in the form

$$Z_n = R_n(1 + iQ_n\gamma_n) \,,$$ (4.73)

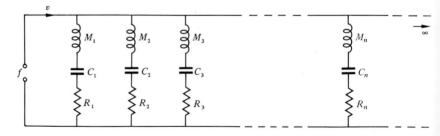

Figure 4.13. Analogous circuit for a mechanical system with continuously distributed mass and compliance (force–voltage analogy).

where

$$Q_n = \frac{\omega_n M_n}{R_n}$$

and

$$\gamma_n = \frac{\omega}{\omega_n} - \frac{\omega_n}{\omega} \ .$$

4.5.6.1 Mode parameters for longitudinal vibrations of a free rod

The normal modes have the form

$$\xi_n = |\xi_n| \cos k_n x \ ,$$

where $k_n = \omega_n/c$, $\omega_n = n\pi c/l$; l is the length of the rod and c the longitudinal phase velocity.

The mode constant q_n for a given reference point A is

$$q_n = \frac{\langle \xi_n^2 \rangle}{\xi_n^2(A)} = \frac{1}{2 \cos^2 k_n x_A} \ .$$

If the point A coincides with a point of maximum displacement, such as a free end of a rod, $q_n = \frac{1}{2}$.

If a point force is applied at the point $x = F$ and the motion is observed at $x = A$, then application of equation (4.65) leads to

$$\kappa_n(F, A) = \frac{\xi_n(F)}{\xi_n(A)} \ .$$

For this case we can write

$$\kappa_n(F, A) = \frac{\cos k_n x_F}{\cos k_n x_A} \ .$$

If F is at one end of the rod and A is at the other end, $\kappa_n = \pm 1$ according as n is even or odd. If F and A coincide, $\kappa_n = 1$. In table 4.2 are summarised the mode parameters required for longitudinal wave propagation in a rod, together with those for torsional and flexural waves and the corresponding values of the characteristic impedances.

As an illustration of the relationships shown in table 4.2, the following parameters have been computed for the first longitudinal and first torsional mode of the brass rod quoted in Experiment E4.2.

Longitudinal mode: $\nu_1 = 1994$ Hz, $\eta_1 = Q^{-1} = 8 \cdot 40 \times 10^{-5}$
  $M_1 = 0 \cdot 229$ kg
  $R_1 = 0 \cdot 241$ Ω
  $C_1 = 27 \cdot 8 \times 10^{-9}$ m N^{-1}

Torsional mode: $\nu_1 = 1516$ Hz, $\eta_1 = Q^{-1} = 5 \cdot 24 \times 10^{-5}$
  $I_1 = 5 \cdot 30 \times 10^{-6}$ kg m^2
  $R_1 = 0 \cdot 264 \times 10^{-6}$ Ω
  $C_1 = 20 \cdot 8 \times 10^{-4}$ N^{-1}.

E

Table 4.2. Resonant mode parameters for a rod.

Parameter	F and A at same end of rod	F and A at opposite ends of rod
q_n	$\frac{1}{2}$	$\frac{1}{2}$
κ_n	1	± 1
Longitudinal waves		
M_n	$\frac{1}{2}M$	$\pm\frac{1}{2}M$
R_n	$\omega_n M_n \eta_n$	$\pm\omega_n M_n \eta_n$
C_n	$\dfrac{1}{\omega_n^2 M_n}$	$\pm\dfrac{1}{\omega_n^2 M_n}$
$Z_0 = \rho c_L S$		
Torsional waves		
I_n	$\frac{1}{2}I$	$\pm\frac{1}{2}I$
R_n	$\omega_n I_n \eta_n$	$\pm\omega_n I_n \eta_n$
C_n	$\dfrac{1}{\omega_n^2 I_n}$	$\pm\dfrac{1}{\omega_n^2 I_n}$
$Z_0 = \rho c_T S$		

Flexural waves
M_n, R_n, C_n as for longitudinal waves
$$Z_0 = \frac{1+\mathrm{i}}{l}(\omega_n c_F K)^{\frac{1}{2}}M$$

The + sign applies if the mode number n is even, the − sign if n is odd. The sign governs the phase relations and may be omitted if magnitudes only are relevant. $\eta = 1/Q_n$ is the loss factor. For torsional waves, I is the moment of inertia about the axis of the rod and $C_n = \theta/\tau$ is an angular compliance where θ is the angular displacement and τ is the torque. K is the radius of gyration of cross-section about an axis perpendicular to the displacement.

4.5.6.2 *Comparison of transmission line and normal mode analysis*
For a half-wavelength transmission line the resistances R_{11} and R_{12} are both given by

$$R_{11} = R_{12} = \frac{m\pi Z_0}{4Q_n} \ .$$

Since for longitudinal waves in a rod

$$Z_0 = \rho c S \ , \qquad \rho = \frac{M}{lS} \ , \qquad c = \frac{\omega_n l}{n\pi} \ , \qquad m = 2n \ ,$$

then

$$R_{11} = R_{12} = \frac{\omega_n M}{2Q_n} \ ,$$

which is identical with the value of R_n in equation (4.67). The values of M_n and C_n can also be made to agree with the equivalent inductance, L_n, and capacitance, C_n, of a half-wavelength transmission line, since (Mason, 1948)

$$L_n = \tfrac{1}{2}Ll, \qquad C_n = \frac{2}{\pi^2}Cl,$$

where L and C are the inductance and capacitance per unit length of line respectively and l is the length of the line. As a check, the geometric mean of the two values of R_{11} given by equations (4.60) and (4.61) is Z_0, in agreement with the corresponding result from normal-mode theory.

Hence, normal mode theory and transmission line theory produce the same results for systems whose lengths are multiples of $\tfrac{1}{4}\lambda$ or $\tfrac{1}{2}\lambda$. Transmission line theory is the more general, since it may be used to find the response at any point along a line, whether resonant or not. In many acoustical problems, where the lengths involved are usually multiples of $\tfrac{1}{2}\lambda$, a discussion in terms of the mode parameters is often sufficient.

4.5.7 Pulse response of a finite rod

When a stress pulse is applied to one end of a finite rod, a series of pulses is developed by multiple reflection from the ends. If the duration of the initial pulse is less than the transit time in the specimen, the series of pulses will be clearly separated. On the other hand, the separate pulses may overlap so that the response of the end faces of the rod may be represented by a series.

If the ends of the rod are taken at $x = 0$ and $x = l$ and, for simplicity, the pulse $f(x, t)$ is assumed to be in the form of a gated continuous wave so that

$$f(x, t) = v_0\exp(-\alpha x)\exp[i(\omega t - kx)] \qquad \text{for } 0 < (t - x/c) < \tau,$$
$$= 0 \qquad\qquad\qquad\qquad\qquad \text{elsewhere},$$

where v_0 represents the input particle velocity amplitude, then the total response at the input end of the rod may be found as follows.

Suppressing the time factor, we can write the velocity at $x = l$ as

$$\hat{v}_l = \hat{v}_0\exp(-i\hat{k}l),$$

where $\hat{k} = k - i\alpha$. The pulse is then reflected from the free end and on return to the input end has the value

$$\hat{v}_2 = \hat{v}_0\exp(-2i\hat{k}l).$$

But we have shown that at a free end of a rod the particle velocity is twice its value when travelling along the rod. Hence the total response at the input end of the rod is

$$\hat{v}_1 = \hat{v}_0 + 2\hat{v}_0\exp(-2i\hat{k}l) + 2\hat{v}_0\exp(-4i\hat{k}l) + \dots = \hat{v}_0\coth i\hat{k}l, \qquad (4.74)$$

if the separate pulse responses are assumed to be in phase (corresponding to a resonant condition). If $\hat{f}_1$ is the force applied at $x = 0$, then

$$\hat{Z}_{11} = \frac{\hat{f}_1}{\hat{v}_1} = \hat{Z}_0 \tanh i\hat{k}l , \tag{4.75}$$

which is the same as equation (4.48) obtained from transmission line theory. In equation (4.75) the ratio $\hat{f}_1/\hat{v}_0$ has been put equal to $\hat{Z}_0$, the impedance for an infinite rod, since, to the first pulse entering the rod, the rod appears to be infinitely long. It is not until the second reflected pulse adds further information to the vibrations of the input end that the vibration begins to assume a periodic character.

The total response at the second end face of the rod may similarly be found, since

$$\hat{v}_2 = 2\hat{v}_0 \exp(-i\hat{k}l) + 2\hat{v}_0 \exp(-3i\hat{k}l) + 2\hat{v}_0 \exp(-5i\hat{k}l) + \dots .$$

Dividing each side by $\hat{f}_1$ we obtain

$$\frac{1}{\hat{Z}_{12}} = \frac{1}{\hat{Z}_0} 2 \exp(-i\hat{k}l)[1 + \exp(-2i\hat{k}l) + \exp(-4i\hat{k}l) + \dots] .$$

In this case it is instructive to consider only the effect of the first two signals; then

$$\hat{Z}_{12} = \tfrac{1}{2}\hat{Z}_0 \exp(i\hat{k}l)[1 - \exp(-2i\hat{k}l)] = \hat{Z}_0 \sinh i\hat{k}l . \tag{4.76}$$

Equation (4.76) is the same as equation (4.51) for the transfer impedance of a short-circuited transmission line. Thus, the periodic response of the system is established provided the end under consideration is still vibrating when the second reflected pulse arrives. The preceding discussion indicates that, when the series of reflected pulses are additive, the pulse response becomes identical with the response under continuous wave excitation.

4.5.8 Composite systems

In many experimental arrangements two or more resonant elements are connected together to form a composite oscillator. Consider a double system consisting of two bars of different material cemented together and having the same cross-section. It will be assumed that the cement is perfect, that is that there is no energy loss at the junction. The behaviour of the system may be conveniently represented by a transmission line (line 1) to which the driving force is applied terminated by a second transmission line (line 2) representing the second bar.

The input impedance of line 2 alone is given by equation (4.48) for short-circuit conditions: $\hat{Z}_{22} = \hat{Z}_{02} \tanh i\hat{k}_2 l_2$, where $\hat{Z}_{02}$ is the characteristic impedance of line 2. If this impedance now acts as a termination to line 1, then from equation (4.47) we have

$$\hat{Z}_{11} = \hat{Z}_{01} \frac{\hat{Z}_{01} \sinh i\hat{k}_1 l_1 + \hat{Z}_{22} \cosh i\hat{k}_1 l_1}{\hat{Z}_{01} \cosh i\hat{k}_1 l_1 + \hat{Z}_{22} \sinh i\hat{k}_1 l_1}$$

$$= \hat{Z}_{01} \frac{\hat{Z}_{01} \tanh i\hat{k}_1 l_1 + \hat{Z}_{02} \tanh i\hat{k}_2 l_2}{\hat{Z}_{01} + \hat{Z}_{02} \tanh i\hat{k}_1 l_1 \tanh i\hat{k}_2 l_2} \ . \tag{4.77}$$

If attenuation within each line is neglected, then the $\hat{k}$'s are real and $\tanh kl$ may be replaced by $i \tan kl$. Equation (4.77) then becomes

$$Z_{11} = Z_{01} \frac{iZ_{01} \tan k_1 l_1 + iZ_{02} \tan k_2 l_2}{Z_{01} - Z_{02} \tan k_1 l_1 \tan k_2 l_2} \ .$$

The composite system will resonate at frequencies ν_c for which the impedance Z_{11} is a minimum (in this case, zero). Thus,

$$Z_{01} \tan k_1 l_1 + Z_{02} \tan k_2 l_2 = 0 \ , \tag{4.78}$$

or, since $k = \omega/c = 2\pi\nu_c/c$,

$$\rho_1 c_1 \tan\frac{2\pi\nu_c l_1}{c_1} + \rho_2 c_2 \tan\frac{2\pi\nu_c l_2}{c_2} = 0 \ . \tag{4.79}$$

The velocities c_1 and c_2 may be expressed in terms of the resonant frequencies of each line when determined separately. Thus, c_1 and c_2 may be written

$$c_1 = \frac{2l_1 \nu_1}{n_1} \ , \qquad c_2 = \frac{2l_2 \nu_2}{n_2} \ ,$$

where ν_1 is the fundamental resonant frequency of a length l_1 of line 1, ν_2 is the fundamental resonant frequency of a length l_2 of line 2, and n_1 and n_2 are the harmonic mode numbers of each line. Hence

$$\rho_1 c_1 \tan\frac{n_1 \pi\nu_c}{\nu_1} + \rho_2 c_2 \tan\frac{n_2 \pi\nu_c}{\nu_2} = 0 \ . \tag{4.80}$$

A common practical case occurs when $n_1 = n_2 = 1$, for which equation (4.80) becomes

$$M_1 \nu_1 \tan\frac{\pi\nu_c}{\nu_1} + M_2 \nu_2 \tan\frac{\pi\nu_c}{\nu_2} = 0 \ , \tag{4.81}$$

where M_1 and M_2 are the masses of the two bars respectively.

4.5.8.1 *Equal transit times*
When the transit times in each bar are approximately equal, that is the frequencies ν_1 and ν_2 are approximately equal, it may be shown (Harris, 1963) that equation (4.81) reduces to

$$\nu_c = \frac{M_1 \nu_1 + M_2 \nu_2}{M_1 + M_2} \ . \tag{4.82}$$

One reason for adopting this case is to make the cemented junction coincide with a stress (strain) node, as shown in figure 4.14, when there will be a minimum damping effect caused by the bond. Also, the system may then be supported at two displacement nodes.

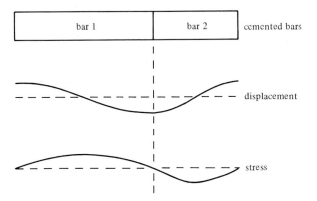

Figure 4.14. Displacement and stress (strain) distributions for two cemented bars $(n_1 = n_2 = 1)$ having equal transit times.

4.5.8.2 Mass loading

When v_2 is greater than v_1 by at least an order of magnitude, $\pi v_c/v_1 \approx \pi$ and $\pi v_c/v_2$ is small. Then equation (4.81) reduces to

$$v_c = \frac{M_1 v_1}{M_1 + M_2} .$$
(4.83)

Thus the difference between v_c and v_1 is due to a mass-loading effect. When a transducer is attached to a specimen it may be necessary to apply a correction to the resonant frequency of the specimen by assuming that the transducer acts as a mass loading.

4.5.8.3 General case

It is not always necessary for each component bar to be a half-wavelength resonator. For instance, if v_2 is required to be measured and the values of M_1, M_2, v_1, v_c are known, then the following method described by Harris (1963; 1966) can be used. Equation (4.81) may be rewritten in the form

$$-\frac{M_1 v_1}{M_2 v_c} \tan \frac{\pi v_c}{v_1} = K \tan \frac{\pi}{K} ,$$
(4.84)

where $K = v_2/v_c$. The left hand side contains only known quantities and the value of v_2 can then be computed from a table of $K \tan(\pi/K)$ versus K (see Experiment E4.4 for an example).

4.5.8.4 *Discrete cross-section change*

In the treatment described in the preceding section it was assumed that the component bars had the same cross section. Redwood and Lamb (1957) have shown that, if there is an abrupt change in cross-section, higher order modes must be generated at the discontinuity in order to satisfy the boundary conditions. Such higher order modes are attenuated rapidly away from the junction and they can be accounted for by placing a shunt capacitance at the discontinuity (Whinnery and Jamieson, 1944). Applying the method of this section, we may regard line 1 to be terminated by a parallel connection of a shunt capacitance C and the short-circuited line 2. Using equation (4.47) we obtain the resonant condition

$$Z_{01} \tan k_1 l_1 + Z_{02} \tan k_2 l_2 - \omega C Z_{01} Z_{02} \tan k_1 l_1 \tan k_2 l_2 = 0 , \tag{4.85}$$

where the characteristic impedances are now written fully as $Z_{01} = \rho_1 c_1 S_1$, etc., S_1 being the area of cross section of bar 1. We can find the value of C for a given cross-sectional change by making $k_1 = k_2 = k$ and $l_1 = l_2 = l$; then from equation (4.85) we obtain

$$C = \frac{Z_{01} + Z_{02}}{\omega Z_{01} Z_{02} \tan kl} . \tag{4.86}$$

4.5.8.5 *Triple composite system*

The transmission line method may be extended to a system consisting of three elements. In practice this may correspond to three bars cemented together, or, in a two-bar system at high frequencies, it may be necessary to represent the bond by a third line. The resonance condition (with damping assumed to be negligible) for this case is

$$Z_{01} \tan k_1 l_1 + Z_{02} \tan k_2 l_2 + Z_{03} \tan k_3 l_3$$

$$+ \frac{Z_{01} Z_{03}}{Z_{02}} \tan k_1 l_1 \tan k_2 l_2 \tan k_3 l_3 = 0 . \tag{4.87}$$

4.5.8.6 *Damping conditions*

If the damping is small but not negligible, the hyperbolic functions in the transmission line equations may be expanded and approximated as follows:

$$\sinh i\hat{k}l \approx \alpha l \cos kl + i \sin kl ,$$

$$\cosh i\hat{k}l \approx \cos kl + i\alpha l \sin kl ,$$

$$\tanh i\hat{k}l \approx \frac{\alpha l + i \sin kl \cos kl}{\cos^2 kl + (\alpha l)^2 \sin^2 kl} .$$

For half-wavelength lines ($l = \frac{1}{2}n\lambda$, n integral):

$$\sinh ikl \approx \alpha l ,$$

$$\cosh ikl \approx 1 ,$$

$$\tanh ikl \approx \alpha l .$$

For instance, the equation corresponding to equation (4.87) is then

$$Z_{11} = Z_{01}\alpha_1 l_1 + Z_{02}\alpha_2 l_2 + Z_{03}\alpha_3 l_3 \ . \tag{4.88}$$

This equation may be related to the attenuation coefficient of the whole system if Z_{11} is written as $Z_{11} = Z_0\alpha l$. Equation (4.88) may also be found by computing the transfer impedance, Z_{13}, of the system.

Experiment E4.4   Composite oscillators
(Harris, R. W., 1963, *Aust. J. Appl. Sci.*, **14**, 213. Harris, R. W., 1971, *Proc. IREE (Aust.)*, **32**, 304.)

Double composite bar
Harris (1963) made a test of the frequency equation (4.81) for a double composite system, using a brass–aluminium composite bar of constant cross-section. The resonant frequency of the aluminium bar, ν_2, was varied by the use of a series of bars of different length, while the frequency of the brass bar, ν_1, remained constant at 9·61 kHz. An electrostatic drive and detection system was employed and the resonant condition found corresponding to each bar containing half a wavelength. As shown in figure E4.7 good agreement was found between the experimental points and the theoretical curve.

The frequency ν_2 of an unknown specimen may be determined from measurement of the lowest composite frequency, corresponding to $n_1 = n_2 = 1$, and application of equation (4.83). As a test of this method, a brass bar ($M_1 = 0·0521$ kg, $\nu_1 = 9·61$ kHz) was cemented to an aluminium bar ($M_2 = 0·0305$ kg) and the composite bar frequency found to be $\nu_c = 6·63$ kHz. The value of $K\tan(\pi/K)$ is therefore 3·628 and from an appropriate table (given in the reference) a value of ν_2 of 33·3 kHz is found. The value of ν_2 found by direct measurement was 33·7 kHz.

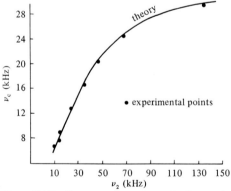

Figure E4.7. Resonant frequency of a brass–aluminium composite bar, ν_c, as a function of the resonant frequency of an aluminium bar, ν_2.

Effect of a small perturbation
The effect produced on the resonant frequency of a bar by a change in mechanical properties at a specific location along the bar has been investigated by the transmission line method. Three approaches may be used: (1) triple transmission line theory may

be applied in which the perturbation is represented by an inserted transmission line having a different characteristic impedance from the original line; (2) if the perturbed part is small, it may be regarded as a shunt reactance placed in the original line; (3) the original line can be subdivided into a large number of segments and one of these replaced by the perturbation.

In examining the latter method, Harris (1971) divided the line into 40 segments. The load impedance for each transmission line segment is the input impedance of the previous segment and hence the input impedance of the perturbed transmission line can be determined and the resonant frequency, corresponding to a zero in the final input impedance, can be computed. The calculation was carried out in a computer, a flow diagram being given in the reference. The characteristic impedance of the modified segment was made γ times, and the propagation constant $1/\gamma$ times, the normal value. Computations were made for values of γ of $1 \cdot 01$, $1 \cdot 05$, and $1 \cdot 10$. In figure E4.8 is shown a plot of ν'_c/ν_c versus discontinuity position, where ν'_c is the resonant frequency of the perturbed line. It is found that a maximum effect is produced at the centre. A comparison of the magnitude of the effect when the perturbation is situated at the centre of the rod for the different values of γ is shown below:

Modification factor	Ratio of frequencies, ν'_c/ν_c
$1 \cdot 01$	$1 \cdot 00049$
$1 \cdot 05$	$1 \cdot 00233$
$1 \cdot 10$	$1 \cdot 00435$

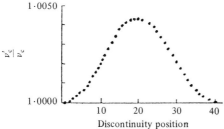

Figure E4.8. Ratio of resonant frequency of perturbed line, ν'_c, to unperturbed line, ν_c, as a function of the position of the modified segment ($\gamma = 1 \cdot 10$).

4.6 Solid plate

4.6.1 Infinite isotropic plate

Propagation of plane sinusoidal waves in an infinite free plate of thickness h can be analysed by a waveguide approach similar to that which has been used for the solid cylinder (section 4.4). The direction of propagation will be the $+x_3$ direction with the x_1 direction normal to the surface of the plate. The possible wave types in such a plate are longitudinal, flexural, and transverse (particle displacement in the x_2 direction). The treatment of Cremer (1948) will be followed which will provide solutions for the longitudinal and flexural waves. Solutions for the transverse waves are given in Redwood (1960).

Solutions are sought for the wave equations

$$\nabla^2 \phi = \frac{1}{c_\ell^2}\frac{\partial^2 \phi}{\partial t^2} , \qquad \nabla^2 \psi_i = \frac{1}{c_t^2}\frac{\partial^2 \psi_i}{\partial t^2} , \qquad i = 1, 2, 3 . \tag{4.89}$$

As before, the displacement is written as

$$u_i = \text{grad}\,\phi + \text{curl}\,\psi_i . \tag{4.90}$$

The boundary conditions require that

$$\sigma_{11} = \sigma_{21} = \sigma_{31} = 0 ,$$

where σ_{21} is automatically zero since there is no motion in the x_2 direction for the assumed longitudinal and flexural waves. On using the method of separation of variables we obtain the solutions

$$\phi = \phi_1(x_1)\exp[i(k_3 x_3 - \omega t)] , \tag{4.91}$$

$$\psi_i = \psi_1(x_1)\exp[i(k_3 x_3 - \omega t)] , \tag{4.92}$$

where

$$\phi_1(x_1) = A_1 \cosh\left\{\left[k_3^2 - \left(\frac{\omega}{c_L}\right)^2\right]^{\frac{1}{2}} x_1\right\} + A_2 \sinh\left\{\left[k_3^2 - \left(\frac{\omega}{c_L}\right)^2\right]^{\frac{1}{2}} x_1\right\} , \tag{4.93}$$

$$\psi_1(x_1) = B_1 \cosh\left\{\left[k_3^2 - \left(\frac{\omega}{c_T}\right)^2\right]^{\frac{1}{2}} x_1\right\} + B_2 \sinh\left\{\left[k_3^2 - \left(\frac{\omega}{c_T}\right)^2\right]^{\frac{1}{2}} x_1\right\} , \tag{4.94}$$

$c_L = (E/\rho)^{\frac{1}{2}}$ is the longitudinal velocity in a thin rod of the same material, and $c_T = (\mu/\rho)^{\frac{1}{2}}$ is the torsional velocity in a rod of the same material.

For flexural waves, the second term of equation (4.93) and the first term of equation (4.94) apply. On combining these with the boundary conditions we obtain the frequency equation

$$\left\{1 - \frac{1}{2}\left(\frac{c}{c_T}\right)^2\right\}^2 \tanh\left\{\frac{\omega h}{2c}\left[1 - \left(\frac{c}{c_L}\right)^2\right]^{\frac{1}{2}}\right\}$$
$$= \left[1 - \left(\frac{c}{c_L}\right)^2\right]^{\frac{1}{2}}\left[1 - \left(\frac{c}{c_T}\right)^2\right]^{\frac{1}{2}} \tanh\left\{\frac{\omega h}{2c}\left[1 - \left(\frac{c}{c_T}\right)^2\right]^{\frac{1}{2}}\right\}. \tag{4.95}$$

At low frequencies and with the assumption that $c < c_L$, $c < c_T$, equation (4.95) reduces to a form in agreement with the simpler classical theory (Skudrzyk, 1968):

$$c_F^4 = \frac{c_T^2 \omega^2 h^2}{6(1-\mu)} , \tag{4.96}$$

where c_F is the phase velocity of flexural waves and μ is Poisson's ratio. It may be shown that, as $\omega \to \infty$, equation (4.95) yields the velocity of Rayleigh surface waves, c_R. The velocity of the flexural waves increases asymptotically to c_R, approaching within 10% when $h = 0.3\lambda_L$, where

λ_L is the wavelength of longitudinal waves in the plate. This distribution of values of c_F gives the first flexural mode for a plate.

For longitudinal waves, the first term of equation (4.93) and the second term of equation (4.94) apply. Together with the boundary conditions this yields the frequency equation

$$\left\{1 - \frac{1}{2}\left(\frac{c}{c_T}\right)^2\right\}^2 \coth\left\{\frac{\omega h}{2c}\left[1 - \left(\frac{c}{c_L}\right)^2\right]^{\frac{1}{2}}\right\}$$

$$= \left[1 - \left(\frac{c}{c_L}\right)^2\right]^{\frac{1}{2}}\left[\left(\frac{c}{c_T}\right)^2 - 1\right]^{\frac{1}{2}} \coth\left\{\frac{\omega h}{2c}\left[\left(\frac{c}{c_T}\right)^2 - 1\right]^{\frac{1}{2}}\right\} \qquad (4.97)$$

If it is assumed that $c > c_T$, then at low frequencies equation (4.97) reduces to the velocity equation given by the simpler theory

$$c_L^2 = \frac{2c_T^2}{1 - \mu} . \qquad (4.98)$$

At higher frequencies

$$c = c_L\left[1 - \frac{\mu(\omega h)^2}{24(1 - \mu)c_L^2}\right] . \qquad (4.99)$$

For a material with $\mu = 0 \cdot 3$, the correction term in equation (4.99) reaches 8% when λ_L falls to a value of $3h$. As $\omega \to \infty$, c reduces asymptotically to the velocity of Rayleigh surface waves. The variation of c with frequency gives the first longitudinal mode for a plate.

In the range $c > c_L$ both equations (4.95) and (4.97) yield many values of ω for equal phase velocities c. Consequently, distributions of phase velocity are obtained which correspond to higher modes of both flexural and longitudinal waves. As $\omega \to \infty$, the phase velocity of the higher modes tends asymptotically to the velocity of transverse waves. These higher wave types are characterised by the occurrence of nodal surfaces parallel to the plate boundaries, beyond which the oscillations have opposite phase. Phase velocity curves for the first few longitudinal and flexural modes are shown in figure 4.15.

An interesting confirmation of the theory for a plate comes from experiments conducted by Sanders (1939). He observed ultrasound transmission maxima at certain combinations of frequency and angles of incidence with foils of brass and nickel immersed in ethyl acetate. Gotz (1943a, 1943b) interpreted these results in terms of the coincidence effect which describes structural resonances occurring as a function of angle of incidence (the resonance-like effect is due to efficient coupling which occurs when the resolved wavenumber for the incident wave is equal to the flexural wavenumber for the material). Since the foils were thin (less than 1 mm thick) and the frequencies high (above 1 MHz), many experimental points were obtained involving a number of different modes. Gotz superimposed Sander's results on the curves of figure 4.15 with good agreement.

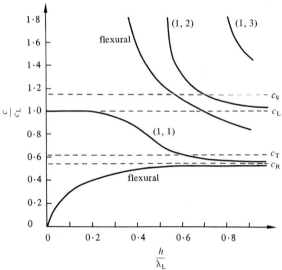

Figure 4.15. Phase velocity curves for longitudinal and flexural waves in a solid plate ($\mu = 0\cdot31$).

4.6.2 Resonant modes of finite isotropic plates

The particular normal modes for a plate depend on its shape and on the method of support. Often in practice a plate is simply supported at its edges. Under these conditions there is no transverse motion but the slope of the plate is not constrained. Agreement between theory and experiment is not good because of effects produced by friction at the edges and the finite mass of the supports.

The eigenfunctions for a freely supported rectangular plate are usually assumed to have the form

$$v(x, y, t) = V_{mn} \sin\frac{m\pi x}{l_1} \sin\frac{n\pi y}{l_2} \cos\omega_{mn}t \,, \tag{4.100}$$

where V_{mn} is the velocity amplitude of the (m, n) mode, and

$$\omega_{mn} = \left[\frac{h^2 E}{12(1-\mu^2)\rho}\right]^{\frac{1}{2}} \pi^2\left(\frac{m^2}{l_1^2} + \frac{n^2}{l_2^2}\right).$$

Somewhat better results may often be obtained by assuming that the velocity is a maximum at the edges and therefore replacing the sines by cosines.

As with the rod, the normal-mode analysis procedure proposed by Skudrzyk (1968) and summarised in section 4.5.6 will be followed. If the plate is excited by a point force at its centre and the velocity measured at the same point, then the excitation constants κ_{mn} are unity. If the plate is driven at its centre and the velocity measured at a corner, the values of κ_{mn} are alternately positive and negative. The mode masses and resistances (incorporating the excitation constants as before) are

$$M_{mn} = \frac{q_{mn}M}{\kappa_{mn}} = q_{mn}M \; ; \qquad R_{mn} = \frac{\eta\omega_{mn}^2 M_{mn}}{\omega} \, , \qquad (4.101)$$

where M is the total mass. The mode constants q_{mn} have the values $\frac{1}{4}$ if $m \neq 0$ or $n \neq 0$, and $\frac{1}{2}$ if either $m = 0$ or $n = 0$. The impedance for a given mode is

$$Z_{mn} = R_{mn} + i\left(\omega M_{mn} - \frac{1}{\omega C_{mn}}\right).$$

Craggs (1969) has shown that when a step input function is applied to a freely supported rectangular plate, the relative displacements at the centre form a series $1/mnr_{mn}^2$, where r is the frequency ratio relative to the fundamental. For the first three modes the relative displacements are $1 : 0 \cdot 025 : 0 \cdot 01$. The first mode carries almost 97% of the total displacement.

Skudrzyk (1968) has given the necessary equations and parameters for centrally symmetrical vibrations of a circular plate. Some of the problems connected with the free vibration of anisotropic rectangular plates are discussed by Whitney (1972).

4.7 Analysis of linear systems
A common method for evaluating the response of a linear system is the use of steady state analysis. If the input is a sinusoidal function of known frequency and constant amplitude, then the output will also be a sinusoidal function of the same frequency but with modified amplitude and phase. The function that describes the way in which the system performs this modification of amplitude and phase is called the **transfer function**, H(iω). In a mechanical system, H(iω) specifies the frequency response of the system to a unit input force at each frequency (the spectrum of a delta function force). If the force is an arbitrary one with an input spectrum, denoted by F(iω), then the output spectrum function, G(iω), of the system can be expressed

$$G(i\omega) = F(i\omega)H(i\omega) . \qquad (4.102)$$

The three quantities in equation (4.102) are in general complex and have the form

$$H(i\omega) = |H(i\omega)| \exp[i\theta(\omega)] , \qquad (4.103)$$

where $|H(i\omega)|$ is the amplitude characteristic of the system and $\theta(\omega)$ is the phase characteristic of the system.

It is sometimes convenient to separate the real and imaginary parts:

$$H(i\omega) = A(\omega) + iB(\omega) , \qquad (4.104)$$

whence

$$|H(i\omega)|^2 = \{A(\omega)\}^2 + \{B(\omega)\}^2 \,,$$

$$\theta(\omega) = \tan^{-1}\frac{B(\omega)}{A(\omega)} \,. \tag{4.105}$$

When the input signal consists of more than one sinusoidal component, the output of the system may be evaluated if it is assumed that the principle of superposition holds, that is that the system is linear.

A linear system can also be examined in the time domain, as shown schematically in figure 4.16. In correspondence with the method used for frequency response, in order to find the response of a linear system at a specific time it is necessary to know the function $h(t)$, called the impulse response function. $h(t)$ measures the response of the system to a delta function input, which is an impulse of amplitude unity that occurs at a particular instant of time (see Appendix 4.2). The output of the system at a particular time for an arbitrary input function $f(t)$ can then be written

$$g(t) = f(t)h(t) \,. \tag{4.106}$$

When $f(t)$ is a delta function, $g(t) = h(t)$.

In the time domain it is often required to find the total time response of the system. To do this it is necessary to use the convolution theorem

$$g(t) = \int_{-\infty}^{\infty} f(\tau)h(t-\tau)\,d\tau \,, \tag{4.107}$$

where τ is an auxiliary time variable. This is a difficult equation to evaluate and is further discussed in Appendix 4.3. In practice it is often easier to evaluate $g(t)$ by a Fourier transform method.

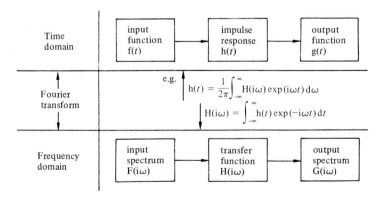

Figure 4.16. Response diagram for a linear system.

4.7.1 Fourier transform method

A time function may be transformed into a frequency function, and vice versa, by means of Fourier transforms (Appendix 4.3). For instance, the input function and input spectrum are related by the following integrals

$$f(t) = \frac{1}{2\pi} \int_{-\infty}^{\infty} F(i\omega) \exp(i\omega t)\, d\omega \,, \tag{4.108}$$

$$F(i\omega) = \int_{-\infty}^{\infty} f(t) \exp(-i\omega t)\, dt \,. \tag{4.109}$$

Analysis in the time domain may be avoided by procedures such as the following which apply Fourier integrals.

(i) Finding $H(i\omega)$ from known functions $f(t)$ and $g(t)$.

The first step is to find the Fourier transform of the input function. This may be written as

$$F(i\omega) = \mathcal{F}[f(t)] \,,$$

where the symbol $\mathcal{F}$ denotes the operation specified by equation (4.109). The Fourier transform of the output function is then

$$G(i\omega) = \mathcal{F}[g(t)] \,.$$

The transfer function is now given by

$$H(i\omega) = \frac{G(i\omega)}{F(i\omega)} = \frac{\mathcal{F}[g(t)]}{\mathcal{F}[f(t)]} \,. \tag{4.110}$$

If required, the impulse response $h(t)$ may now be found as the inverse Fourier transform of $H(i\omega)$:

$$h(t) = \mathcal{F}^{-1}[H(i\omega)] \,,$$

where the symbol $\mathcal{F}^{-1}$ denotes the operation specified by equation (4.108).

(ii) Finding $g(t)$ from known functions $f(t)$ and $H(i\omega)$.

The process to be followed is shown in figure 4.17. First, the Fourier transform of $f(t)$ is found, then the output spectrum, and finally the output function. Thus,

$$g(t) = \mathcal{F}^{-1}[G(i\omega)] = \mathcal{F}^{-1}[F(i\omega)\, H(i\omega)] = \mathcal{F}^{-1}[\mathcal{F}[f(t)] H(i\omega)] \,.$$

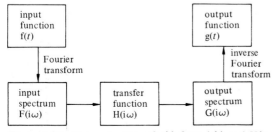

Figure 4.17. Determination of $g(t)$ from $h(t)$ and $H(i\omega)$.

4.7.2 Impulse methods

As suggested by the previous analysis, a quick method for finding the impulse response, $h(t)$, of a linear system is to apply a delta function as input. The Fourier transform of a delta function is unity, $\mathscr{F}[\delta(t)] = 1$, i.e. its spectrum has unit value at all frequencies. Hence, measurement of the output spectrum of a system subjected to a delta function input determines its transfer function directly, since then $G(i\omega) = H(i\omega)$. In practice the actual form of the impulse function is immaterial provided it occurs in a sufficiently short interval of time. In the limit any simple pulse shape will tend towards a delta function.

For instance, consider a rectangular impulse function as shown in figure 4.18 and defined as

$$f(t) \begin{cases} = \dfrac{1}{\tau} & 0 < t < \tau\,, \\ = 0 & \text{elsewhere}\,. \end{cases}$$

Then the Fourier transform of $f(t)$ is

$$F(i\omega) = \int_{-\infty}^{\infty} f(t)\exp(-i\omega t)\,dt = \frac{1}{\tau}\int_{0}^{\tau}\exp(-i\omega t)\,dt$$

$$= \frac{1}{\omega\tau}[\sin\omega\tau + i(\cos\omega\tau - 1)]\,.$$

Hence

$$|F(i\omega)| = \frac{\sin(\tfrac{1}{2}\omega\tau)}{\tfrac{1}{2}\omega\tau}\,,$$

and in the limit as $\tau \to 0$,

$$|F(i\omega)| \to \lim_{\tau \to 0} \frac{\sin(\tfrac{1}{2}\omega\tau)}{\tfrac{1}{2}\omega\tau} \to 1\,.$$

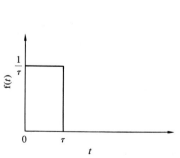

Figure 4.18. Rectangular pulse.

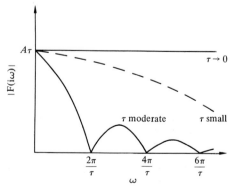

Figure 4.19. Spectrum of rectangular pulse as the duration becomes shorter.

For a rectangular pulse of arbitrary height A, $|F(i\omega)| \to A\tau$, where $A\tau$ is called the strength of the pulse. The effect of shortening the duration of a rectangular pulse on its spectrum is shown in figure 4.19, where it is clear that the briefer the pulse the more it tends towards a delta function. This behaviour is an illustration of the uncertainty principle in the form $\Delta\nu\,\Delta t \approx 1$. Similar results may be obtained from pulses of other shapes such as half-cosine, cosine-squared, triangular, Gaussian, etc.

The transient behaviour of a system may alternatively be described in terms of its response to a unit step function. The two approaches are closely related since the unit step function may be expressed as the integral of the delta function. Conversely, differentiating a unit step function generates a delta function.

The object of impulse methods is to provide a fast analysis of the system. The entire frequency response may be found from a single excitation of the system. An illustration of an impulse method is given in Experiment E5.1. On occasions delta function impulses prove to be too severe on the system or may be difficult to generate. An interesting modified impulse method has been described by White (1971). A swept frequency signal of constant amplitude is generated over a limited frequency range and applied to the system through a suitable transducer. The time-limited and frequency-limited signal then acts as a modified delta function that does not make too severe demands on the system under test.

References

Abramson, H. N., 1957, *J. Acoust. Soc. Am.*, **29**, 42.
Bancroft, D., 1941, *Phys. Rev.*, **59**, 588.
Beranek, L., 1954, *Acoustics* (McGraw-Hill, New York).
Biesterfeldt, H. J., Lange, J. N., Skudrzyk, E. J., 1960, *J. Acoust. Soc. Am.*, **32**, 749.
Cherry, C., 1949, *Pulses and Transients in Communication Circuits* (Chapman and Hall, London).
Chree, C., 1889, *Trans. Camb. Phil. Soc.*, **14**, 250.
Craggs, A., 1969, Ph. D. Thesis, University of Southampton.
Cremer, L., 1948, *The Propagation of Structure-Borne Sound*, Report No.1 (Series B) (DSIR, London).
Davies, R. M., 1948, *Phil. Trans.*, **A240**, 375.
Davies, R. M., 1956, *Survey in Mechanics* (Cambridge University Press, Cambridge).
Dirac, P. A. M., 1935, *The Principles of Quantum Mechanics* (Oxford University Press, London).
Firestone, F. A., 1956, *J. Acoust. Soc. Am.*, **28**, 1117.
Gotz, J., 1943a, *Akust. Zh.*, **8**, 145.
Gotz, J., 1943b, *Z. angew. Math. u. Mech.*, **23**, 294.
Green, W. A., 1960, *Prog. in Solid Mech.*, **1**, 223.
Harris, R. W., 1963, *Aust. J. Appl. Sci.*, **14**, 213.
Harris, R. W., 1966, *Proc. IREE (Australia)*, **27**, 280.
Lanczos, C., 1957, *Applied Analysis* (Pitman, London), chapter 4.
Love, A. E. H., 1927, *The Mathematical Theory of Elasticity* (Cambridge University Press, Cambridge).

Mason, W. P., 1948, *Electromechanical Transducers and Wave Filters* (Van Nostrand, New York).

Paul, R. J. A., 1965, *Fundamental Analogue Techniques* (Blackie, London).

Pearson, J. D., 1953, *Quart. J. Mech. and Appl. Math.,* **6**, 313.

Pochhammer, L., 1876, *J. reine u. angew. Math.,* **81**, 324.

Rayleigh, Lord, 1894, *Theory of Sound* (Macmillan, London).

Redwood, M., 1959, *J. Acoust. Soc. Am.,* **31**, 442.

Redwood, M., 1960, *Mechanical Waveguides* (Pergamon Press, Oxford).

Redwood, H., Lamb, J., 1957, *Proc. Phys. Soc. Lond.,* **B70**, 136.

Sanders, F. H., 1939, *Canad. J. Res.,* **17A**, 179.

Sittig, E., 1957, *Acustica,* **7**, 175, 299.

Skudrzyk, E. J., 1958, *J. Acoust. Soc. Am.,* **30**, 1140.

Skudrzyk, E. J., 1968, *Simple and Complex Vibratory Systems* (Pennsylvania State University Press, University Park, Pa.).

Slater, J. C., 1959, *Microwave Transmission* (Dover Publishing, New York).

Sommerfeld, A., 1950, *Mechanics of Deformable Bodies* (Academic Press, New York).

Whinnery, J. R., Jamieson, H. W., 1944, *Proc. IRE,* **32**, 98.

White, R. G., 1971, *J. Sound Vib.,* **15**, 147.

Whitney, J. M., 1972, *J. Acoust. Soc. Am.,* **52**, 448.

REVIEW QUESTIONS

4.1 Discuss the possible modes of propagation for pressure waves in a fluid plate with free boundaries. (§4.2.2)

4.2 Distinguish between the concepts of phase velocity and group velocity for pressure waves in a fluid plate. What is meant by the term cut-off frequency? (§4.2.3/4)

4.3 Write down expressions for the boundary conditions applicable to waves propagating in a fluid plate with (a) free boundaries, and (b) rigid boundaries. (§4.2/4.3)

4.4 Outline the Pochhammer–Chree method for finding the longitudinal modes of a solid cylinder. (§4.4.1)

4.5 Write down the dual equations for mechanical, electrical, and acoustical oscillations. (§4.5.1)

4.6 Compare the boundary conditions at (a) a free end, and (b) a clamped end of a long rod with those applicable to analogous electrical transmission lines. (§4.5.2)

4.7 What are the variations of current and voltage with distance along a resonant electrical transmission line? (Figure 4.11)

4.8 Find the mode parameters M_n, C_n, R_n, κ_n, q_n for a mechanical oscillator system. (§4.5.6)

4.9 Draw the distributions of displacement and stress for two bars of different materials cemented together for which the transit times in each bar are equal. Draw diagrams for both the fundamental and second harmonic in each bar. (Figure 4.14)

4.10 If a rectangular pulse of strength $A\tau$ and short duration is applied to a linear system, how may the output spectrum be found? (§4.7.2)

PROBLEMS

4.11 Use the wave reflection method to find the allowed values for the wave vector k in a fluid plate with free boundaries.

4.12 Give a proof of equation (4.23) using Rayleigh's energy method. Verify that this equation can be obtained by approximating the Pochhammer–Chree equations.

4.13 Find the force–current and force–voltage analogous circuits for the following mechanical system:

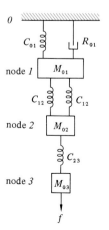

4.14 Draw the equivalent mechanical circuit for the acoustical system shown below and hence find the force-current and force-voltage analogous circuits. (Hint: It is often easier to draw the force-voltage circuit directly for an acoustical system.)

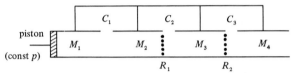

M represents acoustic mass, C acoustic compliance (a function of volume of air), R acoustic resistance.

4.15 Prove by two methods that (a) the particle velocity at the free end of a long rod is doubled, and (b) the stress at a clamped end is doubled.

4.16 Verify equation (4.47) for the input impedance of a finite rod.

4.17 Verify equation (4.55) for the input admittance $|Y_{11}|$ of a rod with free ends and find a similar expression for the transfer admittance $|Y_{12}|$. Comment on the significance of the maxima and minima shown in figure 4.12.

4.18 Verify equations (4.60) to (4.63).

4.19 Compute the transfer impedance Z_{13} of a triple composite transmission line for the case of small damping [compare with equation (4.88)].

4.20 Confirm that equation (4.95) leads to the Rayleigh wave velocity as $\omega \to \infty$.

4.21 Find the Fourier transform of a sawtooth pulse of duration τ for which

$$f(t) \begin{cases} = A\left(1 - \dfrac{t}{\tau}\right) & 0 < t < \tau \\ = 0 & \text{elsewhere .} \end{cases}$$

Show that, in the limit as $\tau \to 0$, this pulse tends to a delta function of strength $\frac{1}{2}A\tau$.

Appendix 4.1   Additional notes on analogous circuits

Firestone's (1956) rules for drawing circuit diagrams

1. *Identify the two terminals of each active and passive element.*
   Discrete variable values (potential or flow) can only be established at the terminals of the elements.
2. *Connect together at a common junction those terminals that move together.*
   Discrete displacements or velocities can only exist at individual masses or at the junctions of two or more elements. These points are called *nodes*. Each mass forms a nodal point.
3. *Connect to the reference junction all terminals that remain stationary with respect to the reference frame.*
   One terminal of each mass is connected to the reference node (the ground) in the force-current analogy. One terminal of each spring is connected to the reference node in the force-voltage analogy.
4. *Mark each source element with an arrow to indicate the positive direction of the force or velocity which it represents.*
5. *Assign a coordinate to each movable junction and indicate by an arrow the direction for positive values of this coordinate.*

Consider the mechanical system shown in figure A4.1. This diagram is often taken to represent a simple motorcar suspension system for a single wheel. M_{01} represents $\frac{1}{4}$ of the mass of the car, M_{02} is the mass of the wheel and axle, C_{12} and R_{12} are constants associated with the main spring and shock absorber, and C_{02} is a linearised

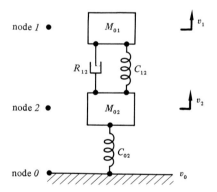

Figure A4.1. Mechanical system representing a motor car suspension for one wheel.

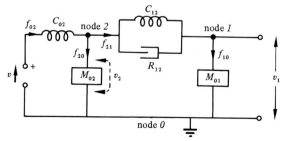

Figure A4.2. Equivalent mechanical network for the system in figure A4.1.

compliance associated with the tyre. In figure A4.2 is shown the equivalent mechanical network using the above rules, and the analogous force-current electrical network is illustrated in figure A4.3. Double subscripts are used on the elements to indicate the nodes to which they are joined.

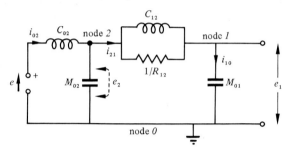

Figure A4.3. Analogous force-current circuit for figure A4.2.

Beranek's (1954) method for converting from the force-current analogy to the force-voltage analogy (the 'dot' method)
Assume that the force-current analogous circuit has been determined and it is required to draw the force-voltage circuit. The following rules may be followed:
1. *Place a dot at the centre of each loop of the circuit and one dot outside all loops. Number the dots consecutively.*
2. *Connect the dots together with lines so that there is a line through each element and so that no line passes through more than one element.*

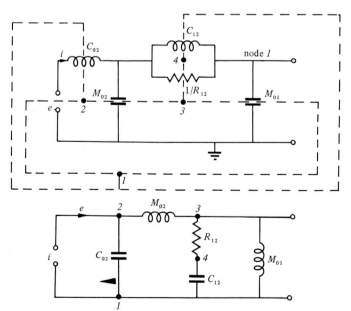

Figure A4.4. Application of the 'dot' method to convert a force-current analogous circuit into a force-voltage circuit.

3. *Draw a new circuit such that each line connecting two dots now contains an element that is the inverse (according to the transformations given in section 4.5.1) of that in the original circuit.*

The method is applied in figure A4.4 to find the dual circuit for figure A4.3. As a final check on the validity of the circuit, allow each element to become either very large or very small and check if the circuit behaves in the same way as the original mechanical system.

Appendix 4.2 The delta function or unit impulse

An impulsive function of considerable value in the transient analysis of linear systems is the **delta function**, $\delta(x)$ or $\delta(t)$, introduced by Dirac (1935) and defined by

$$\delta(t-a) = 0 \qquad \text{for } t \neq a\,,$$

$$\int_{a-\epsilon}^{a+\epsilon} \delta(t-a)\,\mathrm{d}t = 1 \qquad \text{for } \epsilon > 0\,.$$

The first equation shows that $\delta(t-a)$ is zero except at $t = a$, as shown schematically in figure A4.5. The second equation shows that the area under the delta function is unity.

Strictly speaking, the delta function is not a proper mathematical function but it may be used in analysis with the following interpretation. Consider a signal of duration $\Delta\tau$ with constant amplitude $1/\Delta\tau$ in this time interval and zero amplitude outside the interval, as shown in figure A4.6. Now let $\Delta\tau \to 0$ and the amplitude $\to\infty$ so that the area remains equal to unity. In the limit a delta function is formed. In practice, an impulse may have an area differing from unity. The area of such an impulse is called the strength, S, of the impulse.

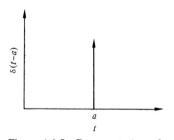

Figure A4.5. Representation of a delta function.

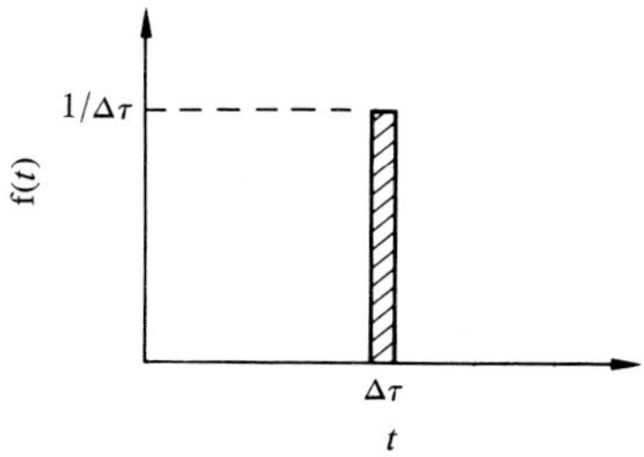

Figure A.4.6. Diagram for defining the delta function.

Some properties of the delta function
(a) If $f(t)$ is a continuous function of t, then

$$\int_a^c f(t)\delta(t-b)\,\mathrm{d}t = f(b) \qquad \text{for } a < b < c\,.$$

The delta function may therefore be regarded as a probe that measures the value of the function when $t = b$. This process is illustrated in figure A4.7.
(b) If the delta function is integrated, a unit step function $u(t)$ is generated (figure A4.8)

$$u(t-a) = \int_0^t \delta(t-a)\,\mathrm{d}t \begin{cases} = 0 & \text{if } t < a\,, \\ = 1 & \text{if } t > a\,. \end{cases}$$

Conversely, differentiating a unit step function generates a delta function:

$$\frac{d}{dt}u(t-a) = \delta(t-a) .$$

(c) Consider a triangular pulse and its derivative, as shown in figure A4.9. Now let $\epsilon \to 0$. Then the triangular pulse becomes a delta function and its derivative becomes the **unit doublet**, denoted by $\delta'(t)$. The integral of $\delta'(t)$ is zero. However, the moment of the unit doublet is

$$\int_{a-\epsilon}^{a+\epsilon} t\delta'(t-a)\,dt = -1 .$$

The doublet is useful in problems such as the application of a couple to a rod or beam or to the charge distribution of a dipole.

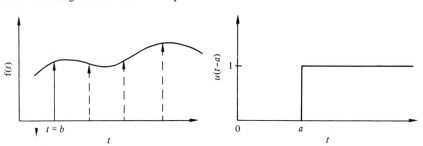

Figure A4.7. Measurement of a given function using a delta function probe.

Figure A4.8. Unit step function.

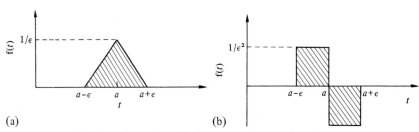

(a)   (b)

Figure A4.9. (a) Triangular pulse; (b) derivative of triangular pulse.

Appendix 4.3 Fourier integrals, convolution, correlation

If $f(t)$ is a periodic function of period T, then $f(t)$ may be expanded in terms of a series of cosine and sine functions (Fourier series) provided $f(t)$ satisfies certain conditions, known as the Dirichlet conditions:

(i) it has at most a finite number of discontinuities in one period;

(ii) it has at most a finite number of maxima and minima in one period;

(iii) the integral $\int_{-T/2}^{T/2} |f(t)|\,dt$ is finite.

Fourier series expansions may be written in three different ways:

(1) *Trigonometric form*

$$f(t) = A_0 + \sum_{n=1}^{\infty} A_n \cos\omega_n t + \sum_{n=1}^{\infty} B_n \sin\omega_n t . \tag{A4.3.1}$$

(2) *Amplitude-phase form*

$$f(t) = C_0 + \sum_{n=1}^{\infty} C_n \cos(\omega_n t + \varphi_n) .$$ (A4.3.2)

(3) *Complex form*

$$f(t) = \sum_{n=-\infty}^{\infty} \hat{D}_n \exp(i\omega_n t)$$ (A4.3.3)

of which the real part is

$$f(t) = D_0 + \sum_{n=1}^{\infty} 2|\hat{D}_n| \cos(\omega_n t + \varphi_n) .$$ (A4.3.4)

As shown in standard textbooks, the equations for evaluating the coefficients in the above equations are

$$A_0 = \frac{1}{T} \int_{-T/2}^{T/2} f(t) \, dt$$ (A4.3.5)

$$A_n = \frac{2}{T} \int_{-T/2}^{T/2} f(t) \cos \omega_n t \, dt$$ (A4.3.6)

$$B_n = \frac{2}{T} \int_{-T/2}^{T/2} f(t) \sin \omega_n t \, dt$$ (A4.3.7)

$$\hat{D}_n = \frac{1}{T} \int_{-T/2}^{T/2} f(t) \exp(-i\omega_n t) \, dt .$$ (A4.3.8)

The relationships between the various coefficients are as follows:

$$A_0 = C_0 = D_0 \qquad\qquad \varphi_n = -\tan^{-1}\frac{B_n}{A_n} = \frac{\operatorname{Im}\hat{D}_n}{\operatorname{Re}\hat{D}_n}$$

$$\hat{D}_n = \tfrac{1}{2}(A_n - iB_n) \qquad\qquad A_n = C_n \cos\varphi_n = 2\operatorname{Re}\hat{D}_n$$

$$|\hat{D}_n| = \tfrac{1}{2}(A_n^2 + B_n^2)^{1/2} \qquad\qquad B_n = -C_n \sin\varphi_n = -2\operatorname{Im}\hat{D}_n$$

$$|C_n| = 2|\hat{D}_n| = (A_n^2 + B_n^2)^{1/2} \qquad \hat{D}_n = \tfrac{1}{2}|C_n| \exp(i\varphi_n) .$$

As remarked by Lanczos (1957), if only positive frequencies are considered, it is necessary to associate with every frequency two functions, $\cos\omega t$ and $\sin\omega t$. If both positive and negative frequencies are introduced, then only the single function $\exp(i\omega t)$ is associated with each frequency. Although negative frequencies have no physical significance, they combine with positive frequencies to give real resultant values. This is necessary since the value of any real function $f(t)$ at any instant of time must be real.

The effect of applying the Fourier method to a periodic function of time is to express the function alternatively in the frequency domain in terms of a spectrum of harmonics whose fundamental frequency ν_1 is the reciprocal of the period T of the original function. For instance, consider a train of rectangular pulses, as shown in figure A4.10. The function of time to be analysed is

$$f(t)\begin{cases} = 0 & -\tfrac{1}{2}T < t < -\tfrac{1}{8}T \\ = A & -\tfrac{1}{8}T < t < \tfrac{1}{8}T \\ = 0 & \tfrac{1}{8}T < t < \tfrac{1}{2}T . \end{cases}$$

Then, it may be confirmed that

$$A_0 = \tfrac{1}{4}A , \qquad A_n = \tfrac{1}{2}A \frac{\sin(\tfrac{1}{4}n\pi)}{\tfrac{1}{4}n\pi} , \qquad B_n = 0 .$$

Note that A_0 gives the average value of the function of time and $B_n = 0$ for a symmetrical function. In figure A4.11 is shown the amplitude spectrum. The time function can therefore be represented by the following Fourier series

$$f(t) = \tfrac{1}{4}A + \frac{2A}{\pi} [\cos\omega_1 t + \tfrac{1}{2}\cos\omega_2 t + \tfrac{1}{3}\cos\omega_3 t + ...] .$$

Consider now a further reduction in the ratio of pulse duration, τ, to period, T. This ratio is often called the mark/space ratio and is denoted by $s = \tau/T$. For the case just considered, $s = \tfrac{1}{4}$. For $s = \tfrac{1}{8}$, repetition of the above computation gives

$$A_0 = \tfrac{1}{8}A , \qquad A_n = \tfrac{1}{4}A \frac{\sin(\tfrac{1}{8}n\pi)}{\tfrac{1}{8}n\pi} .$$

A more general equation for A_n may be written in terms of s:

$$A_n = 2As \frac{\sin(n\pi s)}{n\pi s} . \qquad\qquad (A4.3.9)$$

For the case, $s = \tfrac{1}{8}$, it is noted that the amplitude of each spectral component is half its previous value, and there are now twice as many spectral components in a given interval of the sine function. The spacing between harmonics is equal to the fundamental frequency (and this is equal to $1/T$). Thus, as T increases the fundamental frequency decreases and the spectrum lines move closer together.

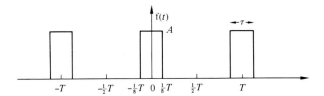

Figure A4.10. Train of rectangular pulses.

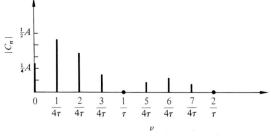

Figure A4.11. Amplitude spectrum for a train of rectangular pulses. The value of $A_0 = \tfrac{1}{4}A$ is shown at zero frequency. τ is the pulse duration.

In all cases, the envelope of the spectra is governed by the function $\sin(n\pi s)/n\pi s$. In communication theory this function is called the **sampling function** and may be written concisely

$$S(x) \equiv \frac{\sin x}{x} , \tag{A4.3.10}$$

where $x = n\pi s = n\pi\tau/T$. A graph of this function is shown in figure A4.12. Note that the first minimum occurs at $|x| = \pi$, that is when $n\pi s = \pi$ or $n = 1/s$. Thus, when $s = \frac{1}{4}$, the first minimum occurs at the fourth harmonic; when $s = \frac{1}{8}$, the first minimum occurs at the eighth harmonic, and so on. Hence, as the spacing between the pulses increases, i.e. as T increases, the spectrum lines crowd together but the envelope function remains the same shape. In the limit, as $T \rightarrow \infty$, we are left with only an isolated pulse and the corresponding spacing between spectral harmonics tends to zero, that is the spectrum becomes continuous.

In the limit as $T \rightarrow \infty$ and $\nu_1 \rightarrow 0$, examination of equations (A4.3.7) and (A4.3.8) shows that $\hat{D}_n \rightarrow 0$ but the ratio $\hat{D}_n/\Delta\nu$ remains finite, where $\Delta\nu = 1/T$ is the spacing between harmonics. In the limit, the discrete quantity $\hat{D}_n$ becomes a continuous function of frequency which will be denoted by $\hat{D}(\omega)$. Also, in place of the discrete frequency of a given harmonic ($\omega_n = n\Delta\omega$), we now have a continuous frequency variable ω. Equation (A4.3.8) may therefore be rewritten

$$\lim_{\Delta\nu \rightarrow 0} \frac{\hat{D}(\omega)}{\Delta\nu} = \int_{-\infty}^{\infty} f(t)\exp(-i\omega t)\,dt .$$

The limits on the integral have been changed since, in the limit, they become $-\infty$ and ∞. The left-hand side of this equation will be denoted by $F(i\omega)$ and is called the **Fourier transform** of $f(t)$. Thus,

$$F(i\omega) = \int_{-\infty}^{\infty} f(t)\exp(-i\omega t)\,dt . \tag{A4.3.11}$$

In a similar way, the limiting form of equation (A4.3.7) may be found by replacing $\hat{D}_n$ by $F(i\omega)\Delta\nu = F(i\omega)\,\Delta\omega/2\pi$, so that, in the limit as $\Delta\omega \rightarrow 0$,

$$f(t) = \frac{1}{2\pi}\int_{-\infty}^{\infty} F(i\omega)\exp(i\omega t)\,d\omega . \tag{A4.3.12}$$

$f(t)$ is sometimes referred to as the **inverse Fourier transform** of $F(i\omega)$. Equations (A4.3.11) and (A4.3.12) tell us that an isolated function of time $f(t)$ can be analysed into a continuous frequency spectrum $F(i\omega)$, and, conversely, $f(t)$ can be synthesised from a knowledge of the amplitude and phase of its spectral components.

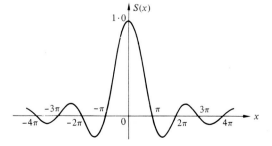

Figure A4.12. The sampling function $S(x) = (\sin x)/x$.

If equations (A4.3.11) and (A4.3.12) are combined, an expression is found for Fourier's double integral theorem:

$$f(t) = \frac{1}{2\pi} \int_{-\infty}^{\infty} \left[\int_{-\infty}^{\infty} f(t) \exp(-i\omega t) \, dt \right] \exp(i\omega t) \, d\omega \ . \tag{A4.3.13}$$

Consider now a linear system for which the input time function is $f(t)$, the impulse response function is $h(t)$ and the output time function is $g(t)$. A frequent practical problem is to find $g(t)$ for a prescribed input function. It will be assumed that spectral analysis has provided the input spectrum function $F(i\omega)$, and the transfer function $H(i\omega)$ of the system. Then,

$$g(t) = \mathscr{F}^{-1}[G(i\omega)] = \mathscr{F}^{-1}[F(i\omega)H(i\omega)] \ ,$$

where $G(i\omega)$ is the output spectrum function and $\mathscr{F}^{-1}$ is the inverse Fourier transform as given by equation (A4.3.12). Therefore

$$g(t) = \frac{1}{2\pi} \int_{-\infty}^{\infty} F(i\omega) H(i\omega) \exp(i\omega t) \, d\omega \ . \tag{A4.3.14}$$

For $F(i\omega)$ write

$$F(i\omega) = \int_{-\infty}^{\infty} f(\tau) \exp(-i\omega\tau) \, d\tau \ ,$$

and substitute in equation (A4.3.14):

$$g(t) = \frac{1}{2\pi} \int_{-\infty}^{\infty} \left[\int_{-\infty}^{\infty} f(\tau) \exp(-i\omega\tau) \, d\tau \right] H(i\omega) \exp(i\omega t) \, d\omega \ .$$

Interchanging the order of the integration, we find

$$g(t) = \int_{-\infty}^{\infty} f(\tau) \left\{ \frac{1}{2\pi} \int_{-\infty}^{\infty} H(i\omega) \exp[i\omega(t-\tau)] \, d\omega \right\} d\tau \ .$$

Consequently $g(t)$ may be written

$$g(t) = \int_{-\infty}^{\infty} f(\tau) h(t-\tau) \, d\tau \ , \tag{A4.3.15}$$

where

$$h(t-\tau) = \frac{1}{2\pi} \int_{-\infty}^{\infty} H(i\omega) \exp[i\omega(t-\tau)] \, d\omega \ ;$$

τ is an auxiliary time function. Equation (A4.3.15) is called the **convolution** (or folding) **integral**. A similar result is obtained if $f(t)$ is treated as the folded function. In place of equation (A4.3.15) we find

$$g(t) = \int_{-\infty}^{\infty} f(t-\tau) h(\tau) \, d\tau \ . \tag{A4.3.16}$$

Equation (A4.3.16) is a little easier to visualise than equation (A4.3.15). The method of evaluating the integral is shown graphically in figure A4.13. The functions are plotted in terms of τ. Here $f(t-\tau)$ represents $f(t)$ folded about the origin; t is then a delay time which shifts $f(\tau)$ along the time axis. For each value of t, the convolution integral is the area formed by the product of $f(\tau)$ and $h(\tau)$. The required function $g(t)$ is found by plotting the areas as a function of t.

A function that looks similar to the convolution integral is the **correlation** function, of which there are two versions. Consider two periodic functions, $f_1(t)$ and $f_2(t)$. If the second function is shifted along the t axis by a time increment τ, and then the average value of the product found, we can write

$$R_{12}(\tau) = \frac{1}{T}\int_0^T f_1(t)f_2(t-\tau)\,dt \,, \tag{A4.3.17}$$

where $R_{12}(\tau)$ is called the **cross-correlation coefficient**.
For signals in the form of a pulse

$$R_{12}(\tau) = \int_{-\infty}^{\infty} f_1(t)f_2(t-\tau)\,dt \,. \tag{A4.3.18}$$

The cross-correlation coefficient therefore gives a measure of the correlation or similarity between two (nonidentical) waveforms as one of the signals, $f_2(t)$, is shifted along the time axis past signal $f_1(t)$. If there is no correlation between the two signals, then $R_{12}(\tau) = 0$. This result may be used as a definition for noncoherent signals.
In a similar way, an **autocorrelation coefficient**, $R_{11}(\tau)$ may be defined as

$$R_{11}(\tau) = \frac{1}{T}\int_0^T f(t)f(t-\tau)\,dt \,. \tag{A4.3.19}$$

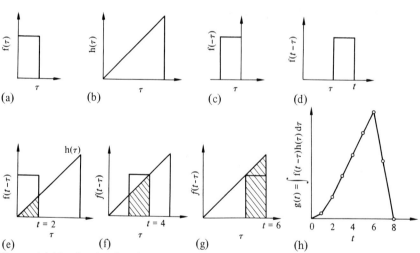

Figure A4.13. Graphical evaluation of the convolution integral: (a) applied force function; (b) system impulse response function; (c) folded force function; (d) force function folded and shifted; (e), (f), (g) three evaluations of the convolution integral; (h) system output function.

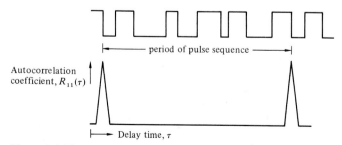

Figure A4.14. Autocorrelation function for a pseudo-random binary sequence.

This function measures the correlation between a given signal $f(t)$ and the same signal at a later time. In figure A4.14 is shown an illustration of an autocorrelation function for a signal in the form of a pseudorandom binary sequence (a repetitive sequence of pulses of random duration). It is noted that the autocorrelation function consists of a series of triangular functions which occur whenever the binary sequence repeats itself.

Some of the properties of the autocorrelation function are listed below.

(a) It is a symmetrical function about the origin, i.e. $R_{11}(\tau) = R_{11}(-\tau)$.

(b) Its value at $\tau = 0$ is equal to the mean square value of the signal, $\overline{[f(t)]^2}$.

(c) The autocorrelation function of any periodic waveform is periodic and has the same period as the waveform itself. Autocorrelation is therefore valuable in revealing periodicity in a signal. However, $R_{11}(\tau)$ does not contain any phase information.

(d) When the autocorrelation function is formed for a random noise signal, $R_{11}(\tau)$ has the form of an impulse at the origin, that is, once the signal has been time-shifted relative to itself, there is no longer any similarity. In general, wide-bandwidth signals produce a narrow $R_{11}(\tau)$ and vice versa.

With the use of the above relationships the following pair of expressions may be proved

$$R_{11}(\tau) = \frac{1}{2\pi}\int_{-\infty}^{\infty} S_{11}(\omega)\exp(i\omega\tau)\,d\omega\,, \qquad\qquad\qquad (A4.3.20)$$

$$S_{11}(\omega) = \int_{-\infty}^{\infty} R_{11}(\tau)\exp(-i\omega\tau)\,d\tau\,, \qquad\qquad\qquad (A4.3.21)$$

where $S_{11}(\omega) = |F(i\omega)|^2$ is called the **power spectral density**.

This theorem, known as the **Wiener–Khintchine theorem**, shows that the autocorrelation function and the power spectral density form a Fourier transform pair. The power spectrum of a signal may therefore be found experimentally by determining the autocorrelation coefficient as a function of time and then taking the Fourier transform. Many analysing instruments find the power or amplitude spectrum in this way, since in computing the autocorrelation function random noise present in the signal tends to average out to zero.

Some applications of correlation are listed below.

(a) *Signal buried in noise.* If the form of the transmitted signal is known, then it may be cross-correlated with the received signal + noise. Since the cross-correlation coefficient for the noise will be zero, there is left only the signal in the form of its autocorrelation function. Since phase information is retained in computing the cross-correlation coefficient, the value of τ for which $R_{12}(\tau)$ has its peak value gives a measure of the transit time of the signal. When the transmitted signal is unknown, it is necessary to use autocorrelation which extracts the signal amplitude information but not the phase information.

(b) *Impulse response.* Using a correlation method it is possible to measure the impulse response $h(t)$ of a system while it is operating normally without disturbing the system in any way. A low-level noise signal is introduced at the input and then the cross-correlation function is formed between the input and the system output function, as shown diagrammatically in figure A4.15. The level of the noise signal can be made very low if a long averaging time is used in computing the correlation function. The test noise signal does not correlate with background noise. In the process the test signal is correlated with itself to give an impulse function. Thus, as far as the system itself is concerned, it has been sampled by an impulse function so that the output contains the impulse response $h(t)$.

(c) *Noise transmission path.* Cross-correlation may be used for determining the delay times taken by a signal following different paths. For instance, in order to investigate the transmission of vibrations from a car wheel into the body at the driver's position, a transducer may be mounted on the wheel axle and a microphone placed near the driver's ear. Cross-correlation of the two signals from the transducer and microphone will reveal peaks corresponding to delayed signals from the wheel. In figure A4.16 are shown two such peaks corresponding to a path up the steering column (*1*) and a path via the floor and driver's seat (*2*). The relative amplitudes are related to the relative amounts of vibration transmitted.

As shown by equation (A4.3.11), a function of time $f(t)$ may be Fourier transformed into a complex spectrum function $F(i\omega)$, which may be written

$$F(i\omega) = \int_{-\infty}^{\infty} f(t)\exp(-i\omega t)\,dt = \int_{-\infty}^{\infty} f(t)\{\cos\omega t - i\sin\omega t\}\,dt\,, \qquad (A4.3.22)$$

where the cosine terms give the in-phase or real components and the sine terms give the out-of-phase (quadrature) or imaginary components. From these components, amplitude and phase can be computed by the usual equations.

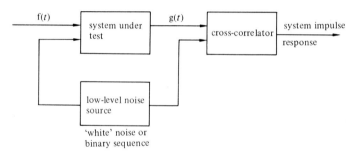

Figure A4.15. Scheme for measuring a system impulse response function.

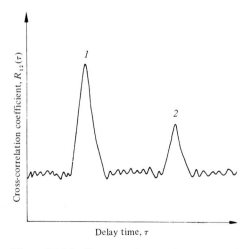

Figure A4.16. Cross-correlogram for two signals travelling from a car wheel to the driver's ear (*1*) via the steering column and (*2*) via the floor and driver's seat.

In order to use a **digital technique of computation** the infinite Fourier integrals must be changed into sums over a finite duration. Thus, the function $f(t)$ to be analysed must be replaced by N samples taken at discrete time intervals Δt over a finite record time T, where $T = N\Delta t$. Instead of a continuous spectrum, one reverts to a type of Fourier series having discrete spectrum lines and a spectral resolution $\Delta \nu$ given by $\Delta \nu = 1/T$. Also, aliasing errors occur (stroboscopic effect due to sampling at equal time intervals) if any components have a frequency greater than the Nyquist frequency: $\nu_{max} = 1/(2\Delta t)$. Frequencies greater than ν_{max} are folded back into frequencies below ν_{max}.

In place of equation (A4.3.22), a **discrete Fourier transform** may be written

$$F(im\Delta\nu) = \frac{1}{N}\sum_{n=0}^{N-1} f(n\Delta t)\exp\left(-\frac{i2\pi mn}{N}\right)$$
$$= \frac{1}{N}\sum_{n=0}^{N-1} f(n\Delta t)\left(\cos\frac{2\pi mn}{N} - i\sin\frac{2\pi mn}{N}\right), \tag{A4.3.23}$$

where ω has been replaced by $2\pi m\Delta\nu$, and t by $n\Delta t$.

One problem with the use of equation (A4.3.23) is that the result is sensitive to the location of the time origin. If the time record is shifted relative to the origin, there will be a relative change in the cosine and the sine components, reflecting the phase shift that has occurred in $f(t)$. An alternative method is to compute the power spectrum, $S_{11}(\omega)$, by multiplying $F(i\omega)$ by its complex conjugate:

$$S_{11}(\omega) = F(i\omega) \, . \, F^*(i\omega)$$
$$= [A(\omega) + iB(\omega)][A(\omega) - iB(\omega)]$$
$$= A^2(\omega) + B^2(\omega) \, . \tag{A4.3.24}$$

It may be noted that there is now no imaginary part, that is phase information has been suppressed. $S_{11}(\omega)$ is therefore independent of the location of the time origin.

The correlation functions may also be written in digital form, assuming N samples are taken at intervals of Δt sec:

$$R_{11}(\tau) = \frac{1}{N}\sum_{n=1}^{N} f(n\Delta t)f(n\Delta t - \tau) \, , \tag{A4.3.25}$$

$$R_{12}(\tau) = \frac{1}{N}\sum_{n=1}^{N} f_1(n\Delta t)f_2(n\Delta t - \tau) \, . \tag{A4.3.26}$$

Since an averaging process is involved, the sampling interval Δt is not related to the delay time τ, and can be much larger. Since the error in determining the coefficients decreases as N increases, the number of samples N is more important than the rate at which they are taken.

5

Ultrasonic experimental techniques

5.1 Discussion and classification of methods

The basic object in making mechanical wave measurements is to determine the velocity and attenuation for the particular type of wave under consideration. These measures may then be related to structural properties of the material. The velocity is primarily a function of the elastic constants and density, while the attenuation is largely determined by dissipative mechanisms peculiar to the type of material, frequency range, grain size, presence of external fields, etc.

Mechanical wave measurements have been made over an enormous range of frequencies, from less than 1 Hz to approximately 10^{11} Hz. The latter frequency is still less than the frequencies of lattice vibrations in crystals which are of the order of 10^{13} Hz. The complete mechanical wave spectrum is compared with the electromagnetic spectrum in figure 5.1. A mechanical wave velocity of 3000 m s^{-1} has been assumed and the spectra are drawn with a common wavelength scale. A comparison on the basis of wavelength is useful, since the detail that may be resolved by both acoustical and optical waves often depends on the relationship between the wavelength and some relevant structural dimension. For instance, mechanical centimetre and millimetre waves are useful in obtaining

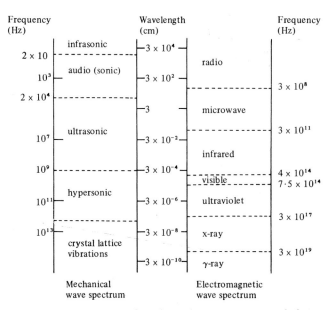

Figure 5.1. Comparison of mechanical wave spectrum and electromagnetic wave spectrum. (A mechanical wave velocity of 3000 m s^{-1} has been assumed in computing the wavelength.)

F

information concerning gross structural effects, such as those produced by the grains in a polycrystalline metal or by the dislocations in a single crystal. Information concerning effects on an atomic scale require the use of high-frequency waves.

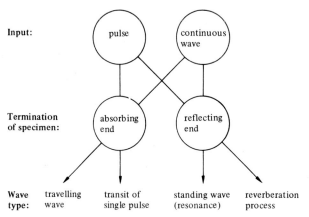

Figure 5.2. Basic methods for measuring velocity and attenuation.

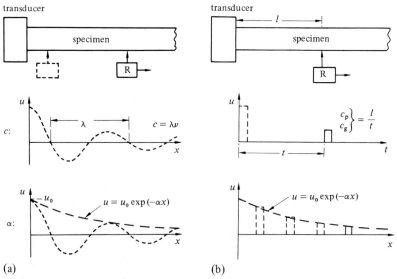

(a)   (b)

Figure 5.3. (a) Travelling wave method. The wavelength is measured by a travelling probe, R. This yields $c = \lambda \nu$ since ν is known. α is determined from the displacement decay curve. (b) Single pulse transit method. The velocity is measured by l/t. For a short pulse the group velocity, c_g, is measured; for a long-pulsed carrier frequency the phase velocity, c_p, is measured. α is determined from the decay in pulse amplitude (provided the pulse shape does not change).

In general a choice may be made between the use of continuous waves or pulses. One of the most common methods is the resonance method which depends on the establishment of a standing wave pattern in the material. Since energy losses occur in all materials, the standing wave pattern will not be perfect and the energy losses may be evaluated from a knowledge of the standing wave ratio in the material. In the resonance method it is important to identify the mode of vibration and to measure accurately the resonant frequency. When the specimen is in the form of a rod, three types of wave are possible: longitudinal, torsional, and flexural. In a plate longitudinal, transverse, and flexural waves may occur. Under certain circumstances surface waves may occur on the surface of a solid specimen. In all cases of wave propagation in a bounded solid the relationship between the wavelength and the dimensions of the specimen is important.

Pulse methods are particularly suited to high frequencies when the specimen size in general becomes too small for the resonance method. Provided the dimensions of the specimen are large compared with the wavelength, bulk propagation conditions apply and there are only two

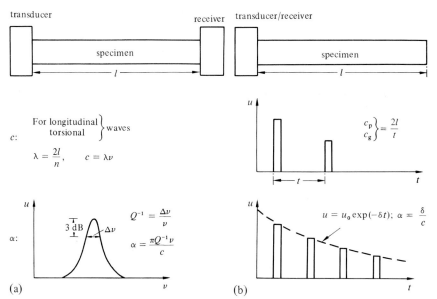

Figure 5.4. (a) Standing wave (resonance) method. λ is determined by the specimen dimensions for a given mode of vibration. The velocity is then found from $c = \lambda \nu$. α is found from the bandwidth of the resonance curve or the decay rate measured at the resonant frequency. (b) Reverberation (pulse–echo) method. If the reflected pulses are discrete, c is determined by the transit time and the length of the specimen. α is found from the decay rate of the echoes. If the pulses overlap, a quasiresonant frequency and wavelength are measured and c found from $c = \lambda \nu$. α is determined from the reverberation decay curve.

possible types of wave: the longitudinal (dilatational), and the transverse (shear or rotational). As long as the effects of the boundary surfaces can be ignored, propagation then occurs as if in an infinite solid. Pulse methods involve the timing of the passage of a wave packet as it traverses the material or as it is reflected multiply between parallel walls of the material. Care must be exercised in the form of the pulse employed in order to be able to assign the correct velocity.

In the case of an arbitrary pulse shape it is difficult to extract the phase velocity corresponding to a given frequency unless Fourier analysis is employed. Even the determination of the group velocity is not reliable if appreciable dispersion is present. Reliable measurements of the phase velocity may be obtained by the use of a wave packet containing a sufficiently large number of sinusoidal waves. Very accurate interference methods may be used for the measurement of small changes in the velocity of propagation. In figure 5.2 are shown, in schematic form, the basic methods for the direct determination of the velocity and attenuation. The procedures for measuring the velocity and attenuation in each case are shown diagrammatically in figures 5.3 and 5.4.

5.2 Travelling wave method
The travelling wave method has been used for measurements on thin strips of materials and fibres. The specimen is driven from one end into either longitudinal or flexural vibration by means of a suitable transducer. At the other end an absorber is frequently used in order to reduce the possibility of standing waves on the specimen. A second transducer in contact with the specimen is used to record the vibrations as it traverses the specimen. The lower frequency limit for the method occurs when the phase changes over the length of the specimen are less than $180°$. The upper frequency limit is reached either when the attenuation becomes excessive or when the wavelength becomes too small to measure accurately (less than 2-3 mm).

Since the stress σ_{xx} at a distance x from the generator for a longitudinal plane wave is given by

$$\sigma_{xx} = \sigma_0 \exp(-\alpha x) \exp[i(\omega t - kx)] , \tag{5.1}$$

where σ_0 is the stress wave amplitude at the generator, then, assuming no reflected wave, the stress amplitude varies along the specimen as

$$\sigma_{xx} = \sigma_0 \exp(-\alpha x) . \tag{5.2}$$

The phase difference, ϕ, between generator and receiver is

$$\phi = kx = \frac{\omega x}{c} , \tag{5.3}$$

where c is the phase velocity. The velocity is found directly from the frequency of the input signal and a measurement of the wavelength.

A plot of the phase difference between the driving voltage and the output from the receiver allows the wavelength to be determined accurately. The attenuation coefficient, α, may be found by recording the level of the receiver output as a function of distance along the specimen. One advantage of using travelling waves is that a continuous range of frequencies may be covered with one specimen. The method has been used successfully over a frequency range from 1 kHz to 100 kHz. Disadvantages of the method include the difficulty of avoiding standing wave effects and complications caused by the generation of flexural waves when longitudinal wave measurements are being made.

5.3 Resonance methods

One of the first methods used for measuring the dynamic elastic properties of solid specimens was that in which a long thin rod or wire was set into resonance at one or more of the normal modes of the system. Resonant systems employing longitudinal, torsional, or flexural vibrations can yield all the constants of the material, such as Young's modulus, shear modulus, Poisson's ratio, and damping measures. This method is especially useful at frequencies below about 200 kHz where the ratio d/λ can be kept small, d being a cross-sectional dimension of the specimen and λ the wavelength. One disadvantage of the resonance method is that coupling between the driving system and the specimen may result in a change in resonant frequency and in the shape of the resonance peak.

The phase velocity for a given mode of vibration may be found from the product of the measured frequency of vibration and the wavelength. The latter is usually calculated from the dimensions of the specimen and knowledge of the particular normal mode that is excited. Measurement of the absorption of energy by a resonant mode may be made by (a) cutting off the driving source suddenly and then recording the decay of the free vibrations of the specimen, or (b) direct measurement of the bandwidth of the resonant peak.

For longitudinal waves in an isotropic rod for which $d/\lambda \ll 1$, the phase velocity, c_L, given by elementary plane wave theory is

$$c_L = \left(\frac{E}{\rho}\right)^{1/2} = \frac{2\nu l}{n} \tag{5.4}$$

where E is Young's modulus, ρ is the density, ν is the frequency, l is the length of the specimen, and n is the mode number. In the elementary theory it is assumed that the specimen vibrates only longitudinally. This is not strictly correct, since longitudinal extension of a rod will cause some lateral contraction, the extent of this effect depending on the ratio d/λ. It is found in practice that, as the thickness increases, the resonant frequency, and hence the computed velocity, decreases (see experiment E4.2). Using Rayleigh's corrected equation (4.23), we obtain the theoretical value of the velocity for a long, thin rod:

$$c_L = c'_L \left[1 + \mu^2 \pi^2 \left(\frac{a}{\lambda_n} \right)^2 \right] \tag{5.5}$$

where μ is Poisson's ratio, a is the radius of the specimen, $c'_L = 2\nu'_n l/n$ is the measured velocity, and ν'_n is the measured frequency. The theoretical value of c_L is required if a value for Young's modulus is to be computed from equation (5.4).

For torsional waves, the phase velocity, c_T, is

$$c_T = \left(\frac{G}{\rho} \right)^{1/2} = \frac{2\nu l}{n} , \tag{5.6}$$

where G is the shear modulus. For flexural waves, the phase velocity, c_F, is

$$c_F = \frac{2\pi K}{\lambda} \left(\frac{E}{\rho} \right)^{1/2} , \tag{5.7}$$

where K is the radius of gyration of cross section of the rod about the neutral axis. As may be noted from these equations, at long wavelengths c_L and c_T are constant for a given material, whereas c_F varies inversely with the wavelength. At high frequencies, or when the ratio d/λ exceeds a value of approximately $0 \cdot 3$, it may be necessary to use the theory developed by Pochhammer and Chree (section 4.4.1) in evaluating experimental results. A number of investigations have been reported in which the constants of the material were determined from the propagation of longitudinal or flexural waves in plates.

5.3.1 Effect of absorption on velocity and modulus
In real materials where absorption is present the values of the elastic constants depend to some extent on the amount of absorption. This effect may be estimated if the appropriate modulus is assumed to be a complex quantity. For instance, Young's modulus may be written

$$\hat{E} = E_1 + iE_2 , \tag{5.8}$$

where E_1 is the elastic part of the complex modulus and E_2 is a measure of the energy loss associated with longitudinal deformation. Writing the modulus in complex form signifies that there is now a phase lag between the strain and the applied stress. Equation (5.8) may also be written

$$\hat{E} = E_1(1 + i\eta) \tag{5.9}$$

where $\eta = E_2/E_1$ is the loss factor. The angle δ by which the strain lags behind the stress is given by $\tan \delta = \eta$.

When a plane wave propagates through an absorptive material, the wave velocity may be written in complex form

$$\hat{c}_L = \left(\frac{\hat{E}}{\rho} \right)^{1/2} = \left(\frac{E_1 + iE_2}{\rho} \right)^{1/2} . \tag{5.10}$$

It is possible now to estimate the effect of absorption on the velocity
and the Young's modulus. The magnitude of the complex modulus is

$$|\hat{E}| = E_1(1+\eta^2)^{\frac{1}{2}} .$$ (5.11)

For absorption to make a 1% change in $|\hat{E}|$, $\eta = 0\cdot4$, that is $Q = 70$.
Hence for specimens with a value of $Q < 70$ correction for absorption
becomes necessary. The correction to the velocity can be similarly
estimated since

$$|\hat{c}_L| = c_L[1+(\tfrac{1}{2}\eta)^2]^{\frac{1}{2}} .$$ (5.12)

For absorption to make a 1% change in $|\hat{c}_L|$, $\eta = 0\cdot28$ and $Q = 3\cdot5$. For
a change of 1 in 10^4 in $|\hat{c}_L|$, $\eta = 0\cdot028$ and $Q = 35$. It can be seen that
the effect of absorption on velocity is small but should be taken into
account for highly absorptive materials.

5.3.2 Electrostatic methods

A block diagram for a basic electrostatic system of measurement is shown
in figure 5.5. The driving unit consists of a fixed metal plate placed close
to one end of the specimen, thus forming a parallel-plate capacitor. When
an alternating voltage is applied between the driving plate and the
specimen, vibrations are induced in the specimen. The electrostatic force
that can be applied in this way is given by

$$F = -\frac{\epsilon A}{2}\frac{V^2}{d^2} ,$$ (5.13)

where ϵ is the permittivity of the medium between the plates, A is the
plate area, V is the applied potential difference, and d is the spacing
between the plates. It is observed that F is proportional to the square of
the applied voltage and inversely proportional to the square of the spacing.
As a consequence, if an alternating voltage is applied to the plates, the
specimen will vibrate at twice the frequency of the applied voltage. This
frequency doubling may be overcome by the application of a steady

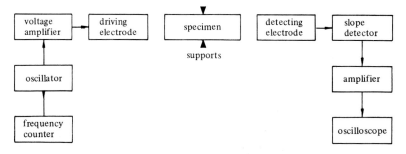

Figure 5.5. Electrostatic drive and detection system.

polarising voltage in addition to the alternating voltage. If $V = V_0 + V_1$, where V_0 is the polarising voltage and $V_1 = V_1 \exp(i\omega t)$ is the applied voltage, then

$$V^2 = V_0^2 + 2V_0 V_1 + V_1^2 \approx V_0^2 + 2V_0 V_1 \, ,$$

if $V_0 \gg V_1$. There are two advantages in this procedure: first, the force now has the same frequency as the voltage; and second, the alternating component of the force is considerably increased in magnitude since in the second term V_1 is multiplied by $2V_0$. The force may further be increased if a thin piece of mica is inserted between the fixed plate and the specimen. It is advisable to cement the mica to the fixed electrode by means of a conducting cement. Although the mica is in contact with the end face of the specimen, the real contact is so poor, because of the normal irregularities of the surfaces, that the specimen is not appreciably damped.

At the receiving end of the specimen a capacitor may again be formed between the end of the specimen and a fixed plate so that changes of capacitance occur owing to the vibrations of the end face of the specimen. The voltage changes across the capacitor may be measured directly by means of a capacitance bridge, or the capacitance changes may be used to alter the tuning of a crystal-controlled oscillator. The oscillator is adjusted to operate slightly off resonance. If there is now a small capacitance change, the operating point will move along the slope of the resonance curve. Large changes in the output current of the oscillator may thus be obtained for very small changes in capacitance. Such a detector is sometimes called a slope detector. Nonconducting specimens may be investigated by this method if the end faces are coated with a thin conducting film.

One of the advantages of the electrostatic method is that the specimen may be freely supported at the centre or some other displacement node with no measuring devices attached directly to the ends. The method may be applied over a frequency range from a few Hz to at least 200 kHz and is particularly suited to measurements at low and high temperatures.

5.3.2.1 Single-electrode method

Bordoni (1947; 1954) employed a single electrode both for driving and detecting. A polarising voltage was not used in order to reduce the possibility of coupling between the driving and detecting circuits. The detector was of the frequency modulation type, isolated from the driving circuits by means of a suitable filter network. The capacitor formed by the external electrode and the end face of the specimen is connected into the tank circuit of a crystal-controlled oscillator operating at a frequency much higher than the driving frequency. The changes in capacitance then produce frequency modulation of the output of the oscillator. The frequency-modulated signal is converted into an amplitude-modulated signal, detected and displayed on a meter or oscilloscope. Bordoni claims that, with this system, displacements amounting to a fraction of an Ångstrom may be detected. The original apparatus was designed to reach frequencies of 50 kHz.

Improvements to this method were subsequently carried out by Pursey and Pyatt (1954) who were able to make measurements up to 250 kHz. Bordoni and Nuovo (1957, 1958) and Bordoni *et al.* (1959) extended the electrostatic method up to nearly 10 MHz, but found it necessary to revert to two electrodes. Also, owing to the small specimen dimensions at high frequencies, it was not convenient to use specimens in the form of rods; instead, thin plates were used. A simpler tuned detector was preferred in the range 0·2 to 10 MHz. The mechanical vibrations are converted into an alternating signal by means of a steady polarising voltage applied through a large impedance, as shown in figure 5.6. If the impedance is an inductor it can be chosen to resonate with the detector capacitance at the vibration frequency. Under these circumstances maximum sensitivity is achieved. The main difference between a frequency-modulated detector and a polarised detector described above is that the former does not remove any appreciable amount of energy from the vibrating system whilst the latter directly converts mechanical into electrical energy. Although the sensitivity increases with the efficiency of energy conversion, at the same time perturbation of the vibrating system caused by the detector also increases.

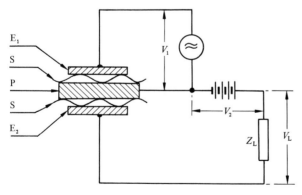

Figure 5.6. Tuned electrostatic detector (Bordoni and Nuovo, 1958). P vibrating plate, E_1 electrostatic actuator, V_1 driving voltage, E_2 probe electrode, V_2 polarising voltage, Z_L load impedance, S dielectric spacers, V_L output voltage.

5.3.3 Electromagnetic methods

5.3.3.1 *Electrodynamic system*

Barone and Giacomini (1954) developed a system in which an end face of a rod or bar is placed between the poles of a permanent magnet and an oscillating electric current passed across the face. With a suitable disposition both longitudinal and torsional waves may be generated, as shown in figure 5.7. The force acting on the specimen is

$$F = Bli_0 \exp(i\omega t) \, ,$$

where B is the magnetic induction, l is the length of the current path, and i_0 is the current amplitude. In the case of a metallic specimen, two current leads are attached at two opposite points on the end face so that the current flows perpendicular to the magnetic field. If the specimen is nonconducting, a thin conducting strip may be deposited or cemented to the end face and, as a matter of convenience, carried along the side faces to the supports where electrical contact is made. A similar system may be used as receiver. This method works satisfactorily over a frequency range from 100 Hz to 200 kHz. The electrodynamic method has been employed for flexural vibrations by Biesterfeldt *et al.* (1960).

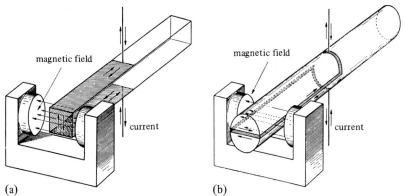

Figure 5.7. Electrodynamic drive system (Barone and Giacomini, 1954) for (a) longitudinal waves, and (b) torsional waves.

5.3.3.2 *Eddy current system*

In the method introduced by Wegel and Walther (1935) the driving end of the specimen is subjected to both a static and an oscillating magnetic field, as shown in figure 5.8. The oscillating field induces eddy currents in the specimen, which in turn are acted on by the stationary inhomogeneous magnetic field giving rise to an oscillating force on the specimen. At the receiving end a similar detecting system may be used. Vibrations induce eddy currents at this end; the field of these eddy currents is then detected by coils surrounding the end. If the specimen is nonferromagnetic, it is necessary to cement small ferromagnetic tips to the ends of the specimen. These may give rise to errors, particularly with damping measurements. The method may be used over a frequency range of 100 Hz to 100 kHz and for temperatures up to 300°C; it is suitable for both longitudinal and torsional waves.

Thompson and Holmes (1956) used the eddy-current method as driving system but added an electrostatic detecting system. The apparatus was arranged to give readings of resonant frequency and internal friction while a copper specimen was being irradiated in a reactor. By means of a small frequency-modulated transmitter signals were relayed to an FM receiver placed outside the reactor. The output voltage of the receiver was then proportional to the displacement amplitude of one end of the specimen. Internal friction values were obtained by measuring the bandwidth of the displacement versus frequency curve when the driving oscillator frequency was swept across the resonance range. Thompson and Glass (1958) describe a regenerative version of this system using the same generator and detector which allows for continuous recording of the resonant frequency and damping. Later, Thompson *et al.* (1967) added an automatic data logger to the system.

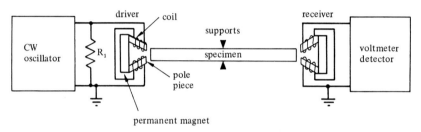

Figure 5.8. Eddy current system (Wegel and Walther, 1935).

5.3.4 Composite bar systems

Composite bar systems have been developed which employ either a 2-bar or a 3-bar resonator. Quimby (1925) investigated a system consisting of a quartz driver bar cemented to specimens of metals and glass in the form of rods. An alternating voltage applied across the appropriate pair of side faces caused the quartz crystal bar to vibrate longitudinally. The velocity of the vibrations induced in the attached specimen was measured with a Rayleigh disk suspended in air. Resonance curves were obtained for the composite system and for the quartz bar alone. Using the equivalent of equation (4.81), Quimby found the resonant frequency, ν_2, for the specimen, and hence the velocity in the material. His measurements were in the frequency range 37 to 60 kHz. A change in resonant frequency of 2 Hz in 50 kHz could be detected.

5.3.4.1 *Marx's three-component system*

In the system described by Marx (1951) and by Marx and Sivertsen (1953), the composite oscillator consists of an $18 \cdot 5°$ X-cut α-quartz driving crystal with full-length electrodes, a nearly identical quartz gauge crystal with electrodes covering only the central one-third of its X faces, and the specimen, as shown in figure 5.9. In the figure the specimen is shown connected to the driver crystal by a dummy quartz bar to allow the specimen to be located in a furnace. The length of the dummy bar can be any multiple of half-wavelengths. In figure 5.10 are shown the displacement and strain distributions for the basic three-component system. Voltmeters record the peak amplitudes of driver and gauge crystals from which the total decrement, Δ_t, and strain amplitude may be determined. Marx shows that

$$\Delta_t = \frac{k_1}{m_t \nu_t^2} \frac{V_d}{V_g} \, , \tag{5.14}$$

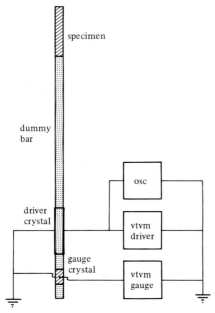

Figure 5.9. Composite oscillator system for longitudinal vibrations (Marx and Sivertsen, 1953).

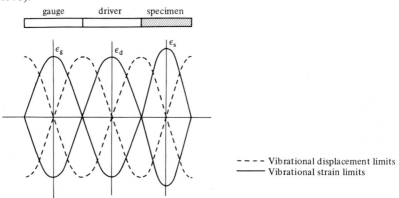

Figure 5.10. Displacement and strain distributions for the basic three-component oscillator (Marx and Sivertsen, 1953).

where m_t is the total mass, ν_t is the resonant frequency of the composite resonator, V_d and V_g are the peak voltages at resonance of driver and gauge crystals respectively. The factor k_1 is found from measurements on the two quartz bars coupled together without the specimen. The vibrational strain amplitude ϵ_0 is proportional to the gauge signal, so

$$\epsilon_0 = k_2 V_g , \tag{5.15}$$

where k_2 is constant for a given system.

It can be shown from equation (4.87) that, provided $\Delta_t < 0\cdot 1$ and the frequencies involved are all approximately the same, the component values of the decrement in a composite system are given by

$$m_t \Delta_t = m_1 \Delta_1 + m_2 \Delta_2 + \dots . \tag{5.16}$$

Thus, if suffix 1 denotes the quartz crystals and suffix 2 the specimen,

$$\Delta_2 = \frac{m_t \Delta_t - m_1 \Delta_1}{m_2} . \tag{5.17}$$

It has been assumed here that the cross-sectional dimensions of each element of the composite system are the same. Marx found that values of the decrements are not appreciably affected by area mismatching at the specimen–driver interface. Area ratios as high as $6:1$ do not produce serious difficulties provided the driver–gauge area is greater than that of the specimen.

If the specimen resonant frequency differs from that of the driver–gauge combination there will be an increase in the measured decrement. If the specimen decrement is less than 10^{-5}, frequency mismatching of $0\cdot5\%$ makes no significant change in the decrement. If the specimen decrement is greater than 10^{-3}, a frequency mismatch up to 2% may be tolerated. A more exact analysis of the Marx system, based on the equivalent electrical circuit, is given by Robinson and Edgar (1974).

5.3.4.2 Bolef and Menes Q-meter method
A resonance method for the MHz frequency range that is simple in principle has been introduced by Bolef and Menes (1960). A thin quartz transducer is cemented to the specimen and then is connected across the terminals of a Q meter. The frequency of the Q meter is adjusted so that a mechanical standing wave pattern near the resonant frequency of the transducer, ν_T, is generated in the composite system. At this frequency a sharp minimum in the Q meter reading is obtained corresponding to a minimum in the effective shunt impedance looking towards the transducer. Further minima occur whenever the standing wave system in the specimen changes by an integral number of half-wavelengths. Application of transmission line theory to the system transducer–bond–specimen shows that it is now necessary to determine the frequencies $\nu_{(n-1)}$, $\nu_{(n-2)}$, ..., $\nu_{(n-m)}$, $\nu_{(n+1)}$, $\nu_{(n+2)}$, ..., $\nu_{(n+p)}$ for which there are respectively $(n-1)$,

$(n-2)$, ..., $(n-m)$, $(n+1)$, $(n+2)$, ..., $(n+p)$ half-wavelengths in the standing wave system. The velocity in the specimen is then found from

$$c = 2l\Delta\nu_{av}\left(1+\frac{m_T}{m_s}\right)$$ (5.18)

where $\Delta\nu_{av}$ is the average difference in frequency between two successive resonances, l is the length of the specimen, m_T is the mass of the transducer and m_s is the mass of the specimen. This method works satisfactorily for high-Q specimens. Bolef and Menes give an alternative method for low-Q specimens.

If a sweep frequency generator is used, acoustical spectra as shown in figure 5.11 are obtained. The envelope is governed by the transducer response, the individual lines corresponding to 'critical' frequencies within the specimen. The damping may be found from the bandwidth of the lines.

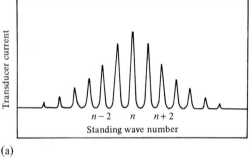

(a)

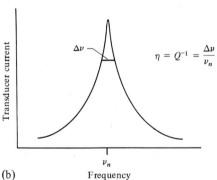

(b)   Frequency

Figure 5.11. Acoustic spectra obtained by the Bolef and Menes (1960) method: (a) a discrete line resonance pattern for longitudinal waves in ruby at 440 MHz and 4 K (Bolef *et al.*, 1962); (b) individual resonance line characterised by the resonant frequency ν_n and loss factor η.

5.3.5 Influence of specimen supports

In making measurements on a resonant system it is inevitable that the supports, and possibly the driving and detecting apparatus, will perturb the vibrations within the specimen. If the specimen is clamped there is usually little effect on any of the normal modes that have the clamping point as a displacement node. For other modes the effect will depend on the impedance offered by the clamp. A disadvantage of clamping is that different clamping positions may be necessary if measurements are required at a number of harmonics of the specimen. When the specimen is clamped at the centre, all the odd-numbered harmonics are relatively unaffected and often provide a sufficient number of frequencies for measurement. A lower impedance than that produced by a clamp may be obtained by suspending the specimen by fine threads or wire loops attached to a rigid support. Again, for highly resonant specimens it is necessary to support the specimen at displacement nodes.

5.4 Pulse methods

The earliest work employing pulses of ultrasonic energy was performed on liquids. Pellam and Galt (1946) describe an interferometer consisting of a liquid column containing a quartz transducer and a polished reflector set up parallel to each other. The X-cut quartz crystal is supplied with a pulse of width 1 μs from a gated oscillator operating at 15 MHz with a pulse repetition rate of 1 kHz. After transmitting the initial pulse the transducer remains quiescent until a series of reflected signals is received. Velocity measurements are made by determining the distance the transducer must be moved to delay the received echoes by a specified increment. Absorption measurements are made by determining the attenuation necessary to keep the receiver signal constant as the transducer is moved. The attenuation coefficient can be measured to an accuracy of about 5% and the sound velocity to about 0·05%.

Early papers dealing with ultrasonic waves in solids include those of Huntingdon (1947), Mason and McSkimin (1947), and Galt (1948). The basic pulse system introduced by Mason and McSkimin is still often used for measurement of the velocity and attenuation coefficient. Their original arrangement is shown in figure 5.12. The signal applied to the driving crystal is a pulsed carrier wave while a receiving crystal at the opposite end of the specimen transmits the signal to an oscilloscope. With improvements in the recovery times of amplifiers, a single transmitter-receiver transducer is often more convenient now. The source of the carrier frequency is a variable-frequency oscillator. The amplifier is turned on and off by a bias signal from the pulse generator whose rate is controlled by a synchroniser which also controls the sweep circuit of the oscilloscope. With a flat-top gating pulse the output of the amplifier is in the form of a wave packet of controllable time duration. Either the

RF output of the amplifier or a rectified output is displayed on the oscilloscope. Examples of rectified received pulses are shown in figure 5.13.

The velocity is measured from the time interval between successive reflections, a separate timing signal being used. The attenuation coefficient is measured from the relative heights of the series of reflected pulses, provided the pulse heights form a reasonably exponential series. Both longitudinal (with an X-cut crystal) and transverse (with a Y-cut crystal) waves may be employed with this method.

The type of pulse used has an important bearing on the significance of such measurements. Three types of pulse have been used by various investigators: (i) a single brief pulse that can be analysed into a Fourier

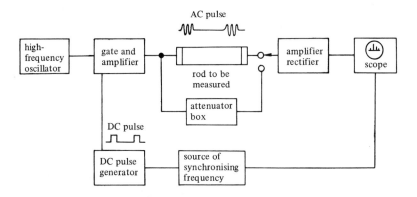

Figure 5.12. Basic pulse system for measuring velocity and attenuation in a solid (Mason and McSkimin, 1947).

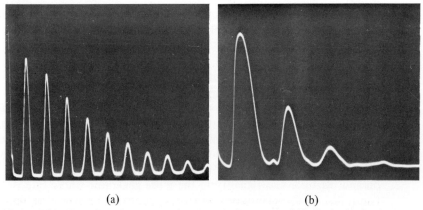

Figure 5.13. Rectified ultrasonic pulses in a single crystal of NaCl obtained using (a) the fundamental mode of a 5 MHz X-cut quartz transducer, (b) the third harmonic (15 MHz) of the X-cut quartz transducer and showing a greater absorption rate.

spectrum containing a continuous range of frequencies; (ii) a short train of decaying vibrations obtained from shock exciting a crystal generator; and (iii) a wave packet consisting of a finite number of sinusoidal waves obtained by gating the output of an oscillator.

Because of the difficulty in assigning a central frequency to the pulse spectrum for types (i) and (ii), type (iii) pulse is now mostly used in physical investigations. As long as the initial pulse contains at least 15 cycles of the carrier frequency, the error in specifying the frequency is estimated to be no greater than 1 in 250.

Many of the early investigators determined the velocity by measuring the transit time of a pulse in the specimen, making use of the leading edge of the pulse for this purpose. However, the shape of the pulse will only remain constant if there is no velocity dispersion in the medium, that is if the velocity is independent of frequency. When dispersion is present the different component frequencies of the pulse will travel with different velocities and the shape of the pulse will change as it progresses. The energy of such a pulse travels with the group velocity c_g:

$$c_g = c_p - \lambda \frac{dc_p}{d\lambda} \; ;$$

c_p is the phase velocity. The precision for the direct measurement of velocity is about $\pm 0 \cdot 2\%$. Much greater precision may be achieved by more sophisticated timing methods and interference techniques. Attempts to measure the damping by determining the heights of pulses of arbitrary shape are unsatisfactory, since in a dispersive medium the peak height of the pulse does not correspond to any particular frequency component. The most satisfactory type of pulse is therefore the wave packet containing a specified number of sinusoidal waves. Provided the packet contains enough cycles, the specimen can be effectively driven into quasi-steady state vibration for the duration of the packet.

Many variations of the basic method have been devised, the aim of most of these being to achieve greater precision in the measurement of the velocity. It is not possible to improve appreciably on the basic method for the measurement of attenuation. Improvements in electronic circuitry have led to more accurate timing methods. These include pulse superposition, sing-around and digital averaging methods. Interference methods based on phase comparison of waves in successive wave packets or with a reference signal lead to an improvement in the precision of velocity measurement by at least one order of magnitude. Brief descriptions of some of these methods will be given below. Further details may be found in the review article by McSkimin (1964). Details concerning the transducers commonly used in ultrasonic measurements may be found in articles by Berlincourt et al. (1964) and Mason (1964).

5.4.1 Pulse superposition method (McSkimin, 1961)

Consider a pulse generator which produces a periodic train of pulses, the repetition period t being determined by the round-trip delay time, δ, in the specimen. If, for instance, the period is equal to 2δ, then the large primary pulse synchronises with all the even echoes and all the odd-numbered echoes are superimposed in between, as shown schematically in figure 5.14. t is adjusted so that the resultant of all the odd echo amplitudes is a maximum. Then

$$t = p\delta - \frac{p\gamma}{360\nu} + \frac{n}{\nu} \quad , \tag{5.19}$$

where p is an integer giving the number of round-trip delays in the specimen, γ is the phase shift occurring at the transducer–cement–specimen interface, ν is the source frequency, and n is an integer. If γ is evaluated from measurements for different cement thicknesses and t determined for the case $n = 0$, then δ can be calculated. t can be measured accurately by means of a frequency counter or can be generated by a frequency synthesiser. Velocities can be measured by the pulse superposition method with a precision of 1 part in 5000. One advantage of this method is that the transducer coupling is taken into account.

A variation on the pulse superposition method is the pulse–echo overlap method in which a repetition rate 100–1000 times slower is used so that the echoes die out between repetitions. The echoes are overlapped in 'scope time' on an oscilloscope. An analysis of the accuracy of both methods has been made by Papadakis (1972).

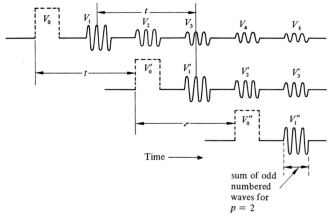

Figure 5.14. Pulse superposition method (McSkimin, 1961). Sequence of received pulses with repetition period t equal to twice the round-trip delay δ.

5.4.2 Sing-around method (Holbrook, 1948)

The stability of modern electronic timing circuity provides the possibility for measuring the time interval between pulses with considerable precision. Holbrook (1948) suggested a synchronous sing-around system, shown in figure 5.15. The pulse generator excites the transmitting transducer and from the resulting series of echoes a particular echo is gated out, amplified, shaped, etc. and then used to trigger off the generator again. By adjusting the combined acoustical–electrical delay until synchronism is obtained, the ultrasonic velocity is found from measurement of the pulse repetition rate. Improvements in the basic technique have been made by Forgacs (1960) to the stage where changes in velocity can be measured with a precision of 1 part in 10^7.

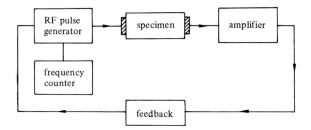

Figure 5.15. Sing-around system (Holbrook, 1948).

5.4.3 Digital averaging method (Lacy and Daniel, 1972)

A standard single-ended pulse–echo system is coupled to a computing counter to give greater accuracy in the measurement of the transit time between any two echoes. The standard amplified RF or video signals may be viewed directly on an oscilloscope while the transit time between any two selected echoes is measured by triggering the counter either from the video signals or from a selected RF cycle within a given echo. In figure 5.16 are shown the main signals involved. In figure 5.16a the counter is

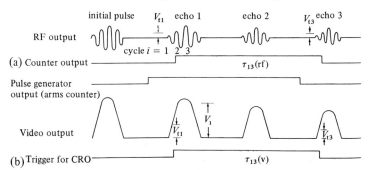

Figure 5.16. Digital averaging method (Lacy and Daniel, 1972). The main signals involved in the measurement of pulse transit time are shown (a) as an RF display, (b) as a video (rectified pulse) display.

shown responding to voltages V_{t1} and V_{t3} arising from echoes 1 and 3 respectively. In figure 5.16b the counter is set to read the time interval between voltage levels V_{t1} and V_{t3} on the video display. The counter may now be set to average digitally the transit times for a number of measurements between 10 and 10^4. In this way transit times may be obtained to an accuracy of $0 \cdot 1$ ns and small changes in transit time may be determined with a sensitivity of 20 ps. The authors discuss in detail the measurement of the longitudinal velocity in an Nb–9%Mo single crystal of length $2 \cdot 50$ cm at 30 MHz and 77 K. A value of the average transit time is given as $10 \cdot 2246 \pm 0 \cdot 0017$ μs. A comparison measurement using the pulse overlap method gave $10 \cdot 2254 \pm 0 \cdot 0018$ μs.

5.4.4 Phase comparison method (McSkimin, 1950)

Another way to improve the basic pulse–echo method is to make use of the interference principle. For this purpose it is necessary to generate phase-coherent wave packets by gating a continuously operating oscillator. A second parallel circuit is provided in which is an adjustable gate and a variable delay line, as shown in figure 5.17. A comparison signal may then be superimposed on any one of the echo signals arising from the specimen. The frequency of the oscillator is first adjusted until the comparison signal and the echo under examination are in phase. If allowance is made for phase changes introduced by the cement, the frequency will now correspond to an integral number of half-wavelengths within the specimen. The phase and amplitude of the comparison signal

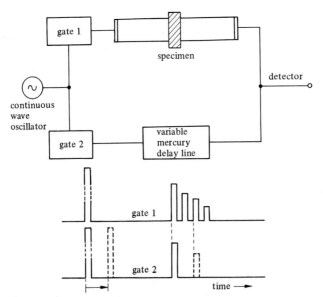

Figure 5.17. Phase comparison system for measuring velocity and attenuation in a small specimen (McSkimin, 1950).

may be adjusted so as to make the two pulses cancel out as in the system of Williams and Lamb (1958). Such a balancing scheme may be made sensitive to 1° of phase shift giving rise to errors as small as 1 part in 10^4. Errors introduced by the cement may be minimised by making the thickness of the layer of cement as nearly as possible an odd number of quarter wavelengths at the frequency of measurement.

An adaptation of this method by McSkimin (1957) has proven useful in making measurements on small crystals. The transducer is attached to a buffer rod of low-loss material, such as fused silica, from which the waves pass into the specimen cemented to the opposite, tapered end of the buffer, as shown in figure 5.18a. A large reflection occurs from the buffer–specimen interface followed by a series of echoes from within the specimen. The pulse duration is now increased so that the echoes overlap, and the frequency is adjusted until an in-phase interference pattern appears, as in figure 5.18b. The velocity is then given by

$$c = \frac{2l\nu_n}{n + \gamma/2\pi} , \qquad (5.20)$$

where ν_n is a frequency corresponding to the in-phase condition, n is an integer which must be determined experimentally, γ is the phase shift occurring at the specimen surfaces and l is the length of the specimen. γ may be determined from the ratio of the magnitudes of the signals E_b/E_t

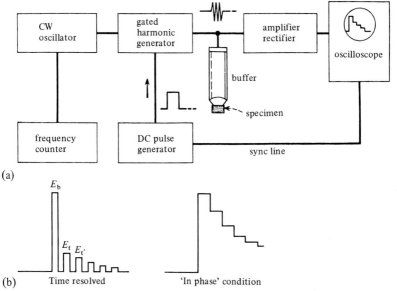

(a)

(b)   Time resolved   'In phase' condition

Figure 5.18. Buffer rod method for measuring velocity and attenuation by phase comparison (McSkimin, 1957). (a) Circuit diagram, (b) video patterns. E_b is the magnitude of signal from buffer-specimen interface; $E_t, E_{t'}$, etc., are the magnitudes of pulse echoes within the specimen.

(see figure 5.18b). Advantages of this method include the direct evaluation of coupling effects; measurements can be carried out on specimens with linear dimensions as small as 2 mm; and it is not necessary to deposit conducting electrodes on the specimen. The velocity can be measured with a precision of about 1 in 10^4.

5.4.5 Phase-locked pulse method

Blume (1963) has combined some of the features of the sing-around system with the interferometer technique. The frequency of the signal from a continuously running generator is controlled by the spacing between echoes through the agency of a gated AFC. A given echo is gated out of the pulse–echo pattern and sent into a phase-sensitive detector together with the reference frequency, as shown schematically in figure 5.19. The output of this stage is adjusted manually to a null condition (phase shift between reference voltage and nth echo $\frac{1}{2}\pi$ or $\frac{3}{2}\pi$) by varying the generator frequency. Subsequently the AFC senses any departure from the null condition and changes the frequency in order to maintain a phase-locked condition. Full circuit details are given by Blume. The resolution of the system is reported to be 2–3 parts in 10^8.

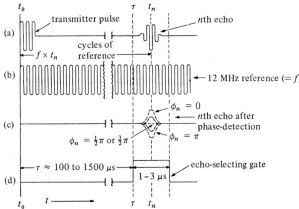

Figure 5.19. Basic timing diagram of the phase-locked pulse interferometer. (After Blume, 1963.)

5.4.6 Coherent-pulse continuous-wave (CW) system

An interesting technique combining some of the advantages of pulse and CW systems has been described by De Klerk (1965). The basis of the system, shown as a block diagram in figure 5.20, is a pulse–echo apparatus consisting of a continuously-operating variable-frequency generator coupled to a gated amplifier. A low-level CW signal derived from the generator is now mixed with the pulse–echo signal via an RF attenuator. Constructive interference between the two signals will occur whenever the phase difference is $r\pi$, where r is an integer. When the CW and pulse signals are out of phase, the pulse–echo envelope shows a sinusoidal modulation.

Frequencies corresponding to $v_{n-m}, ..., v_n, ..., v_{n+p}$ of the Bolef and Menes CW technique may now be determined, for each of which the phase difference is $r\pi$ with r even. Equation (5.18) for the velocity is then applicable. A distinct advantage of this method is that pulse–echo attenuation measurements can be made at the same time. Nonlinearity of the oscilloscope time base and uncertainties in the measurement of transit times due to the presence of the transducer–specimen bond are avoided.

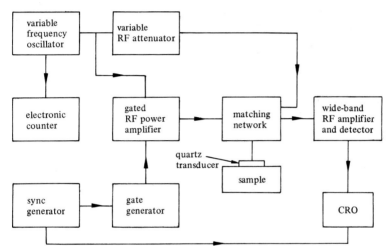

Figure 5.20. Block diagram of pulse/CW system (De Klerk, 1965).

5.4.7 Sources of error

In the pulse–echo method there are numerous sources of error in measuring both the velocity and attenuation. In measuring the velocity the accuracy is affected by problems connected with the pulse shape, triggering, timing, the influence of the bond and the orientation of the specimen axis with respect to its crystal system. Methods for overcoming some of these problems have already been discussed.

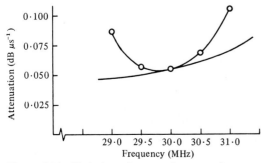

Figure 5.21. Variation of attenuation near the resonant frequency of the transducer (30 MHz). (After Truell *et al.*, 1969.)

The measurement of attenuation is very sensitive to a number of effects related to the propagation of pulses in crystals. Spurious results may be obtained if allowance is not made for diffraction, mode conversion, coupling losses, surface reflection losses, and nonparallelism. In measuring attenuation it is convenient to superimpose on the echo pattern an envelope curve derived from a calibrated exponential generator. The tuning of the oscillator is then adjusted until the minimum value of the attenuation is obtained, as shown in figure 5.21.

5.4.7.1 *Diffraction losses*
In most experiments using the pulse method a small circular or square transducer is cemented to the specimen. If the lateral dimensions of the specimen are greater than those of the transducer, it is usually assumed that the side boundaries have negligible effect and the waves propagate as if in an infinite medium. The manner of radiation of the energy into the specimen is important for the assessment of attenuation. Calculations of radiation patterns from transducers are often made on the assumption of radiation into a perfect fluid.

Seki *et al.* (1956) computed the diffraction field for a circular transducer. In the Fresnel region, adjacent to the transducer, propagation is essentially plane but the pressure shows a series of fluctuations, as illustrated in figure 5.22. In the far field the energy spreads out at an angle determined by the simple diffraction law

$$\sin\beta = \frac{1 \cdot 22\lambda}{D} \tag{5.21}$$

where β is the half-angular width of the beam and D is the diameter of the transducer. The pressure then falls off inversely as the distance as if from a point source.

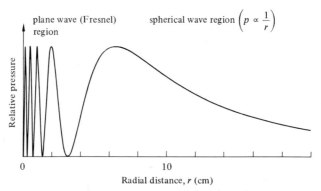

Figure 5.22. Computed pressure variations along the axis of a circular transducer as a function of radial distance from the transducer face. It is assumed that $a = 5\lambda$, where $a = 1 \cdot 3$ cm is the radius of the transducer.

When pulse–echo measurements are made corresponding to various distances in the far field, the beam spreading causes loss of energy and hence an error in the measurement of attenuation. From equation (5.21), the beam spreading is less at high frequencies and for large diameter transducers. Hence, large errors are to be expected at low frequencies and for small transducers.

Seki *et al.* (1956) calculated the diffraction loss in decibels as a function of path length. They found it convenient to introduce a normalised path length s defined by $s = \lambda z/a^2$, where z is the actual path length and a is the radius of the transducer. The actual path length z is then measured in units of a^2/λ. The diffraction loss in decibels was computed from $-20 \log_{10} p_{rms}$, where p_{rms} is the pressure in the diffraction field averaged over the receiving transducer. The diffraction loss as a function of a^2/λ is shown in figure 5.23. Minima occur at $0 \cdot 7$, $1 \cdot 05$, and $2 \cdot 4$ a^2/λ for which the diffraction losses are $1 \cdot 5$, $1 \cdot 8$, and $2 \cdot 4$ dB respectively. Corresponding to these minima will be maxima in the measured decay pattern. The practical effect produced by diffraction is therefore to cause a non-exponential echo decay pattern.

If a mean line is drawn on figure 5.23, the average loss is 1 dB per path length of a^2/λ. For many practical purposes this average value provides sufficient correction. A square transducer gives approximately the same loss as a circular transducer of the same area. Further details of the theory together with a number of photographs of decay patterns are discussed in Truell *et al.* (1969). Seki *et al.* (1956) give an example of the need for making correction for loss by diffraction at low frequencies by quoting measurements on a germanium single crystal with the use of an X-cut quartz transducer at 5 MHz, as shown in figure 5.24.

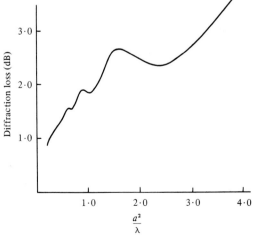

Figure 5.23. Diffraction loss associated with beam spreading as a function of a^2/λ (Seki *et al.*, 1956).

In correcting practical measurements either the average loss figure may be applied, or the more accurate method discussed by Papadakis (1959), also summarised by Truell *et al.* (1969), may be employed. Papadakis (1966) has since extended his calculations to cover the effects of diffraction on both attenuation and velocity in anisotropic media. The previous method has been expanded to use the diffraction loss curves appropriate to the particular crystal system of the specimen.

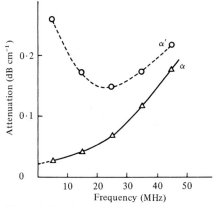

Figure 5.24. Attenuation measurements on a germanium single crystal with the use of the odd harmonics of an X-cut quartz transducer (fundamental frequency 5 MHz). α' represents the measured attenuation values and α the values corrected for diffraction loss (Seki *et al.*, 1956).

5.4.7.2 *Mode conversion*

If the lateral dimensions of the specimen are not great enough (at least five times the transducer diameter) some of the diffracted beam may be reflected from the side walls with possible change of mode. Thus a longitudinal pulse may be reflected partly as a transverse pulse. If such a reflected pulse is recorded by the receiver, a spurious echo will be observed.

One technique that exploits this effect has been described by Redwood (1957). The transducer completely covers one end face of the specimen which is in the form of a ground circular cylinder. The waves then propagate under guided conditions. A large number of reflections are observed with a corresponding improvement in the accuracy of measuring the attenuation. It is important that the cross section of the specimen be accurately circular. A nonexponential decay pattern may be observed if there is interference between the modes.

5.4.7.3 *Other errors*

Provided there is an impedance mismatch between transducer and specimen little energy will be lost on each internal reflection. However, the bonding material may give rise to a phase shift of the reflected pulse. Errors introduced by the transducer–specimen cement may be evaluated by one of the special methods such as that described in section 5.4.4. Commonly used bonding materials at room temperature include oils, stopcock greases, salol, and Eastman 910 cement. Shear wave transducers require a firm bond such as provided by salol or Eastman 910. At low temperatures Nonaq stopcock grease is frequently used.

Errors may also be introduced if the specimen surfaces are not plane and parallel. In the case of a single crystal, where one face of the specimen has been oriented correctly (to within $0 \cdot 5°$ by back reflection x-ray method), the opposite face must now be made flat and parallel, with a wedge angle no greater than 10^{-4} to 10^{-5} radian. For a further extensive discussion of errors of measurement with illustrations see Truell *et al.* (1969).

Although the overall error in measuring attenuation absolutely may well be quite large, of the order of 20%, with the pulse method relative attenuation can be measured fairly accurately. When the interest is on the effects of changing external conditions, the greatest precision is achieved by measuring the change in attenuation for a specific echo rather than measuring the attenuation over the whole pattern before and after the change.

Experiment E5.1 The mechanical impulse method
(Pollard, H. F., 1964, *Aust. J. Phys.*, **17**, 8.)

When an impulse is applied to a mechanical system, the analysis of the response of the system may be carried out in terms of either time or frequency. From the experimental standpoint it is often preferable to work in terms of frequency. Often it is required to find the transfer function, $H(i\omega)$, of the system from measurements made of the output spectrum function, $G(i\omega)$. Provided the input function, $f(t)$, is known, and its Fourier transform, $F(i\omega)$, then $H(i\omega)$ may be found from the relationship

$$G(i\omega) = F(i\omega)H(i\omega) .\tag{E5.1}$$

If $f(t)$ is a delta function, then $F(i\omega) = 1$. In the case of an arbitrary pulse of short duration for which the strength $S = \int_0^\tau f(t)\,dt$, then, as the pulse duration tends to zero, $F(i\omega) \to S$.

Consider a mechanical impulse produced by the impact on the system of a mass m_0 moving with velocity v_0 just prior to impact. If m_0 is brought to rest as a result of the impact, then $S = m_0 v_0$. In practice, the mass will rebound from the specimen, so that

$$S = m_0 v_0' = m_0 v_0(1 + e) ,\tag{E5.2}$$

where e is the coefficient of restitution. In the limit of very short impact time, $F(i\omega) = m_0 v_0'$, and equation (E5.1) becomes

$$G(i\omega) = m_0 v_0' H(i\omega) .\tag{E5.3}$$

If the output velocity function, $v(t)$, of a system is measured, or its Fourier transform, the velocity spectrum $V(i\omega)$, then $G(i\omega)$ in equation (E5.3) is replaced by $V(i\omega)$.

In this case, $H(i\omega)$ is identical with the mechanical admittance of the system, that is, the reciprocal of the mechanical impedance $Z(i\omega)$, since, by definition, $Z(i\omega) = F(i\omega)/V(i\omega)$. Thus, for a mechanical impulse, equation (E5.3) may be written

$$V(i\omega) = \frac{m_0 v_0'}{Z(i\omega)} \, . \tag{E5.4}$$

With the aid of equation (E5.4), the velocity spectrum may be computed for any mechanical system for which $Z(i\omega)$ is known. A convenient method for finding $Z(i\omega)$ is to draw an analogous electrical circuit and then to compute $Z(i\omega)$ by the usual steady-state method.

Semiinfinite elastic rod

Consider the impact of a small elastic sphere on the end face of a semiinfinite elastic rod so that a longitudinal disturbance is produced in the rod. The mechanical system and its force–current analogy are shown in figure E5.1. Analogous mechanical quantities are placed in brackets beside the corresponding electrical quantities. From the circuit

$$\frac{i_1}{e_2} = \frac{1}{R} + i\omega C - \frac{\omega^2 CL}{R} \, .$$

Substituting analogous mechanical quantities, the mechanical transfer impedance is

$$Z_{12}(i\omega) = \frac{f_1}{v_2} = R_0 + i\omega m_0 - \omega^2 m_0 R_0 C_{\mathrm{m}} \, , \tag{E5.5}$$

where $R_0 = S(E\rho)^{1/2}$ is the characteristic impedance, S the area of cross-section, E Young's modulus, ρ the density, and C_{m} the compliance of the rod. Combining equations (E5.4) and (E5.5) gives the velocity spectrum

$$V(i\omega) = \frac{m_0 v_0'}{Z_{12}(i\omega)} = \frac{m_0 v_0'}{R_0 + i\omega m_0 - \omega^2 m_0 R_0 C_{\mathrm{m}}} \, .$$

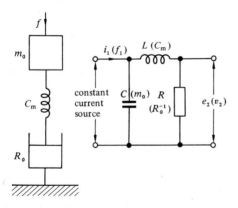

Figure E5.1. Elastic sphere—infinite rod system and its electrical analogy.

This equation may be rearranged in nondimensional form as

$$\frac{R_0}{m_0 v_0'}|V(i\omega)| = \left[1 + \left(\frac{\omega m_0}{R_0}\right)^2 (1+\epsilon)\right]^{-\frac{1}{2}}, \tag{E5.6}$$

where

$$\epsilon = \frac{R_0^2 C_m}{m_0}(\omega^2 m_0 C_m - 2).$$

The velocity spectrum will therefore remain uniform over the range of frequencies for which the right-hand side of equation (E5.6) is unity. At higher frequencies $|V(i\omega)|$ falls off more rapidly as m_0 and C_m are increased. Figure E5.2 shows graphically the velocity spectrum given by equation (E5.6) for a number of different values of C_m. The greater C_m, the 'softer' the impact and the more rapidly does the spectrum fall off at high frequencies.

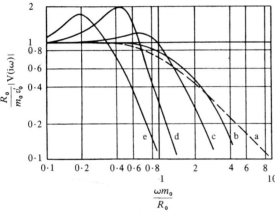

Figure E5.2. Velocity spectrum for elastic sphere–infinite rod system. (a) $C_m = 0$ (limit for impact of rigid sphere), (b) $C_m = 0 \cdot 25 \ m_0/R_0^2$, (c) $C_m = m_0/R_0^2$, (d) $C_m = 4m_0/R_0^2$, (e) $C_m = 10m_0/R_0^2$.

Finite elastic rod

The complete low-frequency resonance response of a finite rod may be computed by the application of transmission line theory. However, near resonant peaks, the theory may be simplified and then agrees with the results of normal mode analysis, as discussed in section 4.5.6. A mode impedance, Z_n, is introduced that governs the response of the nth normal mode of the system and is defined by

$$Z_n = R_n + i\omega M_n + \frac{1}{i\omega C_n}, \tag{E5.7}$$

where R_n, M_n, and C_n are the equivalent resistance, mass, and compliance associated with the nth normal mode. As defined by Skudrzyk (1958), these parameters depend on the type of force distribution (point source, line source, etc.) and on the relative locations of the point of observation and the point of application of the force. At each resonant peak, $Z_n = R_n$. The force–current analogy for such a system at

resonance is shown in figure E5.3. The mechanical transfer impedance of the system
at resonance when excited by a mechanical impulse is then

$$Z_{12}(i\omega) = \frac{f_1}{v_2} = R_n + i\omega m_0 - \omega^2 m_0 C_m R_n \ . \tag{E5.8}$$

As might be expected, equation (E5.8) is similar to equation (E5.5), with R_n replacing
R_0. Substituting equation (E5.7) in equation (E5.4) yields an expression for the
velocity spectrum similar to equation (E5.6), with R_n replacing R_0. In the frequency
region for which the right-hand side of equation (E5.6) is unity,

$$|V(i\omega)| = \frac{m_0 v_0'}{R_n} \ . \tag{E5.9}$$

The particular resonant modes of a system that are excited will depend on the
existence of spectral components of the force at the correct frequencies. In the range
where the spectrum is essentially constant, a mechanical impulse will excite all the
resonant modes of the system. By means of a suitable spectrometer the set of
resonances generated by a mechanical impulse may be recorded.

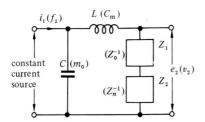

Figure E5.3. Electrical analogy of elastic sphere–finite elastic rod system when vibrating
in a resonant mode. Z_1 is analogous to the characteristic impedance and Z_2 to the
mode impedance.

Impulse spectrometer
An experimental arrangement for determining the response of a rod when excited
longitudinally by a mechanical impulse is shown in figure E5.4. Steel ball bearings of
known mass are released at a controlled rate into the supply tube by means of a
magnetically operated gate. The length and inclination of the tube provides control
over the velocity of impact. The detector consists of the variable capacitor formed
between one end of the rod and a fixed electrode. This capacitor is connected into a
resonant circuit which forms part of a crystal-controlled oscillator operating off-
resonance as a slope detector. In the case of nonmetallic specimens, the end face of
the specimen may be coated with a metallic paint, such as Aquadag.

After amplification the output of the detector is fed into a wave analyser whose
response is conveniently recorded on a logarithmic level recorder. The frequencies of
the resonant modes of the specimen may now be located by driving the wave analyser
slowly through the appropriate frequency range while periodic impulses are applied to
the specimen. For the accurate measurement of individual resonant frequencies the
wave analyser may be tuned manually to the peak response. A stable oscillator is then
switched into the wave analyser and adjusted to give maximum response at the same

setting of the analyser. The frequency setting of the oscillator is read on a frequency counter. An example of a recorded spectrum for a brass rod is shown in figure E5.5, and a slow scan of the fundamental longitudinal mode in figure E5.6a.

The damping associated with each mode is conveniently found by measuring the decay rate at resonance. The specimen is excited with a single impulse and when the decaying signal is examined on the logarithmic recorder, a straight line is observed if the mode has an exponential decay rate. The decay line for the fundamental longitudinal mode of the brass rod is shown in figure E5.6b. When the decay rate becomes too short for accurate measurement it is then more satisfactory to measure the bandwidth of the resonant curve.

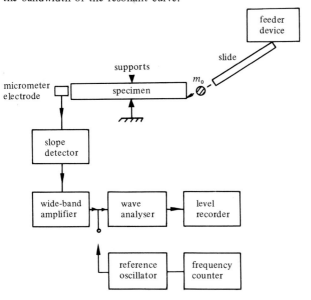

Figure E5.4. Mechanical impulse spectrometer for determining the response of a rod to longitudinal excitation.

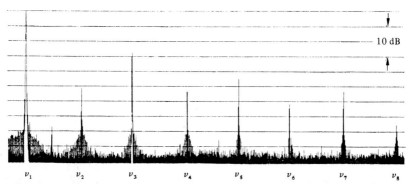

Figure E5.5. Frequency response of a brass rod subjected to mechanical impulses (61% Cu, 36% Zn, 3% Pb; length 75·76 cm; radius 0·477 cm; density $8·48 \times 10^3$ kg m^{-3}, clamped at centre).

Torsional and flexural modes may also be generated with an arrangement similar to that shown in figure E5.7. A steel ball is allowed to fall onto a projecting lug attached to one end of the specimen while the detector electrode is mounted vertically over a similar lug attached either to the same or to the other end of the specimen. With this arrangement both torsional and flexural waves may be generated and detected. The torsional modes are clearly identified since, like the longitudinal modes, the frequency difference between resonant modes is constant whereas for flexural waves the frequency difference increases with frequency.

The impulse spectrometer may also be used to examine the transient time response following an impact on the end face of a specimen. The time base of an oscilloscope may be triggered externally by a signal derived from a photodiode mounted so that the incident ball interrupts a light beam just prior to impact. The resulting waveform may be examined at different times following impact with the aid of a delayed time base. Figure E5.8a shows the initiation of the vibrations in a brass rod, and figure E5.8b the longitudinal vibrations persisting 100 ms after impact.

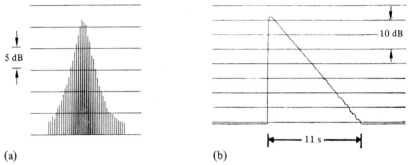

(a)   (b)

Figure E5.6. (a) Slow scan of fundamental longitudinal mode (frequency interval between impulses 1 Hz); (b) decay curve for fundamental longitudinal mode of brass rod (same specimen as in figure E5.5).

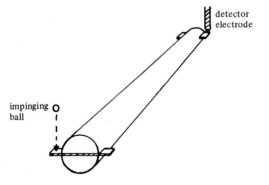

Figure E5.7. Impulse excitation and detection of torsional and flexural waves in a rod.

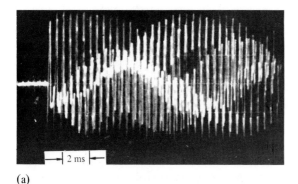

(a)

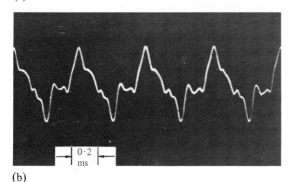

(b)

Figure E5.8. Initiation of vibrations in a brass rod (same specimen as in figure E5.5) following longitudinal impact: (a) total output from slope detector—flexural modes below 1700 Hz attenuated by high-pass filter, ν_1 (longitudinal) = 1994 Hz; (b) as in (a) but with time base of oscilloscope delayed by 100 ms after triggering.

Experiment E5.2 Laser energy deposition method for measurement of elastic constants

(Calder, C. A., Wilcox, W. W., 1974, *Rev. Sci. Instrum.*, **45**, 1557.)

A Q-switched ruby laser pulse is used to generate a very brief stress pulse in one end of a specimen in the form of a rod of circular or rectangular cross section. The sudden deposition of heat energy into the material causes a state of stress due to the material trying to expand against its own inertial forces. The stress field so caused propagates through the rod and is detected at the far end by a 30 MHz transducer. The transducer consists of a 9 mm disc of lithium sulphate backed with acoustically matching material. The backing material damps out stress waves entering it and so minimises reflections within the transducer. The laser provides approximately 1-2 J of energy in about 30 ns with a peak power output of about 50 MW. The 9 mm diameter beam is unfocused giving an irradiance of 80 W cm^{-2}. Part of the beam is deviated by a beam splitter to a photodiode which provides a trigger pulse for the observing oscilloscope. Examination of the output from the transducer enables both the longitudinal and shear velocities of the material to be determined by means of a method of analysis given by Hughes *et al.* (1949).

G

When a short pulse travels along a rod whose length is equal to several diameters (10 or more), the signal received at the far end consists of a number of discrete pulses. The first received pulse is found to travel along the rod with the infinite medium longitudinal (dilatational) velocity, c_ϱ, while subsequent pulses involve one or more mode conversion reflections from the side walls. A typical experimental recording is shown in figure E5.9. The specimen used by Calder and Wilcox in this case was an aluminium alloy bar (6061-T6), 15·30 cm long, cross section 1·27 x 5·08 cm. In explaining the second and later received pulses (trailing echoes), Hughes *et al.* considered a longitudinal pulse launched at near grazing incidence in a cylinder. Since the boundary conditions at a free side wall cannot be satisfied by the presence of a longitudinal wave alone, a transverse disturbance must also be generated at the boundary. As discussed in section 3.4.1, if α is the angle of incidence of the longitudinal wave and γ is the angle of reflection of the transverse wave, then $c_t \sin\alpha = c_\varrho \sin\gamma$. If $\alpha \approx 90°$, then $\sin\gamma = c_t/c_\varrho$. The transverse wave then travels across the rod at angle γ and at the opposite side wall generates a reflected transverse wave and a longitudinal wave which propagates down the rod at near grazing angle. The second received pulse corresponds to the reception of this longitudinal pulse which has been delayed in its transit down the rod by the additional time taken for the transverse wave to cross the rod at the angle γ. The transverse pulse generated at the second side wall then travels back across the rod to the first boundary where further transverse and longitudinal pulses are generated. Thus the third received pulse corresponds to a longitudinal pulse that has been twice delayed in travelling along the rod.

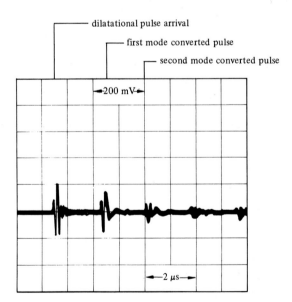

Figure E5.9. Record of received pulses for an aluminium bar, length 15·30 cm, cross section 1·27 x 5·08 cm. Velocity values found are: $c_\varrho = 6300$ m s^{-1}, $c_t = 3070$ m s^{-1}.

Hughes *et al.* give for the arrival time, t, of the various pulses the equation

$$t = m\frac{L}{c_\varrho} + nd\frac{(c_\varrho^2 - c_t^2)^{1/2}}{c_\varrho c_t} \ , \tag{E5.10}$$

where L is the length of the rod, d is the diameter (or cross-sectional dimension), $n = 0, 1, 2, 3, ..., m = 1, 3, 5, ...$. The integers m and n are subject to the condition $mL \geqslant nd\tan\gamma$. The agreement found between the predicted and experimental arrival times is very good.

From the values of c_ϱ and c_t for the material tested, a value for Young's modulus of $7\cdot16 \times 10^{10}$ Pa and Poisson's ratio of $0\cdot335$ were found. The accuracy of measuring elastic constants is mainly limited by the accuracy of the oscilloscope. Using a calibrated time base one can measure elastic constants with an error of $\pm1\%$ using specimens in the form of circular rods or rectangular rods or plates. With a rectangular specimen it is possible to obtain shear-wave information for each lateral direction.

5.5 Optical diffraction methods

Debye and Sears (1932) and Lucas and Biquard (1932) observed independently that, when a beam of monochromatic light passes through a liquid perpendicular to travelling sound waves, the successive compressions and rarefactions of the sound beam act as a phase diffraction grating giving rise to an optical diffraction pattern. It is found that the loss in intensity of the light beam at a given point is proportional to the sound intensity. The basic scheme for observing light diffraction caused by ultrasonic waves is shown in figure 5.25. Monochromatic light from a mercury source is collimated and passes through an aperture into a tank containing a liquid and a quartz transducer, and thence is focused onto a photocell.

Similar phenomena are observed when a transparent solid replaces the liquid tank. In the Schaefer–Bergmann (1934) scheme, described in Experiment E2.3, a circular source-slit is used and as many elastic modes as possible are generated in the solid specimen. In the line diffraction method developed by Hiedemann and Hoesch (1935), a rectangular source-

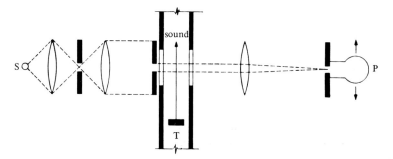

Figure 5.25. Basic experimental arrangement for observing light diffraction caused by ultrasonic waves. The photomultiplier cell scans the various diffraction orders in the plane of the final slit. S—monochromatic light source; T—quartz transducer; P—scanning photomultiplier. (After Hargrove and Achyuthan, 1965.)

slit is used and standing plane waves in one direction in the specimen are generated. In order to do this the specimen size should be large enough to be regarded as an infinite medium. The ultrasonic beam is produced by a quartz transducer set to resonance and the thickness of the specimen is a multiple of half wavelengths.

The theory for the light diffraction effect was deveoped by Raman and Nath (1935; 1936). Their theory predicts that the nth diffraction order propagates in a direction which satisfies the condition (for travelling waves)

$$\sin\theta_n = \frac{n\lambda}{\lambda^*} \qquad n = 0, \pm1, \pm2, \dots, \tag{5.22}$$

where θ_n is the angle of diffraction of the nth order, and λ and λ^* are the wavelengths of light and sound respectively. The normalised intensity is given by

$$I_n = J_n^2(v), \tag{5.23}$$

where $v = 2\pi\mu l/\lambda$, μ is the peak change in refractive index caused by the sound wave, and l is the distance the light travels in the sound beam.

The Hiedemann–Hoesch method has been extended to the measurement of the elastic constants of single crystals by Mayer and Hiedemann (1958). A typical line diffraction pattern is shown in figure 5.26 in which lines due to longitudinal and shear modes in the specimen are visible. Diffraction orders as high as the sixth have been observed for longitudinal waves in glass. It is possible to measure both the spacing and the intensity of the lines accurately allowing the ultrasonic velocity to be determined with a precision of about 1 part in 10^4.

A variation of this technique is known as the visibility method. Instead of the light being focused by the final lens L_3, a microscope is used to view the beam of light. A system of secondary interference fringes is observed with the lines spaced exactly one half acoustic

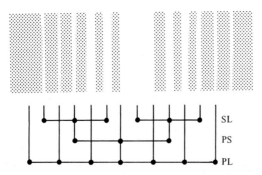

Figure 5.26. Line diffraction pattern. PL—primary longitudinal orders; PS—primary shear orders; SL—secondary longitudinal orders with primary shear orders as zero order. (After Mayer and Hiedemann, 1958.)

wavelength apart. It is then possible to measure the wavelength of the waves, longitudinal or shear, in the specimen accurately by direct observation.

5.6 Ultrasonic spectrum analysis

In the past most ultrasonic measurements have been made with relatively sharply tuned transducers attached directly to the specimen under examination. Under these circumstances the system only responds efficiently near the resonant frequency of the transducer. A number of attempts have been made to develop broad-band transducers suitable for use over a range of frequencies. The object is to obtain information on the propagation of ultrasonic waves in a given system in the form of a response spectrum. More detailed information on the material or on defects present may be obtained when measurements are made at more than one frequency.

5.6.1 Ultrasonic attenuation measurement by a broad-band buffer system

Papadakis *et al.* (1973) show that a system in which a buffer rod (solid rod or liquid column) is introduced between a transducer and a specimen produces more satisfactory results than one in which a heavily damped transducer is bonded directly to the specimen. In the latter case it is found that erroneously high attenuation readings are obtained. The measurement system used is shown in figure 5.27. The first received echo, A, represents part of the incident signal which has been partly reflected at the buffer–specimen interface. Echo B represents part of the incident signal that is transmitted into the specimen and then reflects back into the buffer rod. Echo C is a similar echo involving two internal reflections in the specimen. A particular echo may be gated out and analysed by a spectrum analyser. The result of analysing echoes A, B, and C from a nickel specimen at the end of a water buffer column are shown superimposed in figure 5.28. The amplitude of a given echo at a given frequency may therefore be obtained.

It is now necessary to correct each recorded amplitude because of diffraction (beam spreading). A method for doing so is discussed by Papadakis *et al.*, together with the appropriate equations for computing R, the amplitude reflection coefficient, and α, the attenuation coefficient. In figure 5.29 are shown data relating to measurements on a nickel specimen which had previously been tested by the standard pulse–echo method. The authors conclude that the spectrum analysis method using a buffer rod is as accurate as the standard bonded-transducer method.

The input spectrum available, that is with transducer and buffer rod but without specimen attached, for a buffer rod of fused quartz of length 6·22 cm is shown in figure 5.30. This graph has been corrected for diffraction and for variations in the spectrum of the pulser used to drive the transducer.

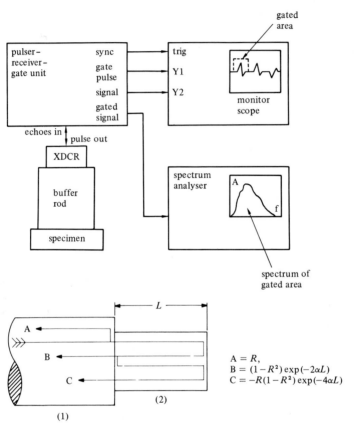

Figure 5.27. Block diagram for a broad-band buffer system, and ultrasonic paths within specimen. (After Papadakis *et al.*, 1973.)

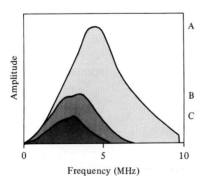

Figure 5.28. Spectra of echoes A, B, C (as in figure 5.27) for a nickel specimen at the end of a water buffer column. Echo A is attenuated 20 dB. (After Papadakis *et al.*, 1973.)

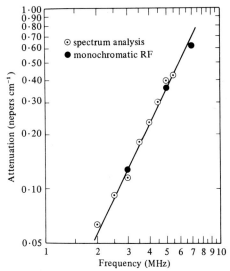

Figure 5.29. Attenuation–frequency spectrum for the specimen for which the spectra are plotted in figure 5.28. Open circles are data obtained by spectrum analysis, solid circles are data using a pulse-echo method with quartz transducers bonded to the specimen. (After Papadakis *et al.*, 1973.)

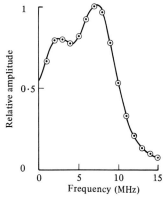

Figure 5.30. Input spectrum available to specimen after correction for diffraction and variations in pulser spectrum by means of a buffer rod of fused quartz of length 6·22 cm. (After Papadakis *et al.*, 1973.)

5.6.2 Detection of defects in solids

If the standard pulse–echo method is used for the detection of a defect in a solid, it is possible to locate the defect by measuring the transit time of the reflected echo. However, if the defect is of the same order of magnitude or smaller than the beam width, it is not possible to obtain further information concerning the nature of the defect by the pulse–echo method. Attempts have been made to conduct pulse–echo measurements at various frequencies. Although such measurements yield information on the defect geometry, they are usually very difficult to carry out.

Gericke (1963) overcame this difficulty by using a highly damped barium titanate transducer ($Q = 2 \cdot 5$) excited by a three-cycle pulse of duration $0 \cdot 6 \ \mu$s. The spectrum of such a pulse is shown in figure 5.31 together with spectra of a rectangular pulse and a single-cycle pulse. A number of tests were conducted on specimens containing known flaws; in each case examination of the spectra with a spectrum analyser with a frequency range of 5 to 14 MHz showed marked differences often when the signal time traces were almost identical. For instance, figure 5.32 shows the time traces and spectra obtained for two aluminium test specimens containing cylindrical holes of $\frac{1}{8}$ and $\frac{1}{32}$ in diameter respectively.

It was also found possible to distinguish between a porosity consisting of a large number of small holes as distinct from one consisting of a small

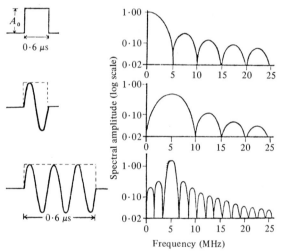

Figure 5.31. Spectra of a rectangular pulse, a single-cycle sine and a three-cycle sine pulse. (After Gericke, 1963.)

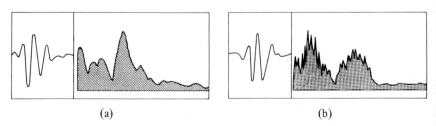

Figure 5.32. Ultrasonic time traces and spectra obtained from (a) $\frac{1}{8}$ in, and (b) $\frac{1}{32}$ in diameter holes in aluminium test specimens by means of a wide-band method. Frequency range: 5–14 MHz (linear). (After Gericke, 1963.)

number of large holes (same reflecting cross section). Again, in detecting
internal cracks with the standard pulse–echo method, serious errors can
occur as the method is insensitive to the orientation of the crack.
Spectrum analysis reveals markedly different spectra for specimens
containing simulated cracks at various orientations, as shown in figure 5.33.

Adler and Whaley (1972) proposed an interference model to explain the
observed spectral behaviour of pulses scattered by a defect. They
suggested that, apart from a specular reflection from the defect, the
observed spectrum was the consequence of interference between sets of
Huygen's secondary spherical wavelets originating from the edges of the
defect. By considering two such sets of waves the following interference
condition is found:

$$\nu_n = \frac{nc}{2d\sin\theta} \,, \tag{5.24}$$

where ν_n are values of the frequency for which the wavelets interfere
constructively, n is an integer, c is the velocity of sound in the medium,
d is the diameter of the defect (assumed to be in the far field of the
transducer) and θ is the angle between the normals to transducer and
defect (assumed plane for simplicity).

In order to verify the model an experiment was set up in which a
transducer was mounted in a water bath at a distance of 6 in from the
specimen. The transducer ($2 \cdot 25$ MHz, lithium sulphate) was shock excited
by a short untuned voltage pulse and acted both as transmitter and
receiver. Being highly damped, the transducer emitted a band of
frequencies ranging from $0 \cdot 5$ to 4 MHz. A given reflected pulse was

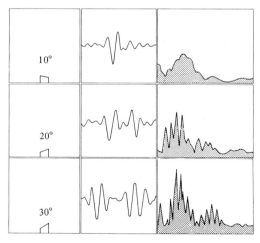

Figure 5.33. Time traces and spectra for specimen with simulated cracks at different
orientations. Same time and frequency scales as in figure 5.32. (After Gericke, 1963.)

selected by means of a gating network and its spectrum displayed on a spectrum analyser. Spectra obtained for various orientations of a 0·125 in diameter reflector are shown in figure 5.34. If the diameter of the reflector is changed, somewhat different spectra are obtained. For each value of θ the frequencies of the interference maxima were measured and compared with values calculated from equation (5.24). A comparison of the values found for the 0·125 in diameter reflector is shown in figure 5.35. The authors comment that the lack of agreement at small angles is due to the difficulty in observing the interference peaks when the strong specular reflection overlaps the interfering waves.

A new explanation for this type of experiment has been given by Simpson (1974). Simpson develops a Fourier-based theory for interfering pulses and shows that it is possible to extract still further information from the observed spectra. A number of experiments were conducted to confirm the theory. The characteristic modulation of the observed spectra was found to be a consequence of the time delay between the two pulses.

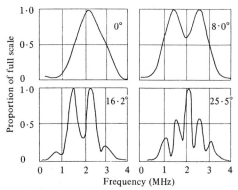

Figure 5.34. Frequency spectra obtained at various angles for a 0·125 in diameter reflector. (After Adler and Whaley, 1972.)

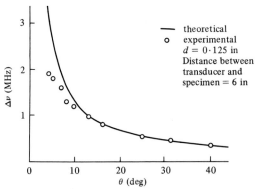

Figure 5.35. The variation of the frequency spacing, $\Delta\nu$, versus angle of incidence for a 0·125 in diameter reflector. (After Adler and Whaley, 1972.)

From the spectra it is possible to extract information on the attenuation as a function of frequency. Also the phase shift of one signal with respect to the other provides information concerning relative acoustic impedances, allowing the detection of nonbonded material and discrimination between cracklike flaws and inclusions. Thus from the interference effects observed between two pulses it is possible to measure simultaneously in a single test the time delay, attenuation, and phase shift in the signals. Further details of Simpson's analysis are given in experiment E5.3.

5.7 Ultrahigh-frequency methods

The pulse–echo method may be used for measurements up to frequencies of the order of 1 GHz for which the higher harmonics of quartz transducer plates are employed. With solid specimens the flatness of the end faces becomes a serious limitation. In order to avoid phase errors due to plane wavefronts travelling through the specimen and impinging on the end faces at an angle, it is necessary for these faces to be flat and parallel to within one-fifth of a wavelength. For a velocity of propagation of 3000 m s^{-1}, the wavelength at a frequency of 1 GHz is 3×10^{-6} m. Since the latter value is approaching that for optical wavelengths, the end faces must be prepared to optical standards. In addition, the ultrasonic attenuation due to thermal vibrations increases with frequency, so it becomes necessary to conduct such high-frequency experiments at low temperatures. Alternative methods for generating and detecting ultrasonic waves at frequencies above 1 GHz include the use of microwave cavities with evaporated thin film transducers, Brillouin scattering methods (see chapter 8) and continuous wave spectrometers.

5.7.1 Microwave cavity method

Bömmel and Dransfield (1958) describe a method for generating high-frequency ultrasonic waves (1–2·5 GHz) in a quartz rod one end of which is placed in a coaxial resonant microwave cavity, as shown in figure 5.36. The effect of the electromagnetic field on the end of the piezoelectric rod is to produce a sharp gradient of piezoelectric stress. The surface then

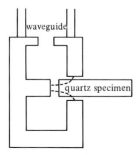

Figure 5.36. Coaxial resonant microwave cavity for the generation of longitudinal waves in a quartz rod.

acts as a transducer, generating an ultrasonic wave which travels along the axis of the rod. The piezoelectric rod thus acts both as a transducer and as a transmission line. In their original experiments Bömmel and Dransfield used an optical diffraction method for detection. Later they employed a similar microwave cavity at the other end of the rod for detection.

Measurements may be made on a nonpiezoelectric specimen by bonding it to a quartz rod. Problems associated with bonding may be avoided if an evaporated thin film is used as the transducer. Cadmium sulphide is mostly used for this purpose, but piezoelectric thin films may be made from zinc sulphide, zinc oxide, aluminium nitride, lithium niobate, etc. The thin film may be evaporated or sputtered directly onto the specimen or onto a thin metallic interface. The crystallographic orientation of the thin film must be appropriate for the type of wave to be generated. Thin films may be constructed for producing both longitudinal and shear waves in the specimen. Further details concerning the making of piezoelectric thin-film transducers are given by De Klerk (1966) and a fuller discussion concerning the generation and detection of ultrahigh-frequency waves may be found in the text by Tucker and Rampton (1972).

5.7.2 High-frequency continuous-wave spectrometers

The Bolef and Menes Q-meter method (see section 5.3.4.2), originally developed for the megaherz range, has been extended for use in the gigahertz range with substantial improvements in sensitivity. A number of new continuous-wave spectrometers are described by Bolef and Miller (1971) who claim that it is possible with these methods to measure changes in phase velocity of 1 part in 10^7 and changes in attenuation coefficient of 10^{-6} cm^{-1}. A basic CW transmission spectrometer is shown in figure 5.37. The measuring system consists of three parts: a transmitter section which includes either a stable oscillator or a swept-frequency generator, a resonator assembly, and a receiver section. Evaporated cadmium sulphide thin-film transducers are used for very high frequencies. They have a number of advantages over bonded transducers, including avoidance of bonding and parallelism errors, and they have a very broad bandwidth allowing continuous operation over a wide frequency range. As in the Q-meter method, the specimen is driven through a series of mechanical resonances. The velocity is determined from the length of the specimen and the number of half-wavelengths in the specimen. As before, the attenuation may be obtained from the bandwidth of the resonance curves. Very small changes in the acoustic phase velocity may be measured by tuning the system to a particular resonant frequency and then recording the shift in frequency with changes in the relevent external parameter. Small changes in attenuation may be measured by recording changes in the amplitude of the signal with change in external parameter.

Further improvements in sensitivity may be obtained by adopting a marginal oscillator as detector. The principle of the marginal oscillator is illustrated in figure 5.38. The condition for oscillation is that the tank circuit represents a zero of reactance and a large impedance. If each mechanical resonance is represented by a series LRC circuit, the coupling network then serves to convert the impedance minimum at the centre of a mechanical resonance to an impedance maximum, as shown in figure 5.38. This impedance maximum can control the RF oscillation level and frequency provided the mechanical Q is sufficiently high. This change in level is amplified and detected. It transpires that the out-of-phase component of the acoustic signal determines the frequency of oscillation whereas the in-phase component determines the level of oscillation.

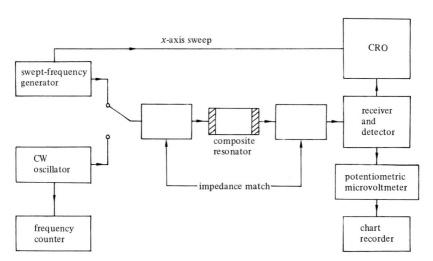

Figure 5.37. Basic CW transmission spectrometer for measuring ultrasonic phase velocity and attenuation. (After Bolef and Miller, 1971.)

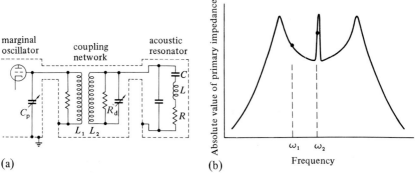

Figure 5.38. (a) Marginal oscillator ultrasonic spectrometer; (b) impedance presented to the marginal oscillator by the coupling network. (After Bolef and Miller, 1971.)

Bolef and Miller (1971) describe a new sampled CW technique that is designed to incorporate the advantages of both CW and pulse–echo methods. The pulse–echo method suffers from at least two disadvantages: there is a lack of monochromaticity in the transmitted signal and the method is limited when dealing with thin specimens. On the other hand, the pulse–echo method has the important advantage that it minimises cross talk between the input and output signals. In the sampled CW technique freedom from cross talk is incorporated while the other advantages of a CW system are retained. Figure 5.39a shows a simplified

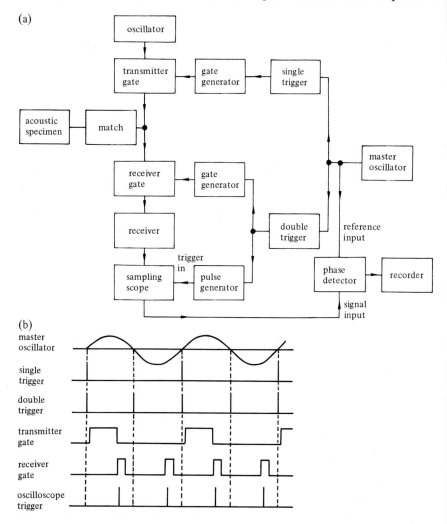

Figure 5.39. (a) Simplified block diagram of sampled CW spectrometer; (b) timing sequence of sampled CW spectrometer. (After Bolef and Miller, 1971.)

sampled CW spectrometer, while figure 5.39b shows the timing sequence of the signals. The CW oscillator is gated on for a time t_d sufficiently long to achieve steady state conditions in the resonator. At time $t = t_d$ the transmitter is gated off and the receiver gated on. The receiver is therefore off during the operation of the transmitter and then proceeds to sample the decaying ultrasonic signal. Only one transducer is necessary with this method. Cross talk is eliminated since the output of the transducer is proportional to the instantaneous acoustic particle velocity at the input face with the transmitter signal off. The output may be examined either in the frequency domain, in which case the mechanical resonance frequency range is slowly scanned, or in the time domain, when the decay of the signal is measured at the fixed resonant frequency.

Ultrahigh-frequency CW methods have been used for a number of applications in the field of solid state physics. In the range 100 MHz to 1 GHz, the resonant interaction of ultrasound with electron spins associated with paramagnetic impurities in diamagnetic crystals has been studied. CW nuclear acoustic resonance studies have been extended to nuclei in metals and in magnetically ordered crystals. Paramagnetic crystals have been studied at frequencies between 0·5 and 10 GHz using a CW acoustic paramagnetic resonance system. CW systems are now sufficiently sensitive to enable measurement of small dispersive velocity changes which accompany resonant coupling of ultrasound to nuclear and electron spins. Small changes in elastic and magnetoelastic properties of crystals which accompany phase changes in the material may be measured. It is also possible to use the method to study nonresonant magnetic-field-dependent effects in semiconductors and metals.

Experiment E5.3 Ultrasonic spectroscopy
(Simpson, W. A., 1974, *J. Acoust. Soc. Am.,* **56**, 1776.)

This paper explores the broadband ultrasonic spectrum analysis technique in terms of the Fourier analysis of the pulses involved. The Fourier transform, $F(i\omega)$, of an input pulse, $f(t)$, is given by equation (4.109)

$$F(i\omega) = \int_{-\infty}^{\infty} f(t) \exp(-i\omega t)\, dt = \mathscr{F}[f(t)] , \tag{E5.11}$$

where $\mathscr{F}$ indicates the Fourier transform. Although theoretically the limits of integration go to infinity, in practice the integral would only be evaluated over the time duration of the pulse. If the input pulse, $f(t)$, suffers a time shift, then, according to the shift theorem of Fourier analysis

$$\mathscr{F}[f(t - t_0)] = \exp(-i\omega t_0) F(i\omega) \tag{E5.12}$$

where $f(t - t_0)$ is identical to $f(t)$ except that it is now centred about $t = t_0$.

Consider now a second identical pulse with a time delay of $2t_0$ between the two pulses. The two pulses may be separated in time or may overlap. Such a signal, consisting of two time-separated pulses, when analysed leads to the spectrum

$$|\mathscr{F}[f(t+t_0)+f(t-t_0)]| = |\exp(i\omega t_0)F(i\omega) + \exp(-i\omega t_0)F(i\omega)|$$

$$= |2\cos\omega t_0||F(i\omega)| \, . \tag{E5.13}$$

This spectrum has the shape of either pulse alone but is modulated by the factor $|2\cos\omega t_0|$. Maxima therefore occur in the resultant spectrum whenever $\omega t_0 = n\pi$, $n = 0, 1, 2, ...,$ or

$$\nu_n = \frac{n}{2t_0} = \frac{n}{\Delta t} \, . \tag{E5.14}$$

The separation between maxima is therefore

$$\Delta\nu = \frac{1}{2t_0} = \frac{1}{\Delta t} \, . \tag{E5.15}$$

If for Δt in equation (E5.14) is substituted the total path-length difference between the two pulses divided by the ultrasonic velocity, equation (5.24), derived by Adler and Whaley, is confirmed. This analysis applies to any two pulses separated in time but the interest here is in the case of two pulses arising as a result of scattering by a defect.

More precisely, the maxima are given by the maxima of $|\cos\omega t_0||F(i\omega)|$. Thus, the condition for the maxima is

$$\tan\omega t_0 - \frac{1}{t_0}\frac{F'(i\omega)}{|F(i\omega)|} = 0 \, , \tag{E5.16}$$

where $F'(i\omega) = d|F(i\omega)|/d\omega$. Thus, strictly, the maxima given by equation (E5.14) only occur when the slope of the envelope is zero $[F'(i\omega) = 0]$. When the slope is not zero the maxima are shifted. Simpson suggests that where readings must be taken with the slope not zero the minimum values of equation (E5.13) be used.

Attenuation measurements
To allow for attenuation, assume that the frequency domain signal is given by $k(\omega)F(i\omega)$, where $k(\omega)$ is a real function: $0 \leqslant k(\omega) \leqslant 1$. Consider the case of an ultrasonic pulse incident normally on a planar medium. Let $f_1(t)$ be the front-surface signal and $f_2(t)$ the back-surface reflection. Ignoring any phase change on reflection, if these two signals are electronically gated to a spectrum analyser, we can write the resultant spectrum as

$$|\mathscr{F}[f_1(t+t_0)+f_2(t-t_0)]| = |\exp(i\omega t_0)F(i\omega) + k(\omega)\exp(-i\omega t_0)F(i\omega)|$$

$$= \{[1-k(\omega)]^2 + 4k(\omega)\cos^2\omega t_0\}^{1/2}|F(i\omega)| \, . \tag{E5.17}$$

If $k(\omega) = 1$ (no attenuation), equation (E5.17) reduces to equation (E5.13). If $0 < k(\omega) < 1$, the spectrum is similar to that described by equation (E5.13) except that the minima do not extend all the way to the axis. The attenuation as a function of frequency is found from the ratio of minimum to maximum for each spectral peak.

Phase measurements
Suppose that the two input signals are identical, except that the second has been shifted in phase by a constant amount ϕ with respect to the first. With the aid of the

shift theorem, we obtain an expression for the spectrum

$$|\mathscr{F}[f(t+t_0)+\exp(\pm i\phi)f(t-t_0)]| = |\exp(i\omega t_0)+\exp(\pm i\phi)\exp(-i\omega t_0)||F(i\omega)|$$

$$= |2\cos(\omega t_0 \mp \tfrac{1}{2}\phi)||F(i\omega)|. \qquad (E5.18)$$

The resulting spectrum is thus identical to that given by equation (E5.13) except for a frequency shift equal to half the phase shift between the two input signals. The equation for the frequency maxima is now

$$\nu_n = \frac{n \pm \phi/2\pi}{\Delta t}, \qquad n = 0, 1, 2, \dots . \qquad (E5.19)$$

The analysis may be extended to the case of two completely arbitrary signals $f(t)$ and $g(t)$ separated in time by $2t_0$ and the case of any number of input signals with arbitrary variable delay between each pulse. The case of several identical signals with constant delay between pulses is then analogous to a diffraction grating producing a spectrum with sharp maxima.

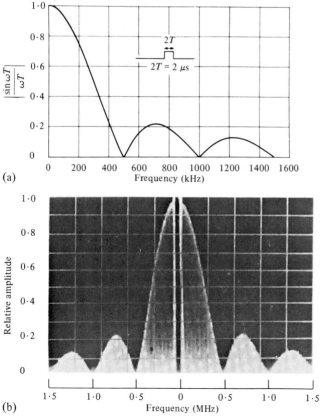

(a)

(b)

Figure E5.10. Spectrum of rectangular pulse, unit height, duration 2 μs (a) calculated, (b) measured. The break at the centre of the measured spectrum is a zero-frequency marker.

Experimental confirmation of theory

A rectangular pulse was chosen for the source. The spectrum calculated for a pulse of unit height and $2T$ duration is shown in figure E5.10a. In figure E5.10b is shown the measured spectrum for a pulse of duration 2 μs. A second identical pulse was then introduced with a delay of 10 μs. The resulting modulated spectrum is shown in figure E5.11. Measurements made on the spectrum were in excellent agreement with equation (E5.13) and the observed frequency shifts agreed with equation (E5.16).

To investigate the effect of attenuation, the second pulse was reduced in amplitude to 20% of that of the first pulse. For this case k is independent of frequency. The resulting spectrum is shown in figure E5.12, and as predicted by equation (E5.17) the attenuation is a function of the modulation depth. Also investigated was the case involving a phase shift of one of the pulses and the case involving interference between two signals of different durations. The method of investigation emphasises the close connection between the observed acoustic phenomena and the results obtained for multiple-slit diffraction in optics.

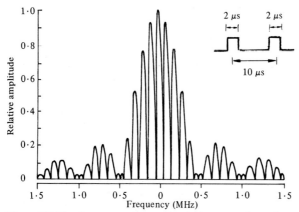

Figure E5.11. Spectrum of two identical rectangular pulses with time separation of 10 μs.

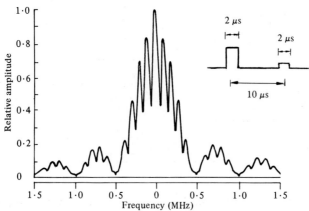

Figure E5.12. Spectrum of two rectangular pulses with 5 : 1 amplitude ratio.

Experiment E5.4 Microwave phonon attenuation measurements in garnet crystals

(Dutoit, M., 1974, *J. Appl. Phys.*, **45**, 2836, and 1971, *Phys. Rev.*, **B3**, 453.)

Synthetic garnet crystals have very low acoustic energy losses at high frequencies and therefore make excellent media in which to propagate high-frequency ultrasonic waves. They are of considerable technological importance and are used in lasers, and in microwave, ultrasonic, magnetic, and optical devices. The measurements described in this paper were made by the pulse–echo technique involving only one microwave cavity. The experimental system used, described more fully in Dutoit (1971), is shown in figure E5.13.

A 2J51 magnetron provides high-power microwave pulses of $0 \cdot 25$ μs duration at a repetition rate of 500 pulses per second. About 500 mW average power is fed to a reentrant cavity. The high electric field across its gap excites a thin-film transducer attached to the sample. A superheterodyne receiver is used for detecting the small return signal. The double-gated integrator averages over many cycles the difference between two short samples of the signal taken at any two instants within a period. One gate is set in the noise long after all echoes have died out and the other gate is positioned on the first echo to obtain a maximum in the DC output. Nonlinearities in the system are bypassed by always reducing the signal to a fixed level with a calibrated attenuator. The relative attenuation is thus directly read in decibels.

Sample dimensions were 5 mm × 5 mm × 10 mm and were cut parallel to a [111] direction. The transducers were sputtered thin films of ZnO sandwiched between chromium/gold electrodes. Absolute values of attenuation were evaluated by matching

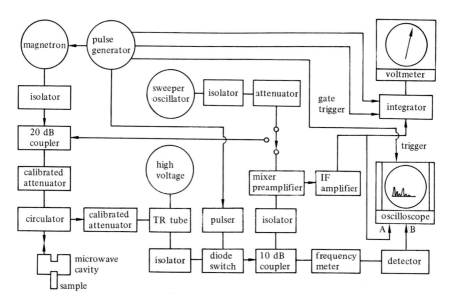

Figure E5.13. Experimental system for the measurement of microwave ultrasonic attenuation.

a calibrated exponential to the echo train at 300 K and at low temperatures. The
temperature dependence was deduced from the change in amplitude of several echoes.
For the low temperature measurements a conventional double Dewar cryogenic system
was used. Figure E5.14 shows the attenuation results obtained as a function of
frequency for longitudinal waves in yttrium aluminium garnet ($Y_3Al_5O_{12}$). Figure
E5.15 shows the attenuation data as a function of temperature for the same material.
Similar data on gadolinium gallium garnet are also reported.

In good dielectric crystals, interactions between microwave phonons and thermal
phonons are responsible for most of the observed loss. At temperatures below about
50 K, the phonon lifetime is long enough so that individual phonon–phonon collisions
need to be considered. A model based on three-phonon processes, which takes into
account the uncertainty in energy of thermal phonons, gives reasonable agreement
with experiment. The attenuation predicted is proportional to ωT^4, where ω is the
acoustic frequency and T is the absolute temperature.

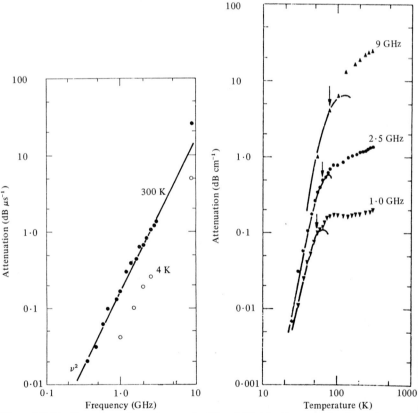

Figure E5.14. Frequency dependence of
attenuation of longitudinal waves in
yttrium aluminium garnet. Attenuation
is plotted as a function of (frequency)^2.

Figure E5.15. Temperature dependence of
attenuation of longitudinal waves in
yttrium aluminium garnet. Arrows indicate
temperatures at which $\omega\tau = 1$.

At temperatures near room temperature, the attenuation may be explained in terms of the Akhiezer mechanism. Ultrasonic waves are treated as a perturbation on the energy of thermal phonons. This entails a relaxation loss since it takes a finite time to reestablish equilibrium. The Akhiezer model predicts an attenuation varying as $\omega^2\tau$, where τ is the thermal-phonon relaxation time. It was found that the room temperature attenuation data agreed with the predictions of the Akhiezer model to within a factor of 2, with no adjustable parameters.

At very low temperatures, the contribution of phonon–phonon interactions to ultrasonic attenuation becomes negligible. In this region, a residual attenuation, α_0, is observed. α_0 is difficult to evaluate since it depends critically on the measurement technique, parallelism of sample faces, coupling losses, and the presence of defects and impurities. The value of α_0 found for yttrium aluminium garnet was typically 0·04 dB μs^{-1}. This value is about 20% of the value at room temperature.

References

Adler, L., Whaley, H. L., 1972, *J. Acoust. Soc. Am.*, **51**, 881;
Balamuth, L., 1934, *Phys. Rev.*, **45**, 715.
Barone, A., Giacomini, A., 1954, *Acustica*, **4**, 182.
Berlincourt, D. A., Curran, D. R., Jaffe, H., 1964, in *Physical Acoustics*, **1A**,
    Ed. W. P. Mason (Academic Press, New York).
Biesterfeldt, H. J., Lange, J. N., Skudrzyk, E. J., 1960, *J. Acoust. Soc. Am.*, **32**, 749.
Blume, R. J., 1963, *Rev. Sci. Instr.*, **34**, 1400.
Bolef, D. I., Menes, M., 1960, *J. Appl. Phys.*, **31**, 1010.
Bolef, D. I., De Klerk, J., Gosser, R. B., 1962, *Rev. Sci. Instr.*, **33**, 631.
Bolef, D. I., Miller, J. G., 1971, in *Physical Acoustics*, Ed. W. P. Mason, R. N. Thurston,
    Volume 8 (Academic Press, New York).
Bömmel, H. E., Dransfield, K., 1958, *Phys. Rev. Letters*, **1**, 234.
Bordoni, P. G., 1947, *Nuovo Cimento*, **4**, 177.
Bordoni, P. G., 1954, *J. Acoust. Soc. Am.*, **26**, 495.
Bordoni, P. G., Nuovo, M., 1957, *Acustica*, **7**, 1.
Bordoni, P. G., Nuovo, M., 1958, *Acustica*, **8**, 351.
Bordoni, P. G., Nuovo, M., Verdini, L., 1959, *Nuovo Cimento*, **14**, 273.
Bradfield, G., 1970, *Ultrasonics*, **8**, 112.
Brown, A. F., Weight, J. P., 1974, *Ultrasonics*, **12**, 161.
Debye, P., Sears, F. W., 1932, *Proc. Nat. Acad. Sci.*, **18**, 409.
De Klerk, J., 1965, *Rev. Sci. Instr.*, **36**, 1540.
De Klerk, J., 1966, in *Physical Acoustics*, Ed. W. P. Mason, **4A** (Academic Press,
    London).
Dutoit, M., 1971, *Phys. Rev.*, **B3**, 453.
Forgacs, R. L., 1960, *J. Acoust. Soc. Am.*, **32**, 1697.
Galt, J. K., 1948, *Phys. Rev.*, **73**, 1460.
Gericke, O. R., 1963, *J. Acoust. Soc. Am.*, **35**, 364.
Hargrove, L. E., Achyuthan, K., 1965, in *Physical Acoustics*, **IIB**, Ed. W. P. Mason
    (Academic Press, New York).
Hiedemann, E., Hoesch, K. H., 1935, *Naturwissenschaften*, **23**, 705.
Holbrook, R. D., 1948, *J. Acoust. Soc. Am.*, **20**, 590A.
Hughes, D. S., Pondrom, W. L., Mims, R. L., 1949, *Phys. Rev.*, **75**, 1552.
Huntingdon, H. B., 1947, *Phys. Rev.*, **72**, 321.
Lacy, L. L., Daniel, A. C., 1972, *J. Acoust. Soc. Am.*, **52**, 189.
Lucas, R., Biquard, P., 1932, *J. Phys. Radium*, **3**, 464.
Mason, W. P., 1964, in *Physical Acoustics*, **IA**, Ed. W. P. Mason (Academic Press, New
    York).

Mason, W. P., McSkimin, H. J., 1947, *J. Acoust. Soc. Am.,* **19**, 464.

Marx, J. W., 1951, *Rev. Sci. Instr.,* **22**, 503.

Marx, J. W., Sivertsen, J. M., 1953, *J. Appl. Phys.,* **24**, 81.

Mayer, W. G., Hiedemann, E. A., 1958, *J. Acoust. Soc. Am.,* **30**, 756.

McSkimin, H. J., 1950, *J. Acoust. Soc. Am.,* **22**, 413.

McSkimin, H. J., 1957, *IRE Trans. Ultrasonic Eng.,* **PGUE-5**, 25.

McSkimin, H. J., 1961, *J. Acoust. Soc. Am.,* **33**, 12.

McSkimin, H. J., 1964, in *Physical Acoustics,* **1A**, Ed. W. P. Mason (Academic Press, New York).

Papadakis, E. P., 1959, *J. Acoust. Soc. Am.,* **31**, 150.

Papadakis, E. P., 1966, *J. Acoust. Soc. Am.,* **40**, 863.

Papadakis, E. P., 1972, *J. Acoust. Soc. Am.,* **52**, 843.

Papadakis, E. P., Fowler, K. A., Lynnworth, L. C., 1973, *J. Acoust. Soc. Am.,* **53**, 1336.

Pellam, J. R., Galt, J. K., 1946, *J. Chem. Phys.,* **14**, 608.

Pursey, H., Pyatt, E. C., 1954, *J. Sci. Instr.,* **31**, 248.

Quimby, S. L., 1925, *Phys. Rev.,* **25**, 558.

Raman, C. V., Nath, N. S., 1935, *Proc. Indian Acad. Sci.,* **A2**, 406.

Raman, C. V., Nath, N. S., 1936, *Proc. Indian Acad. Sci.,* **A3**, 75.

Redwood, M., 1957, *Proc. Phys. Soc.,* **70**, 721.

Robinson, W. H., Edgar, A., 1974, *IEEE Trans. Sonics and Ultrasonics,* **SU-21**, 98.

Schaefer, C., Bergmann, L., 1934, *Naturwissenschaften,* **22**, 685.

Seki, H., Granato, A., Truell, R., 1956, *J. Acoust. Soc. Am.,* **28**, 230.

Simpson, W. A., 1974, *J. Acoust. Soc. Am.,* **56**, 1776.

Skudrzyk, E. J., 1958, *J. Acoust. Soc. Am.,* **30**, 1140.

Thompson, D. O., Holmes, D. K., 1956, *J. Appl. Phys.,* **27**, 713.

Thompson, D. O., Glass, F. M., 1958, *Rev. Sci. Instr.,* **29**, 1034.

Thompson, D. O., Buck, O., Barnes, R. S., Huntington, H. B., 1967, *J. Appl. Phys.,* **38**, 3051.

Truell, R., Elbaum, C., Chick, B. B., 1969, *Ultrasonic Methods in Solid State Physics* (Academic Press, New York).

Tucker, J. W., Rampton, V. W., 1972, *Microwave Ultrasonics in Solid State Physics* (North-Holland, Amsterdam).

Wegel, R. L., Walther, H., 1935, *Physics,* **6**, 141.

Williams, J., Lamb, J., 1958, *J. Acoust. Soc. Am.,* **30**, 308.

REVIEW QUESTIONS

5.1 What is the basis for the resonance method for determining ultrasonic velocity and attenuation? What velocity corrections must be applied (a) for thick specimens, (b) for high absorption? (§5.3)

5.2 In the electrostatic method of measurement, what is the purpose of applying a polarising voltage? (§5.3.2)

5.3 What conditions must be complied with in using a three-component composite resonator? (§5.3.4.1)

5.4 In the pulse–echo method for measuring ultrasonic velocity is the measured velocity a phase or a group velocity? (§5.4)

5.5 What is the principle of the pulse superposition method? (§5.4.1)

5.6 Discuss possible sources of error in using the pulse–echo method. (§5.4.7)

5.7 What are the basic principles involved in the optical diffraction method for measuring elastic constants? (§5.5)

5.8 Explain how broad-band interference effects may be used to determine the presence of a defect in a solid. (§5.6.2 and E5.3)

5.9 Explain how the impact of a small sphere on the end of a cylindrical rod may be used to determine the spectral response of the rod. (E5.1)

Appendix 5.1   Ultrasonic transducers

This appendix is not intended as a thorough review of a rather extensive subject but represents some brief notes that may prove to be supplementary to the material presented in this chapter.   Further details may be found in review articles such as Berlincourt *et al.* (1964) and Bradfield (1970).

Piezoelectric transducers

The **direct piezoelectric effect** refers to the electric polarisation produced when a mechanical stress is applied to certain crystals.   An electric dipole moment is developed that is proportional to the applied stress and whose sign depends on the direction of the stress.   Symbolically this effect may be stated

$$P_i = d_{ijk}\sigma_{jk} \tag{A5.1.1}$$

where P_i is the polarisation per unit area (or dipole moment per unit volume), σ_{jk} is the applied stress, and d_{ijk} are the piezoelectric moduli.   In matrix notation, equation (A5.1.1) may be stated

$$P_i = d_{ij}\sigma_j \,, \tag{A5.1.2}$$

where $i = 1, 2, 3$ and $j = 1, 2, 3, 4, 5, 6$.

The **converse piezoelectric effect** refers to the case when an electric field is applied to a piezoelectric crystal and it becomes strained by an amount directly proportional to the electric field strength:

$$\epsilon_{jk} = d_{ijk}E_i \,, \tag{A5.1.3}$$

where ϵ_{jk} is the mechanical strain and E_i is the electric field strength.   In matrix notation,

$$\epsilon_j = d_{ij}E_i \,. \tag{A5.1.4}$$

It may be proved that the values of d_{ijk} in equations (A5.1.1) and (A5.1.3) are the same.   The reversible nature of the piezoelectric effect is of great practical importance.   In ultrasonic measurements, using for instance a quartz crystal as transducer, the same crystal may be used both for transmission and reception.

Although d_{ijk} has a possible 27 independent coefficients, crystal symmetry reduces this number to manageable proportions.   For instance, for quartz at room temperature there are only 5 independent coefficients as follows (using matrix notation)

$$\begin{bmatrix} d_{11} & -d_{11} & 0 & d_{14} & 0 & 0 \\ 0 & 0 & 0 & 0 & -d_{14} & -2d_{11} \\ 0 & 0 & 0 & 0 & 0 & 0 \end{bmatrix}$$

To illustrate the use of these coefficients, consider the section of a quartz crystal shown in figure A5.1.   If a compressive or tensile stress σ_1 is applied parallel to the x_1 axis, then from equation (A5.1.2),

$$P_1 = d_{11}\sigma_1; \qquad P_2 = 0; \qquad P_3 = 0.$$

Hence the electric polarisation is also along the x_1 axis.   In a similar way it may be confirmed that a tensile stress applied in the x_2 direction also produces a polarisation in the x_1 direction.

As an example of the converse effect, consider an electric field applied in the x_1 direction.   Then, using equation (A5.1.4), the resulting strains are

$$\epsilon_1 = d_{11}E_1; \qquad \epsilon_2 = -d_{11}E_1; \qquad \epsilon_3 = 0; \qquad \epsilon_4 = d_{14}E_1; \qquad \epsilon_5 = \epsilon_6 = 0.$$

Thus, longitudinal strains of opposite sign may appear in either the x_1 or x_2 direction together with a shear strain in the $x_1 x_2$ plane.

If a quartz bar is cut with the orientation shown in figure A5.2, i.e. with its thickness in the x_1 direction (**X-cut**), and an alternating electric field is applied in the x_1 direction, as in the last example, a practical transducer is obtained for longitudinal vibrations either in the x_2 direction or in the x_1 direction. The latter are usually called thickness vibrations. Longitudinal vibrations (x_2 direction) are useful in the frequency range of approximately 10–200 kHz. The fundamental mode for thickness vibrations may be used up to about 30 MHz while harmonics have been used up to 500 MHz.

The fundamental frequency of vibration for longitudinal vibrations is given by (half-wavelength resonator, assuming free ends)

$$\nu_1 = \frac{1}{2y}\left(\frac{s_{22}}{\rho}\right)^{1/2} , \qquad \nu_1 \text{ (Hz)} . y \text{ (m)} = 2728 ,$$

and for thickness vibrations by

$$\nu_1 = \frac{1}{2x}\left(\frac{c_{11}}{\rho}\right)^{1/2} , \qquad \nu_1 \text{ (Hz)} . x\text{(m)} = 2875 ,$$

where x is the thickness of the bar, y is the length in the x_2 direction, c_{11} is an elastic constant and s_{22} is an elastic compliance coefficient. Harmonics may be generated but only for odd multiples of ν_1. Since the sign of the strain changes each half wavelength, the strains cancel out for the even harmonics. The piezoelectric effect can only occur when opposite charges appear on the electrodes. X-cut quartz crystals operating in the thickness mode are usually in the form of thin circular or rectangular plates. For frequencies above about 30 MHz the thickness is such that the crystal becomes fragile and it is then necessary to use the odd harmonics in order to generate higher frequencies. Other cuts are often used for specific purposes, such as the $-18°$ X-cut which produces a purer thickness vibration.

When a quartz plate is cut with its thickness in the x_2 direction (**Y-cut**) application of an electric field to the large surfaces will generate shear vibrations. The fundamental frequency of vibration for a Y-cut crystal operating in the shear mode is given by

$$\nu_1 = \frac{1}{2y}\left(\frac{G}{\rho}\right)^{1/2} , \qquad \nu_1 \text{ (Hz)} . y \text{ (m)} = 1975 .$$

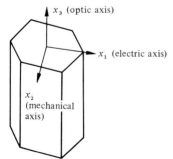

Figure A5.1. Section of a quartz crystal showing the orientation of the reference axes.

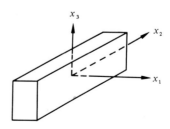

Figure A5.2. X-cut quartz longitudinal vibrator. The electric field is in the x_1 direction, longitudinal (bar) vibrations are in the x_2 direction, thickness vibrations are in the x_1 direction.

Other cuts, such as the AT-cut and BT-cut, are often used. For instance, an AT-cut crystal produces shear vibrations and has a very low temperature coefficient of frequency near room temperature.

Torsional vibrations may be induced if a quartz rod is cut with its length in the x_1 direction and four strip electrodes are connected in pairs. The fundamental frequency is given by the same equation as for shear vibrations above.

A number of other crystalline materials have been used including Rochelle salt (sodium potassium tartrate), ADP (ammonium dihydrogen phosphate), EDT (ethylene diamine tartrate), and DKT (dipotassium tartrate).

One of the first ceramic transducers in use was barium titanate ($BaTiO_3$). This polycrystalline material is electrostrictive, that is, application of an alternating electric field produces a mechanical strain that is proportional to the square of the electric field strength. As discussed in section 5.3.2, this behaviour results in frequency doubling which can be removed by the addition of a large polarising field in addition to the alternating field. The variations in the strain will now occur at the same frequency as the electric field variations.

Permanent polarisation may be introduced in the manufacturing process by applying a large DC field at a temperature above the Curie point and maintaining the field during gradual cooling. The permanently polarised material then behaves in a similar way to a piezoelectric material. Other substances in this class include PZT (lead zirconate titanate), lead metaniobate and sodium metaniobate.

Being polycrystalline, ceramic transducers can be made in a variety of different shapes, including bars, tubes, and focusing forms. The fundamental frequency equations for barium titanate for some commonly used modes of vibration are shown below.

Longitudinal vibrator—bar polarised along its length with the AC field applied to opposite side faces:

$$\nu_1 l = \frac{1}{2}\left(\frac{E}{\rho}\right)^{\frac{1}{2}} = 2260 \; ;$$

Thickness compressional vibrator—ceramic plate with AC field in the same thickness direction and polarised in the same direction:

$$\nu_1 l = \frac{1}{2}\left(\frac{\lambda + 2G}{\rho}\right)^{\frac{1}{2}} = 2540 \; ;$$

Thickness shear vibrator—AC field applied in direction perpendicular to the direction of polarisation, the plane of shear including the directions of the AC and polarising fields

$$\nu_1 l = \frac{1}{2}\left(\frac{G}{\rho}\right)^{\frac{1}{2}} = 1430 \; ;$$

Radial disc vibrator

$$\nu_1 r = \frac{2 \cdot 03}{2\pi}\left[\frac{E}{\rho(1 - \mu^2)}\right]^{\frac{1}{2}} = 1520 \; ,$$

where r is the radius of the disc.

Broadband transducers

In the usual pulse methods the transducer is tuned to a natural resonance for which Q can be very high. In the case of a quartz transducer plate Q may well be at least 2×10^4 and as high as 10^6 for a carefully mounted plate in a vacuum. For ceramic transducers, such as barium titanate, Q is of the order of 100 or more. For ultrasonic spectroscopy the transducer frequency characteristic should be as flat as possible over a frequency range of at least $0 \cdot 5 - 20$ MHz.

In a transducer plate with free surfaces, the wave emission occurs at each surface (where there is the greatest gradient of piezoelectric stress) and results in both a forward-going and a backward-going wave from each surface. The aim of broadband transducer design is to eliminate the wave originating at the back surface.

Suppose that the transducer is backed with a material that completely absorbs any backward-going wave. The forward output of the transducer will then consist of two pulses of opposite polarity time-separated by the transit time of the pulse from the back surface through the transducer to the front surface. If a wave of frequency ν leaves the back surface, its phase will be changed by $2\pi/t$ by the time it reaches the front surface, where t is the transit time. If the transducer thickness is equal to a whole number of wavelengths at frequency ν, then this frequency will be cancelled out of the emitted spectrum. The output spectrum from the transducer will therefore be modulated with minima at frequency intervals $\Delta\nu = 1/t$. A suitable broadband transmitter could therefore be a thick one for which this interval is less than the frequency resolution of the system. Again, the transmitter could be made very thin so that $\Delta\nu$ becomes acceptably high. While a thick transducer is satisfactory as a transmitter, a thin transducer is required for satisfactory reception.

A number of systems have been devised for producing a broadband spectrum including transducers with specially designed backing structures; multiple transducers with elements whose thicknesses form a logarithmic series; transducers based on the inter-digital principle; and transducers having variable piezoelectric properties across the transducer. The generation of broadband ultrasonic signals and methods for detection have been discussed by Brown and Weight (1974).

Use of surface waves

Surface waves of the Rayleigh type may be generated by placing either an X-cut quartz plate or a Y-cut plate on the surface of a specimen. Some of the energy radiated by the plates is converted into Rayleigh waves but the efficiency is not very high. A method that has been widely used is the **wedge method** in which an X-cut quartz thickness plate radiates into a wedge of suitable material (often plastic) placed on the surface, as shown in figure A5.3. The angle of incidence of the longitudinal waves

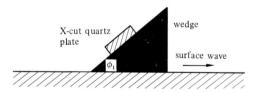

Figure A5.3. Wedge transducer for generation of surface waves.

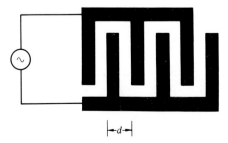

Figure A5.4. Interdigital Rayleigh wave transducer. The Rayleigh wavelength is $2d$.

from the quartz plate, ϕ_1, is chosen so that the angle of refraction, ϕ_2, is 90°. Then

$$\sin\phi_1 = \frac{c_w}{c_s} \ ,$$

where c_w is the velocity in the wedge material and c_s is the surface wave velocity. At this angle it is found that a strong surface wave is propagated in the forward direction only.

A further method that has been used employs an X-cut quartz plate placed on top of an aluminium comb consisting of a set of equidistant teeth and gaps each of width a. The Rayleigh wave generated then has a wavelength of $2a$.

More recently, efficient surface wave production has been made possible by the introduction of interdigital electrodes deposited on the surface of a piezoelectric crystal. When an alternating electric field is applied to the fingers of this device, as shown in figure A5.4, piezoelectric coupling induces a periodic mechanical stress in the crystal which gives rise to a Rayleigh wave. If d is the spacing between neighbouring fingers, the frequency of the electric field is chosen so that the Rayleigh wavelength is equal to $2d$.

6

Internal friction in solids

6.1 Introduction

According to the classical theory of elasticity based on Hooke's law, stress waves are transmitted by a perfectly elastic solid with the following characteristics:

(a) the strain at any point in the body is a function only of the instantaneous stress at the point;

(b) on removal of the stress the elastic strain is instantaneously and fully recoverable;

(c) since the classical process of elastic deformation is reversible, there is no dissipation of energy. Such a process can only occur if the deformation is performed quasi-statically, that is equilibrium is established in the body at every instant of time. A travelling pulse or sine wave would then have constant energy and an isolated perfectly elastic solid, when set into vibration, would continue to vibrate indefinitely.

Experimental measurements on real solids reveal the presence of energy dissipation even at very small strains. Real motion involves finite velocities and therefore the solid cannot be in equilibrium at every instant. Internal processes which attempt to re-establish equilibrium have the effect of rendering the motion irreversible so that energy is lost from the system, ultimately being converted into heat. Thus, dissipation of energy is associated with nonelastic or anelastic behaviour of the body.

The specific type of anelastic behaviour which predominates in a given experiment depends on the imposed experimental conditions. The effects observed include resonance and relaxation phenomena due to the structural details of the material. These effects are often only significant in limited ranges of frequency and temperature.

In general, when an elastic wave travels through a crystalline solid, energy may be lost from the wave by

(i) redirection of the energy through

    (a) a scattering process if the wave impedance of the medium changes,

    (b) diffraction if the wave is incident on an obstacle or void,

    (c) mode conversion at a surface;

(ii) conversion of the energy

    (a) into thermal motion;

    (b) into motion of some other constituent of the lattice, such as a dislocation, an electron, a nucleus, etc.

Conversion of energy can be interpreted fundamentally in terms of interaction processes involving phonons, point defects, dislocations, electrons and nuclei.

Two approaches may be used in studying damping mechanisms. From a phenomenological point of view modifications may be made to Hooke's law in order to allow for dissipation of energy. Such theories involve the motion of a continuous medium and do not attempt to grapple with the causes of dissipation on an atomic scale.

In the second approach an attempt is made to explain specific experimental results by means of atomic models which may not always be amenable to detailed computation. In this chapter we shall adopt mainly the first approach and examine some of the 'classical' damping mechanisms. Most materials, whether classified as liquid or solid, exhibit some degree of flow in addition to elasticity. Newton defined a perfectly viscous fluid as one in which the applied stress is linearly related to the rate of strain, that is,

$$\sigma = \eta \dot{\epsilon}$$

where η is the coefficient of viscosity. Theoretical phenomenological models make use of various combinations of elastic and flow elements. In equivalent mechanical models elastic behaviour is represented by a spring and viscous behaviour by a dashpot.

6.2 Maxwell's material

One of the earliest models introduced to explain the behaviour of real fluids is that of Maxwell (1867) in which the material is assumed to respond to an applied stress with both elasticity and flow. The total deformation of the medium under an applied shear stress is then regarded as made up of an elastic deformation plus deformation due to flow as a Newtonian fluid.

From Hooke's law

$$\epsilon_{12} = \frac{\sigma_{12}}{G} \,, \tag{6.1}$$

where ϵ_{12} is the shear strain resulting from the application of a shear stress σ_{12}, and G is the shear modulus of the material.

From Newton's law for a viscous fluid

$$\dot{\epsilon}_{12} = \frac{\sigma_{12}}{\eta} \,, \tag{6.2}$$

where η is the **shear viscosity coefficient**.

Combining equations (6.1) and (6.2) we obtain

$$\dot{\epsilon}_{12} = \frac{\dot{\sigma}_{12}}{G} + \frac{\sigma_{12}}{\eta} \,. \tag{6.3}$$

6.2.1 Stress relaxation at constant strain

Suppose the medium is suddenly deformed and then constrained so that it retains its new shape. Equation (6.3) then becomes

$$\frac{\dot{\sigma}_{12}}{G} + \frac{\sigma_{12}}{\eta} = 0 \tag{6.4}$$

for which the solution may be written in the form

$$\sigma_{12} = \sigma_0 \exp\left(-\frac{t}{\tau}\right),$$ (6.5)

where σ_0 is the maximum value of the stress determined by Hooke's law and $\tau = \eta/G$ is called the **relaxation time**. Under constant strain, the stress therefore relaxes with time because of the assumed flow mechanism. This behaviour is illustrated in figure 6.1.

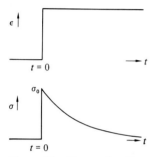

Figure 6.1. Stress relaxation at constant strain for Maxwell material.

6.2.2 Behaviour under constant stress

Consider the application of a stress pulse as shown in figure 6.2a. During the time when the applied stress has a constant value σ_0, $\dot{\sigma}_0 = 0$ and equation (6.3) becomes

$$\dot{\epsilon}_{12} = \frac{\sigma_0}{\eta}.$$ (6.6)

Its solution is

$$\epsilon_{12} = \frac{\sigma_0}{\eta}t + \epsilon_0,$$ (6.7)

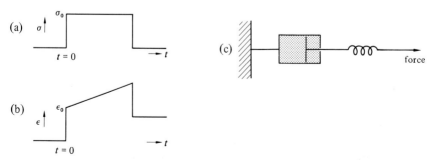

Figure 6.2. Stress pulse of a constant height applied to Maxwell material: (a) stress pulse; (b) resulting strain; (c) equivalent mechanical model (a spring and dashpot in series).

where ϵ_0 is the Hookean strain at time $t = 0$. The behaviour of the body under these conditions is shown in figure 6.2b. There is an instantaneous elastic strain followed by a linear flow region, the two becoming equal at $t = \tau$. In figure 6.2c is shown an equivalent mechanical model, consisting of a spring and a dashpot in series, that has an equation of motion of the same form as equation (6.3), when G is equated to a spring constant and η to a resistive constant.

Maxwell's model has been applied to substances such as pitch and, with some success, to the shear properties of simple liquids and polymers. Values of τ of the order of 1 μs are readily measured experimentally, whereas for liquids having low viscosity τ becomes too short. For example, τ can be measured for polymerised castor oil for which $G \approx 1 \cdot 2 \times 10^6$ N m^{-2}, $\eta \approx 1 \cdot 8$ N s m^{-2}, and $\tau \approx 1 \cdot 5$ μs.

If the medium is made to vibrate sinusoidally, then solution of equation (6.3) under these conditions shows that corresponding to τ there is a relaxation frequency, ν_0, equal to $1/2\pi\tau$.

6.3 Viscoelastic solid (Kelvin–Voigt solid)

Kelvin (1875) and Voigt (1892) independently suggested a modification of Hooke's law in which the stress is not only related to the strain but also to the rate of strain. For an applied shear stress the equation of state has the form

$$\sigma = G\epsilon + \eta\dot{\epsilon} \ . \tag{6.8}$$

In general, for an anisotropic solid

$$\sigma_{ik} = c_{iklm}\,\epsilon_{lm} + \eta_{iklm}\,\dot{\epsilon}_{lm} \ , \tag{6.9}$$

where η_{iklm} is the **viscosity tensor**.

6.3.1 Behaviour under constant stress

If a stress pulse of constant height σ_0 is applied at $t = 0$, then the solution of equation (6.8) is

$$\epsilon(t) = \frac{\sigma_0}{G}\left[1 - \exp\left(-\frac{t}{\tau}\right)\right] \ . \tag{6.10}$$

There is no instantaneous value of the strain; instead it gradually approaches the asymptotic value $\epsilon_0 = \sigma_0/G$.

Similarly, when the stress suddenly returns to zero, the strain decreases accordingly to

$$\epsilon(t) = \epsilon_0\exp\left(-\frac{t}{\tau}\right) \ . \tag{6.11}$$

This behaviour is shown in figure 6.3 together with an equivalent mechanical model.

If a mass m (per unit area) is connected to a viscoelastic material, the equation of motion may be written

$$m\ddot{\epsilon} + \eta\dot{\epsilon} + G\epsilon = \sigma . \tag{6.12}$$

The transient behaviour (with $\sigma = 0$) is that of a damped simple oscillator described by

$$\epsilon(t) = A \exp(-\delta t) \cos(\omega_d t + \varphi) , \tag{6.13}$$

where $\delta = \eta/2m$ is a **damping coefficient**, ω_d is the damped circular (angular) frequency, and φ is a phase constant. Thus, the viscoelastic model involves energy loss; it is the simplest model for representing the internal friction of solids.

There has been some success in applying the model to materials such as cork and rubber. As regards the interaction between elastic waves and conduction electrons at very low temperatures many metals behave like a viscoelastic body.

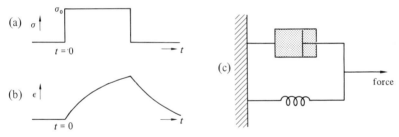

Figure 6.3. Stress pulse of constant height applied to a viscoelastic solid: (a) stress pulse; (b) resulting strain; (c) equivalent mechanical model (a spring and dashpot in parallel).

6.3.2 Wave propagation in a viscoelastic medium

If isotropic conditions are applied to equation (6.9), the equation of state for an isotropic viscoelastic body may be written

$$\sigma_{ik} = \lambda\theta\delta_{ik} + 2G\epsilon_{ik} + \chi\dot{\theta}\delta_{ik} + 2\eta\dot{\epsilon}_{ik} , \tag{6.14}$$

where χ is the **compressional viscosity coefficient**. On combining this equation with Newton's law we obtain the equations of motion for an isotropic viscoelastic body (omitting body forces)

$$\rho\ddot{u}_i = \partial_k\sigma_{ik} = (\lambda + G)\partial_i\theta + G\partial_k\partial_k u_i + (\chi + \eta)\partial_i\dot{\theta} + \eta\partial_k\partial_k\dot{u}_i . \tag{6.15}$$

Consider now a longitudinal disturbance in the x_1 direction. From equation (6.15) we have

$$\rho\ddot{u}_1 = (\lambda + 2G)\frac{\partial^2 u_1}{\partial x_1^2} + (\chi + 2\eta)\frac{\partial^3 u_1}{\partial x_1^2 \partial t} . \tag{6.16}$$

H

If the disturbance is a plane harmonic wave of the form

$$\hat{u}_1 = \hat{A} \exp[i(\hat{k}x_1 - \omega t)], \tag{6.17}$$

where $\hat{k} = k + i\alpha = \omega/c_\varrho + i\alpha$, α is the attenuation coefficient and c_ϱ is the longitudinal wave velocity, then substitution of equation (6.17) in equation (6.16) and separation of real and imaginary parts shows that

$$c_\varrho = \left(\frac{\lambda + 2G}{\rho}\right)^{\frac{1}{2}} \tag{6.18}$$

and

$$\alpha \approx \frac{\omega^2(\chi + 2\eta)}{2\rho c_\varrho^3}, \tag{6.19}$$

where higher order terms are neglected.

It is observed that the velocity of plane longitudinal waves given by equation (6.18), is governed solely by the elastic constants, whereas the attenuation coefficient, described by equation (6.19), is directly proportional to the viscosity coefficients.

A similar procedure may be used to investigate the propagation of plane transverse waves. If the wave is assumed to have the form

$$\hat{u}_2 = \hat{B} \exp[i(\hat{k}x_1 - \omega t)], \tag{6.20}$$

then substitution in equation (6.16), with u_2 replacing u_1, shows that

$$c_t = \left(\frac{G}{\rho}\right)^{\frac{1}{2}}, \tag{6.21}$$

$$\alpha \approx \frac{\omega^2 \eta}{2\rho c_t^3}. \tag{6.22}$$

6.4 Anelastic solid (Zener solid)
Zener (1948) introduced a modification of the Kelvin–Voigt model of a solid in the form of a first-degree equation containing the time derivations of both stress and strain:

$$a_1 \sigma + a_2 \dot{\sigma} = b_1 \epsilon + b_2 \dot{\epsilon}. \tag{6.23}$$

There are three independent constants involved which may be written for convenience

$$\tau_1 = \frac{a_2}{a_1}, \qquad \tau_2 = \frac{b_2}{b_1}, \qquad M_R = \frac{b_1}{a_1}, \tag{6.24}$$

where τ_1 is the relaxation time of the stress under constant strain, τ_2 is the relaxation time of the strain under constant stress, and M_R is the ratio of stress to strain after the system has completely relaxed. M_R is called the **relaxed elastic modulus**.

Equation (6.23) may therefore be written

$$\sigma + \tau_1 \dot{\sigma} = M_R(\epsilon + \tau_2 \dot{\epsilon}) . \tag{6.25}$$

Zener's model takes into account the fact that for many materials, particularly metals, there is an instantaneous strain on sudden application of stress and an instantaneous recovery on removal of the stress followed by time-dependent effects.

6.4.1 Application of a constant stress

If a constant stress σ_0 is suddenly applied at $t = 0$, then solution of equation (6.25) shows that

$$\epsilon(t) = \frac{\sigma_0}{M_R} + \left(\epsilon_0 - \frac{\sigma_0}{M_R} \right) \exp \left(-\frac{t}{\tau_2} \right) . \tag{6.26}$$

This equation is often a better approximation to the behaviour of real materials than the corresponding equation (6.10) for the Kelvin–Voigt solid. The behaviour described by equation (6.26) is shown graphically in figure 6.4 together with an equivalent mechanical model. It may be observed that the strain rises instantaneously to a value ϵ_0 and then gradually rises to a value σ_0/M_R. On removing the load σ_0 the strain drops suddenly by ϵ_0 and then gradually decreases to zero.

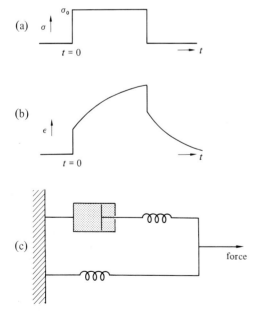

Figure 6.4. Stress pulse of a constant height applied to a Zener solid: (a) applied stress pulse; (b) resulting strain; (c) equivalent mechanical model.

6.4.1.1 *Unrelaxed elastic modulus*

Suppose that the stress increases suddenly by $\Delta\sigma$ in a very short time interval Δt. If equation (6.25) is integrated with respect to time, the first term on each side tends to zero as Δt tends to zero, leaving

$$\tau_1 \Delta\sigma = M_R \tau_2 \Delta\epsilon .$$

We now define the **unrelaxed elastic modulus** M_U as

$$M_U = \frac{\Delta\sigma}{\Delta\epsilon} \; ; \tag{6.27}$$

M_U is thus the ratio of stress to strain in a time interval too short for relaxation to take place. Therefore

$$\frac{M_U}{M_R} = \frac{\tau_2}{\tau_1} . \tag{6.28}$$

The deviation of this ratio from unity gives a measure of the relative change in stress or in strain which may occur through relaxation.

6.4.2 Application of a constant strain

For a constant strain ϵ_0 suddenly applied at $t = 0$ the solution of equation (6.25) is

$$\sigma(t) = M_R\epsilon_0 + (\sigma_0 - M_R\epsilon_0)\exp\left(-\frac{t}{\tau_1}\right) . \tag{6.29}$$

There is an instantaneous stress σ_0 followed by a gradual decrease to the value $M_R\epsilon_0$, as shown in figure 6.5.

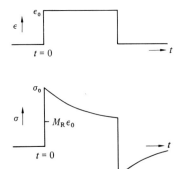

Figure 6.5. Stress-time relationship at constant strain for a Zener solid.

6.4.3 Energy dissipation or internal friction

The viscous elements in the theoretical models described in the preceding section give rise to energy dissipation or internal friction, as the loss is frequently called. If the stress and strain are periodic functions of time, we may write

$$\sigma = \sigma_0 \exp(i\omega t),$$

$$\epsilon = \epsilon_0 \exp[i(\omega t - \delta)], \qquad (6.30)$$

where, in general, there will be a phase difference, δ, between stress and strain. Substitution of equations (6.30) in equation (6.25) gives

$$(1 + i\omega\tau_1)\sigma = M(1 + i\omega\tau_2)\epsilon.$$

We now write this expression as

$$\sigma = M^*\epsilon, \qquad (6.31)$$

where

$$M^* = \frac{1 + i\omega\tau_2}{1 + i\omega\tau_1} M_R \qquad (6.32)$$

and is called the **complex modulus**.

The phase angle δ is given by

$$\tan\delta = \frac{\text{Im}\,M^*}{\text{Re}\,M^*} = \frac{\omega(\tau_2 - \tau_1)}{1 + \omega^2(\tau_1\tau_2)},$$

or

$$\tan\delta = \frac{\omega(\tau_2 - \tau_1)}{1 + \omega^2\tau^2}, \qquad (6.33)$$

where

$$\tau = (\tau_1\tau_2)^{\frac{1}{2}} \qquad (6.34)$$

is the geometrical mean of the two relaxation times.

Also, a **mean modulus**, $\bar{M}$, may be defined by the relation

$$\bar{M} = (M_U M_R)^{\frac{1}{2}}. \qquad (6.35)$$

Equation (6.33) may now be written with the aid of equation (6.28):

$$\tan\delta = \frac{M_U - M_R}{\bar{M}} \frac{\omega\tau}{1 + (\omega\tau)^2}. \qquad (6.36)$$

This form of the equation for $\tan\delta$ has the same form as the Debye relaxation equation in the theory of dielectric relaxation. Hence $\tan\delta$ is a convenient measure for the energy loss in analogy with the corresponding quantity used in electric circuit theory; $\tan\delta$ will be a maximum when $\omega\tau = 1$.

Case A. When $\omega \ll 1/\tau$, then $|M^*| \to M_R$; during each cycle of the relaxation process there is now sufficient time for the process to take place nearly completely and no phase lag occurs.

Case B. When $\omega \gg 1/\tau$, then $|M^*| \to (\tau_2/\tau_1)M_R$; during each cycle there is now insufficient time for relaxation to take place.

The behaviour of $\tan\delta$ and $|M^*|$ with frequency is shown in figure 6.6.

Another way to show the connection between time-dependent behaviour and energy dissipation is to calculate the energy dissipated per cycle, ΔE:

$$\Delta E = \int_0^{2\pi/\omega} \sigma\dot\epsilon \, dt = \pi\sigma_0\epsilon_0 \sin\delta \; . \tag{6.37}$$

The energy loss is often expressed in terms of the loss factor, Q^{-1}, where

$$Q^{-1} = \frac{1}{2\pi} \frac{\Delta E}{E} = \sin\delta \approx \tan\delta \tag{6.38}$$

if δ is small. The logarithmic decrement, Δ, may then be found from $\Delta = \pi Q^{-1}$.

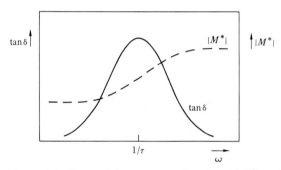

Figure 6.6. Form of the variation of $\tan\delta$ and $|M^*|$ with frequency.

6.4.4 Wave propagation in a Zener solid

The propagation of plane stress waves in a Zener solid may be investigated by combining the equation of state (6.25) with Newton's law and the equation defining strain. Newton's law may be written

$$\frac{\partial^2 \sigma_{ik}}{\partial x_k^2} = \rho \frac{\partial^2 \epsilon_{ik}}{\partial t^2} \tag{6.39}$$

since

$$\rho \ddot u_i = \partial_k \sigma_{ik} \quad \text{and} \quad \epsilon_{ik} = \tfrac{1}{2}(\partial_k u_i + \partial_i u_k) \; .$$

Combining equations (6.39) and (6.25) and considering plane waves travelling in the x_1 direction, we obtain the appropriate wave equation

$$\frac{\partial^2 \varphi}{\partial t^2} + \tau_1 \frac{\partial^3 \varphi}{\partial t^3} = \frac{M_R}{\rho} \left(\frac{\partial^2 \varphi}{\partial x_1^2} + \tau_2 \frac{\partial^3 \varphi}{\partial x_1^2 \partial t} \right) , \tag{6.40}$$

where φ may represent either σ or ϵ. If harmonic waves are assumed to be of the form

$$\hat\varphi = \hat C \exp[i(\hat k x - \omega t)] ,$$

then substitution in equation (6.40) shows that the phase velocity, c, of the stress (strain) waves is

$$c \equiv \frac{\omega}{k} = \left[\frac{M_R}{\rho}(1 + 2\tau_2 \alpha c)\right]^{1/2} = \left[\frac{M_U}{\rho}\left(1 - \frac{\Delta M/M_U}{1 + \omega^2 \tau^2}\right)\right]^{1/2}, \qquad (6.41)$$

where

$$\frac{\Delta M}{M_U} = \frac{M_U - M_R}{M_U} \ll 1 \ ,$$

and the attenuation coefficient, α, is

$$\alpha = \frac{\omega^2(\tau_2 - \tau_1)}{2c(1 + \omega^2 \tau^2)} \approx \frac{1}{2\tau}\frac{\Delta M}{M_U}\frac{\omega^2 \tau^2}{1 + \omega^2 \tau^2}\left(\frac{\rho}{M_U}\right)^{1/2}. \qquad (6.42)$$

Equation (6.42) agrees with equation (6.33) since $\alpha = \omega Q^{-1}/2c$. Thus, α varies as ω^2 whereas $\tan \delta$ (Δ and Q^{-1}) varies as ω.

6.4.5 Temperature dependence

Although a relaxation curve resembles a resonance curve, there are some important differences. The relaxation curve is usually broader than a resonance curve, often covering several decades of frequency, and the maximum value of α is dependent on the temperature. When the relaxation involves atomic movement by diffusion, the relaxation time depends on temperature according to

$$\tau = \tau_0 \exp\frac{U}{kT} \ , \qquad (6.43)$$

where U is an **activation energy**. The frequency at which maximum attenuation occurs is then also dependent on the temperature

$$\omega = \omega_0 \exp\left(-\frac{U}{kT}\right). \qquad (6.44)$$

In principle, a relaxation curve can be determined by measuring α over a range of frequencies at constant temperature, or by measuring α at one frequency over a range of temperatures. Such frequency–temperature dependence and the absence of any amplitude dependence are special features of this type of internal friction.

The Zener model with only one characteristic relaxation time has been successfully applied to systems involving the relaxation of interstitial atoms. However, the relaxation peak, as measured, is frequently broader indicating either the presence of two or more component peaks or the possibility of τ fluctuating in different parts of the material owing to local strains or defects. In the latter case considerable attention has been devoted to the computation of a statistical distribution of relaxation times designed to match the experimental data.

In anisotropic crystals the elastic moduli are functions of orientation so that relaxation phenomena must also be determined as a function of orientation. If measurements are made by means of stress waves then both velocity of propagation and attenuation will depend on orientation. This factor will be referred to again in the next chapter.

6.4.6 Spectrum of relaxation processes

In practice the measured relaxation peak is often broader than that given by equation (6.36). In some cases the curve may be resolved into two or more simpler relaxation peaks. In other cases there is evidence of a distribution of relaxation processes in the material leading to a distribution of values of τ. Berry and Nowick (1966) showed that $\tan \delta$ is then no longer a simple function but is a ratio of two integrals.

6.5 Stress-induced ordering

An interesting example of a relaxation process produced by point defects that leads to internal friction is that known as stress-induced ordering. A point defect in a crystal will produce local elastic strain. Under the action of an applied stress the equilibrium state of a collection of defects in a crystal will change over a period of time to a new equilibrium state. To give rise to internal friction the defects must cause local strains in the lattice with a lower symmetry than that of the lattice. Such defects are sometimes called asymmetric defects or elastic dipoles.

The theory of elastic dipoles is discussed at length by Berry and Nowick (1966). The elastic dipole is defined in terms of a tensor, λ_{ij}, representing the components of strain per unit concentration of defects (all oriented in the same direction) when the concentration is expressed as a mole fraction. If the defect of interest has n crystallographically equivalent orientations, then λ_{ij} may take on a different form for each orientation. In this case the defining equation may be written

$$\lambda_{ij}^{(p)} = \frac{\partial \epsilon_{ij}}{\partial C_p} , \tag{6.45}$$

where ϵ_{ij} is the component of strain, C_p is the mole fraction of defects of orientation p, and $p = 1, 2, ..., n$.

Point defects that may cause relaxation effects include
(a) an interstitial solute atom, denoted by i;
(b) a substitutional solute atom, s;
(c) a vacancy, v;
(d) fcc lattice only: a split or dumb-bell self-interstitial, d;
(e) interacting combinations, such as i-s, i-i, s-v, etc.

Many mechanisms of interest involve relaxation times of the order of 1 s. Measurements in this region are usually conducted with a torsional pendulum technique. As an example, in figure 6.7 are shown internal friction peaks in an Fe-18·5%Mn steel containing various amounts of carbon, as measured by Kê and Tsien (1956) at $2 \cdot 2$ Hz. Measurements made at different frequencies give a value of 34 kcal mol^{-1} for the activation energy. A study of the temperature dependence of the activation energy indicates that the relaxation is a rate process involving the jumping of carbon atoms. The authors propose that the relaxation centres are basically of the i–s type.

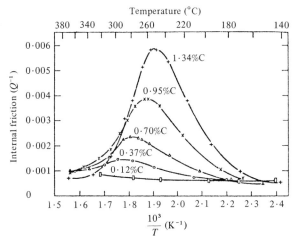

Figure 6.7. Internal friction peaks in Fe–18·5% Mn steel with different carbon contents. Frequency of vibration: $2 \cdot 2$ Hz. (From Kê and Tsien, 1956.)

Experiment E6.1   Oxygen relaxation peak in silicon
(Southgate, P. D., 1960, *Proc. Phys. Soc. (Lond.)*, **76**, 385.)

Southgate (1960) reports ultrasonic measurements made on silicon and germanium at 100 and 300 kHz in a temperature range between room temperature and the melting point. In both materials an electronic relaxation peak appears at a temperature $0 \cdot 56$ of the melting temperature. A relatively large relaxation peak occurs in silicon at higher temperatures, which Southgate attributes to impurity oxygen atoms. The apparatus employed electrostatic drive and FM detection. Longitudinal resonance was induced in the specimen which was suspended by fine tungsten wires. The specimen was held in a high vacuum during measurement and was heated in a molybdenum tube furnace. The decrement was obtained by direct measurement of the decay curve when the drive voltage was switched off. The single crystal specimens were measured with their axes in the [111] direction. The recorded damping-temperature curves showed three components: the electronic peak, the oxygen peak, and a constant mounting loss.

The measurements representing the oxygen relaxation peak were found to be in good agreement with the standard Debye-type equation:

$$\Delta = \Delta_0 \frac{\omega \tau}{1 + \omega^2 \tau^2} , \tag{E6.1}$$

where Δ_0 is the **relaxation strength**. A single relaxation process with an activation energy of 2·55 eV was found to be involved. A comparison of the recorded data and the theoretical equation is shown in figure E6.1.

In determining the nature of the mechanism responsible for the relaxation two specimens were grown under identical ambient conditions, one having a [111] axis and the other a [100] axis. A relaxation peak appeared in the first but not the second, demonstrating that the sites between which oxygen can diffuse are identically located with respect to a [100] axis but not with respect to [111]. Since oxygen was the only principal impurity present, it is presumed that the relaxation is caused by oxygen atoms located in asymmetric positions where they can bond preferentially with just two silicon atoms. Such asymmetry will render the oxygen atoms susceptible to stress-induced ordering.

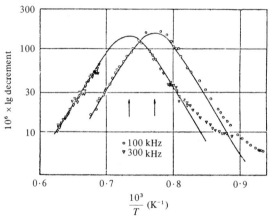

Figure E6.1. Oxygen relaxation peak in silicon. The experimental points are compared with theoretical Debye curves.

6.6 Thermoelasticity

When a solid is compressed, elastic energy is converted into thermal energy and the compressed regions become hotter than the extended regions. There may then be heat flow between the hot and cold regions. As a consequence of the thermal conduction there may be a loss of energy when elastic waves pass through the solid. This effect is frequency dependent. At very high frequencies the process is virtually adiabatic and no losses occur. At very low frequencies, the body is always in thermal equilibrium and the process is isothermal and reversible. If the period of the applied stress is comparable with the time taken for heat transfer there will be an irreversible conversion of elastic energy into heat. The phenomenon therefore has the characteristics of a relaxation process. Thermoelastic losses do not occur with transverse waves since no change in volume is involved.

It will be assumed that when a solid experiences thermal expansion, a thermal strain, ϵ_{ik}^{th}, will develop which adds to the elastic strain, ϵ_{ik}^{el}, already present to give the total strain, so that

$$\epsilon_{ik} = \epsilon_{ik}^{el} + \epsilon_{ik}^{th} .$$ (6.46)

If it is assumed that the total stress is given by a relationship having the form of Hooke's law, then for an anisotropic material

$$\sigma_{ik} = c_{iklm} \epsilon_{lm}^{el} - \gamma_{ik} \Delta T ,$$ (6.47)

where

$$\gamma_{ik} = \left(\frac{\partial \sigma_{ik}}{\partial T} \right)_\epsilon$$

are the components of a second rank tensor representing the temperature coefficients of stress at constant strain (Nye, 1960). Wave propagation problems may be solved by combining equation (6.47) with the equations of motion and the thermal conductivity equation

$$C_\epsilon \frac{\partial T}{\partial t} = K_{ik} \frac{\partial^2 T}{\partial x_i \partial x_k}$$ (6.48)

where K_{ik} is the **thermal conductivity tensor** and C_ϵ is the specific heat at constant elastic strain.

In the case of an isotropic solid, equation (6.47) is replaced by

$$\sigma_{ik} = \lambda \epsilon_{ll}^{el} \delta_{ik} + 2G \epsilon_{ik}^{el} - \gamma \Delta T \delta_{ik} .$$ (6.49)

6.6.1 Relations between elastic constants

When an elastic deformation of an anisotropic body is accompanied by a change in temperature, the change in entropy, ΔS, is

$$\Delta S = \gamma_{ik} \epsilon_{ik} + \frac{C_\epsilon}{T} \Delta T .$$ (6.50)

This equation follows since $S = f(\epsilon_{ik}, T)$ and hence, on expanding (first order)

$$dS = \left(\frac{\partial S}{\partial \epsilon_{ik}} \right)_T d\epsilon_{ik} + \left(\frac{\partial S}{\partial T} \right)_{\epsilon_{ik}} dT ;$$

integration then yields equation (6.50) where the definitions

$$\gamma_{ik} = \left(\frac{\partial S}{\partial \epsilon_{ik}} \right)_T , \qquad C_\epsilon = T \left(\frac{\partial S}{\partial T} \right)_{\epsilon_{ik}}$$

are used.

For isothermal conditions ($T = $ constant) we have

$$\sigma_{ik} = c_{iklm}^T \epsilon_{lm} ,$$

where c_{iklm}^T represents the isothermal elastic constant tensor.

For adiabatic conditions (S = constant) equations (6.47) and (6.50) become

$$\sigma_{ik} = c^S_{iklm}\epsilon_{lm} - \gamma_{ik}\Delta T \,,$$

$$0 = \gamma_{ik}\epsilon_{ik} + \frac{C_\epsilon}{T}\Delta T \,.$$

Eliminating ΔT between these two equations gives

$$c^S_{iklm} = c^T_{iklm} - \frac{\gamma_{ik}\gamma_{lm}T}{C_\epsilon} \,. \tag{6.51}$$

In a similar way it may be shown that

$$s^S_{iklm} = s^T_{iklm} - \frac{\beta_{ik}\beta_{lm}T}{C_\sigma} \,, \tag{6.52}$$

where C_σ is the specific heat at constant stress, β_{ik} are linear thermal expansion coefficients, s_{iklm} is the elastic compliance tensor, and

$$\alpha_{ik} = \left(\frac{\partial S}{\partial \sigma_{ik}}\right)_T = \left(\frac{\partial \epsilon_{ik}}{\partial T}\right)_\sigma \,.$$

Also, the relations for the Lamé constants are

$$\lambda^S = \lambda^T + \frac{(3\lambda+2G)^2\alpha^2 T}{C_\epsilon} \,; \qquad G^S = G^T \,. \tag{6.53}$$

6.6.2 Thermoelastic losses in a Zener solid

At high frequencies ($\omega \gg 1/\tau$) the deformation process will be assumed to be adiabatic and the appropriate elastic modulus will be written M_{ad}. For isothermal conditions ($\omega \ll 1/\tau$) the modulus M_R may be used. For the adiabatic case we may write

$$\frac{\Delta M}{M} \equiv \frac{M_{ad} - M_R}{M_R} \,, \tag{6.54}$$

where $\Delta M/M$ is called the fractional modulus change or **modulus defect**. In this case it is the relative difference between the adiabatic and isothermal moduli. The problem now is to calculate values for $\Delta M/M$ and τ for the particular damping mechanism under examination, such as the thermoelastic effect. Substitution into equations (6.41) and (6.42) then yields the velocity and attenuation of plane stress waves in a viscoelastic solid, with thermoelastic losses taken into account.

As shown by Lücke (1956), a suitable value for τ for the heat flow between a compression and a rarefaction is

$$\tau = \frac{D}{c^2} = \frac{K}{\rho C_\sigma c^2} \,, \tag{6.55}$$

where D is the **thermal diffusivity**, c is the phase velocity, K is the thermal conductivity, and C_σ is the specific heat at constant stress.

6.6.2.1 Case 1. *Low-frequency approximation*

Following Lücke (1956), a value will be found for $\Delta M/M$ for the case of an isotropic rod with a stress wave propagating along its axis (x_1 direction), the dimensions in the x_2 and x_3 directions being small compared with the wavelength (low-frequency approximation).

The total strain at any point along the rod will be a function both of the stress and of the temperature. Hence,

$$d\epsilon_{11} = \frac{d\sigma_{11}}{E_T} + \beta \, dT \, , \tag{6.56}$$

$$d\epsilon_{22} = d\epsilon_{33} = -\frac{\mu \, d\sigma_{11}}{E_T} + \beta \, dT \, , \tag{6.57}$$

where E_T is the **isothermal Young's modulus**, μ is Poisson's ratio and β is the **linear thermal expansion coefficient**.

The work done in producing this strain is

$$dW = \sigma_{11} \, d\epsilon_{11} \, . \tag{6.58}$$

The strains $d\epsilon_{22}$ and $d\epsilon_{33}$ do not contribute to the work since the corresponding stresses are zero.

Combining equation (6.58) with the first and second laws of thermodynamics yields

$$\frac{\rho C_\sigma}{T} \, dT = \left(\frac{\partial \epsilon_{11}}{\partial T}\right)_\sigma d\sigma_{11} = \beta \, d\sigma_{11} \, , \tag{6.59}$$

where C_σ is the specific heat at constant stress (equal to the specific heat at constant pressure, C_p, for small stresses).

On combining equations (6.56) and (6.59) we obtain

$$d\epsilon_{11} = \frac{d\sigma_{11}}{E_{ad}} = \frac{1}{E_T} - \frac{\beta^2 T}{C_p} \, d\sigma_{11} \, , \tag{6.60}$$

where E_{ad} is the **adiabatic Young's modulus**. Hence

$$\frac{\Delta M}{M} = \frac{E_{ad} - E_T}{E_T} = \frac{\beta^2 T}{\rho C_p} E_{ad} \approx \frac{\beta^2 T}{\rho C_p} E_T \, . \tag{6.61}$$

6.6.2.2 Case 2. *High-frequency approximation*

Consider a stress wave travelling in the x_1 direction in a solid whose dimensions in the x_2 and x_3 directions are large compared with the wavelength (high-frequency approximation). Therefore lateral contraction need not be considered, so that $\epsilon_{22} = \epsilon_{33} = 0$. The changes in elastic strain then just compensate for the changes due to thermal expansion.

For an isotropic solid under these conditions

$$d\epsilon_{22}^{el} = d\epsilon_{33}^{el} = -\beta \, dT \,,$$

$$d\epsilon_{11} = d\epsilon_{11}^{el} + \beta \, dT \,. \tag{6.62}$$

Then it may be shown (with the use of Hooke's law) that

$$d\epsilon_{11} = \frac{(1+\mu)(1-2\mu)}{E_T(1-\mu)} \, d\sigma_{11} + \left(\frac{1+\mu}{1-\mu}\right)\beta \, dT = \frac{1}{E_T'} \, d\sigma_{11} + \beta' \, dT \,, \tag{6.63}$$

where the effective values E_T', β' replace E_T and β in equation (6.56), and

$$\frac{\Delta M}{M} = \frac{1+\mu}{(1-2\mu)(1-\mu)} \frac{\beta^2 T E_T}{\rho C_p} \,. \tag{6.64}$$

6.6.2.3 Case 3. Cubic crystal

In the case of anisotropic materials a similar treatment will yield the effective values E_T' and β' for the different orientations. For instance, for propagation along the [100] axis of a cubic crystal we have

$$\sigma_{11} = c_{11}\epsilon_{11}^{el} + c_{12}(\epsilon_{22}^{el} + \epsilon_{33}^{el}) \,.$$

Combining this with equation (6.62) we obtain

$$d\epsilon_{11} = \frac{d\sigma_{11}}{c_{11}} + \frac{c_{11} + 2c_{12}}{c_{11}}\beta \, dT \,, \tag{6.65}$$

where the effective values are

$$E_T' = c_{11}E_T \,; \qquad \beta' = \frac{c_{11} + 2c_{12}}{c_{11}}\beta \,.$$

The appropriate expression for $\Delta M/M$ then follows.

In Lücke (1956) are given further expressions for the [110] and [111] directions in cubic crystals, expressions for hexagonal crystals and tables of computed values of $\Delta M/M$, which are generally of the order of 10^{-2}. The value of τ for a good thermal conductor is of the order of 10^{-11} s.

Lücke quotes some measurements made by Waterman in zinc crystals and shown in figure 6.8, in which comparison is made between losses caused by dislocation damping and the thermoelastic effect. When a longitudinal wave propagates along the hexagonal axis in zinc there is no shear stress component in the slip planes; its attenuation is then given accurately by the computed thermoelastic loss. On the other hand, a shear wave propagating along the hexagonal axis will not suffer any attenuation caused by the thermoelastic effect; its attenuation is assumed to be largely due to dislocation damping. This experiment provides one of the few examples in which the absolute attenuation of a stress wave may be computed accurately.

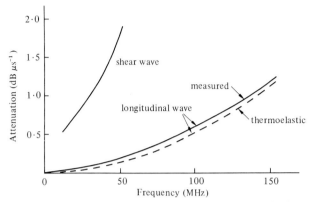

Figure 6.8. Ultrasonic attenuation in zinc single crystals along the hexagonal axis as a function of frequency. The attenuation of a longitudinal wave is governed by thermoelastic losses, whereas for a shear wave dislocation damping is predominant. (After Waterman; as quoted by Lücke, 1956.)

6.6.3 Zener effect in polycrystalline materials

A polycrystalline material consists of an assembly of small single crystals (grains) oriented more or less at random. When a wave propagates through such a material the varying orientations of the grains result in some grains becoming more compressed than others with a consequent heat flow between grains. Zener (1948) shows that the loss factor, η, due to this effect is given by

$$\eta = Q^{-1} = \frac{C_p - C_v}{C_v} R \frac{\nu \nu_0}{\nu^2 + \nu_0^2} \, , \tag{6.66}$$

where R is the fraction of the total strain energy associated with dilatational changes, and ν_0 is the intergrain relaxation frequency

$$\nu_0 = \frac{D}{L_c^2} = \frac{K}{\rho C_p L_c^2} \; ;$$

where L_c is the mean diameter of the grains, and D is the thermal diffusivity. Randall et al. (1939) studied this effect experimentally in brass specimens having varying grain sizes. Their results are shown in figure 6.9 together with a plot of equation (6.66), measurements having been made at 6, 12 and 36 kHz. The effect was found to be small compared with losses caused by dislocations.

Note. When a stress wave passes through a polycrystalline material, some of the energy is scattered by the grains because of the abrupt change in elastic constants between adjoining grain boundaries. Mason and McSkimin (1947) have shown that when $\lambda \geqslant 3L_c$ the attenuation coefficient due to Rayleigh-type scattering is proportional to the third power of the mean grain diameter and inversely proportional to the fourth power of the wavelength:

$$\alpha_{sc} = \frac{8\pi^4 L_c^3 S}{9\lambda^4}$$

where S is a scattering factor dependent on the anisotropy of the material. At shorter wavelengths the attenuation changes less rapidly with frequency and when $\lambda \ll L_c$ the loss is independent of the frequency.

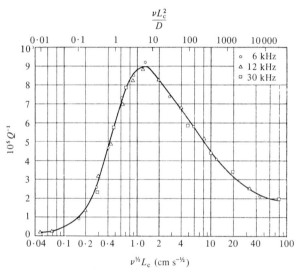

Figure 6.9. Comparison of theoretical curve and measurements of internal friction for polycrystalline brass as a function of frequency, ν, and grain diameter, L_c. (After Randall *et al.*, 1939.)

Experiment E6.2   Ultrasonic hysteresis absorption in polymers
(Hartmann, B., Jarzynski, J., 1972, *J. Appl. Phys.*, **43**, 4304.)

The object of this investigation is to measure the longitudinal and shear velocities and absorptions as a function of frequency, temperature, and strain for the following materials: polymethylmethacrylate (PMMA, density $1 \cdot 191$ kg m^{-3}, weight-average molecular weight $4 \cdot 4 \times 10^3$ kg mol^{-1}), polyethylene (PE, density $0 \cdot 957$ kg m^{-3}, weight-average molecular weight 200 kg mol^{-1}), and polyethylene oxide (PEO, density $1 \cdot 208$ kg m^{-3}, weight-average molecular weight $2 \cdot 9 \times 10^3$ kg mol^{-1}).

An ultrasonic pulse method was used in which the specimen and transducers were immersed in a bath of silicone liquid. Measurements were made both with and without the specimen in the path of the sound beam. Strain in the specimen was estimated from the measured voltage applied to the transducer and the manufacturer's values of the piezoelectric constants.

Ultrasonic measurements

Ultrasonic measurements were made in the immersion apparatus on four PMMA specimens of different thicknesses at $0 \cdot 6$, $2 \cdot 2$, and $5 \cdot 5$ MHz. Graphs of both longitudinal absorption and shear absorption as a function of frequency produced good straight lines. Thus, both sets of data have the form

$$\alpha = A + B\nu ,$$

where α is the absorption coefficient, A and B are constants. The type of absorption observed, for which the major component is $\alpha = B\nu$, or $\alpha\lambda = $ constant, is referred to as **hysteresis absorption**. As shown in figure E6.2, the product $\alpha\lambda$ plotted against temperature is a linear function for PMMA. The absorption is therefore of the hysteresis type over this temperature range.

Measurements were also made to determine whether or not the absorption is strain dependent. For both longitudinal and shear waves in PMMA at room temperature and at a frequency of $2 \cdot 2$ MHz, the absorption was found to have no measurable strain dependence for strains in the range from 10^{-9} to 10^{-6}.

Measurements were made on three PE specimens in the immersion apparatus at frequencies of $0 \cdot 6$, $2 \cdot 2$ and $5 \cdot 5$ MHz and also in an aluminium delay-rod apparatus at 3, 5, 7, $8 \cdot 8$, and 11 MHz. The results of these measurements are shown in figure E6.3 and again indicate a hysteresis type of absorption. In this case, when $\alpha\lambda$ was

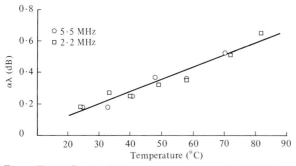

Figure E6.2. Graph of $\alpha\lambda$ versus temperature for PMMA.

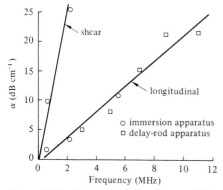

Figure E6.3. Graph of attenuation coefficient versus frequency for PE, measured in an immersion apparatus and also in an aluminium delay-rod apparatus.

plotted as a function of temperature for the measurements at $2 \cdot 2$ and $5 \cdot 5$ MHz, there was agreement at room temperature and near the melting point, but in between there was a deviation from hysteresis behaviour owing to a relaxation process. Again, no dependence on strain was found.

Similar longitudinal measurements were made on PEO, which yielded a straight-line relationship between α and frequency. Shear-wave measurements could not be made because of the high attenuation. When $\alpha\lambda$ was plotted versus temperature, hysteresis behaviour was observed at room temperature and near the melting point with a deviation from hysteresis inbetween. Again no strain dependence was observed.

Comparisons were made for all three materials with torsion pendulum measurements made at 1 Hz. Although the comparison is not entirely conclusive, PMMA appears to have a constant value of $\alpha\lambda$ over a frequency range of 1 to 10^7 Hz, except in the vicinity of a relaxation. A similar result is found for PE, while for PEO $\alpha\lambda$ appears to be constant over the frequency range from 10^{-3} to 10^7 at room temperature.

Existing theories

(1) *Relaxation absorption.* This type of absorption cannot explain the experimental results since α is proportional to ν^2 for a single relaxation, as shown in the text in equations (6.19) and (6.22). A distribution of relaxation times was also considered unlikely to explain the experimental results.

(2) *Scattering at voids or spherulites.* Scattering effects depend on the ratio of the wavelength to the size of the scatterer. This mechanism does not apply to PMMA since it is amorphous and has no appreciable voids. For both PE and PEO the void size is such that scattering would be in the long wavelength limit for which α is dependent on ν^4. Scattering by spherulites is known to have ν^2 dependence.

(3) *Phonon–phonon interaction.* If $\omega\tau \ll 1$, the absorption could be of thermoelastic type for which α depends on ν^2, or the Akhiezer mechanism (the strain modulates the frequencies of the thermal phonons) which also has a ν^2 dependence. If $\omega\tau \gg 1$, it is possible for phonon–phonon interaction to produce hysteresis absorption. However, estimates for the experiments reported show that $\omega\tau \ll 1$. A further point against this mechanism is that it is inherently anharmonic and thus strain dependent, but, as observed, there was no strain dependence. Also the temperature dependence for this type of loss was not observed experimentally.

(4) *Dislocation damping.* As will be discussed further in chapter 7, dislocations may produce absorption in four general ways: (i) a resonance absorption of the dislocation length which has ν^2 dependence, (ii) a dislocation relaxation mechanism (Bordoni loss), also with ν^2 dependence, (iii) a strain-dependent hysteresis loss caused by dislocations breaking away from pinning points, (iv) a strain-independent hysteresis loss caused by the motion of kinks in the dislocation length. The fourth type would be a possibility, but is considered unsatisfactory since no specific molecular mechanism has been worked out.

Polymer absorption model

In general, hystersis can be attributed to the existence of a large number of independent domains, at least some of which can exhibit metastability. The term domain implies that some element of the system exists in one or other of two states. In polymers there are many independent configurations possible for a polymer chain. Further, there are many irregularities in the potential energy leading to many positions of metastability. The suggested model involves rotation about the covalent bonds in the main chain such that the polymer becomes trapped in one of its many local metastable potential energy minima. This mechanism should result in a hysteresis absorption from very low frequencies to 10^{13} Hz. For the crystalline polymers at least, this suggests a volume dependence of the absorption, which in turn means that the absorption is directly related to the degree of crystallinity.

The above investigation shows that the sound absorption in high polymers has a hysteresis background throughout the temperature–frequency plane. Superimposed on this background are various relaxation processes. The relatively complicated shape of the hysteresis α versus T plot is reduced to a straight line by plotting $\alpha\lambda$ versus V, where V is the volume. The temperature, frequency and strain dependence of the hysteresis absorption in partially crystalline polymers can be expressed as

$$\alpha(T, \nu, \epsilon) = \frac{a - bV(T)}{c(T)/\nu} \quad,$$

where a and b are constants for a given polymer, and c is the velocity of sound.

6.7 Wave propagation in piezoelectric semiconductors
When an ultrasonic wave passes through a piezoelectric crystal, the applied periodic stress causes a periodic deformation of the crystal lattice. Since a piezoelectric material does not possess a centre of symmetry, the deformation will cause an additional electrical polarisation. This effect is illustrated in figure 6.10 for the wurtzite (CdS, ZnO) structure which consists of two interpenetrating close-packed hexagonal lattices. If a compressional stress is applied along the hexagonal axis, the resulting strain will cause the centres of positive and negative charge to move apart along the hexagonal axis. A stress wave (longitudinal or transverse) with particle motion along the hexagonal axis will therefore produce a

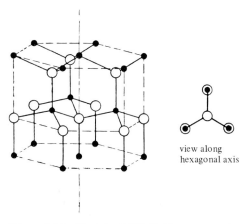

view along
hexagonal axis

Figure 6.10. Wurtzite (CdS, ZnO) structure showing two interpenetrating close-packed hexagonal lattices. Each atom of each type is at the centre of a tetrahedron whose corners are occupied by atoms of the other type. (After McFee, 1966.)

longitudinal piezoelectric field. As shown by McFee (1966), longitudinal electrostatic fields produced in this way are much stronger than any transverse electric fields. Acoustic plane waves which are accompanied by strong longitudinal piezoelectric fields are termed **piezoelectrically active waves**.

If the material possesses both piezoelectricity and conductivity, there will be a further interaction of the stress wave with the charge carriers since the motion of the latter depends on the lattice potentials. The effects produced by this coupling depend on the relationship between the ultrasonic frequency and the value of the conductivity relaxation frequency $\omega_c = \sigma_c/\epsilon$, where σ_c is the electrical conductivity and ϵ is the permittivity. If $\omega \ll \omega_c$, the charge carriers redistribute themselves quickly enough so that their field cancels the piezoelectric field caused by the ultrasonic wave. If $\omega \gtrsim \omega_c$, the charge carriers cannot respond quickly enough and a strong piezoelectric field is the result as in an insulator. The ultrasonic wave then suffers dispersion and attenuation and a number of interesting interaction effects appear, such as acoustic amplification. The latter will be discussed further in section 6.7.2.

In a piezoelectric medium there may be coupling between any of the three possible acoustic modes (one longitudinal and two transverse) and the two possible transverse electromagnetic waves. An electromagnetic wave will be accompanied by an acoustic wave and vice versa. However, the accompanying waves in each case are usually of small magnitude. It is the stronger electrostatic field produced by the piezoelectric polarisation that has a more marked coupling effect on an ultrasonic wave.

6.7.1 One-dimensional macroscopic wave theory

The basic equations of state for a piezoelectric crystal are

$$\sigma_{ij} = c_{ijkl}\epsilon_{kl} - d_{lij}E_l , \tag{6.67}$$

$$D_k = d_{klm}\epsilon_{lm} + \kappa_{ik}E_i , \tag{6.68}$$

where σ_{ij} are the stress components, c_{ijkl} is the elastic constant tensor, ϵ_{kl} are the strain components, d_{lij} is the piezoelectric constant tensor, E_l is the electric field, D_k is the electric displacement, and κ_{ik} are the components of the dielectric tensor. Equation (6.67) shows that an applied stress causes both an elastic strain and an electric field to appear. Equation (6.68) shows that the electric displacement field is composed of both a piezoelectric and a dielectric part.

Since the longitudinal piezoelectric fields are the only significant ones, the subscripts may be omitted and the above equations written in one-dimensional form:

$$\sigma = c\epsilon - dE , \tag{6.69}$$

$$D = d\epsilon + \kappa E . \tag{6.70}$$

It will be assumed that the semiconductor is n-type and extrinsic and that it contains charge carriers which behave like electrons with an isotropic effective mass m^*. Completely nondegenerate (Maxwell–Boltzmann) statistics will be assumed for the electron distribution. The theory outlined below follows Hutson and White (1962) and White (1962).

Consider a plane acoustic wave (longitudinal or transverse) propagating in the x direction with displacement u given by

$$u = u_0 \exp[i(k_s x - \omega t)] \tag{6.71}$$

where $k_s = \omega/c_s + i\alpha$ is the complex wave vector, c_s is the acoustic wave velocity, and α is the attenuation coefficient. It is valid to write k_s this way provided the coupling between the acoustic wave and the electrons is small in an absolute sense.

Putting $\epsilon = \partial u/\partial x$ for the strain, we can write the relevant wave equation in the form

$$\rho \frac{\partial^2 u}{\partial t^2} = \frac{\partial \sigma}{\partial x} = c \frac{\partial^2 u}{\partial x^2} - d \frac{\partial E}{\partial x} \ . \tag{6.72}$$

To solve equation (6.72), an expression is required for E in terms of the displacement u. The equations required to do this are

$$\frac{\partial D}{\partial x} = -qn \ , \qquad \text{(Poisson's equation);} \tag{6.73}$$

$$\frac{\partial J}{\partial x} = q \frac{\partial n}{\partial t} \ , \qquad \text{(continuity equation);} \tag{6.74}$$

$$J = q(n_0 + n)\mu E + qD_n \frac{\partial n}{\partial x} \ , \qquad \text{(current density);} \tag{6.75}$$

q is the magnitude of the electronic charge, μ and D_n are the electronic mobility and diffusion constants respectively, n_0 is the equilibrium density of conduction electrons in the absence of an acoustic wave, and n is the variation in the density of conduction electrons caused by the acoustic wave.

It is assumed that the electric field and displacement have plane-wave space and time dependences

$$E = E_0 + E_1 \exp[i(k_s x - \omega t)] \ , \tag{6.76}$$

$$D = D_0 + D_1 \exp[i(k_s x - \omega t)] \ , \tag{6.77}$$

where E_0 is an external DC electric field applied parallel to the x axis to produce a steady electronic drift.

A linear small-signal theory is obtained if terms containing the product of two wave amplitudes are neglected. The term nE_1 in equation (6.75) is therefore neglected, which will be valid provided $n \ll n_0$. Eliminating E

from equation (6.72) then gives

$$\rho\frac{\partial^2 u}{\partial t^2} = c'\frac{\partial^2 u}{\partial x^2} ,$$

(6.78)

where c' is an effective elastic constant that takes into account the effects of piezoelectricity and conductivity on the acoustic wave propagation:

$$c' = c\left\langle 1 + \frac{d^2}{c\epsilon}\left\{1 + \frac{i\sigma}{\epsilon\omega}\left[1 + \frac{\mu k_s}{\omega}E_0 + i\left(\frac{k_s^2}{\omega}\right)D_n\right]^{-1}\right\}^{-1}\right\rangle .$$

(6.79)

Assuming that the coupling between the acoustic wave and the electrons is small so that we may write

$$k_s = \frac{\omega}{c_s} + i\alpha$$

(6.80)

and that $|\alpha| \ll \omega/c_s$, we can solve equation (6.78) on the basis of the plane wave solution (6.71). The acoustic wave velocity c_s and the attenuation coefficient (per unit length) α are then given by

$$c_s = \rho^{\frac{1}{2}}\text{Re}[(c')^{\frac{1}{2}}] ,$$

(6.81)

$$\alpha = \omega\rho^{\frac{1}{2}}\text{Im}[(c')^{-\frac{1}{2}}] .$$

(6.82)

The small coupling approximation also implies that the velocity dispersion will be small, hence c_s will not differ significantly from $c_0 = (c/\rho)^{\frac{1}{2}}$. Thus, k_s may be replaced in equation (6.79) by ω/c_0, giving

$$c_s = c_0\left[1 + \frac{K^2}{2}\left(1 - \frac{\omega_c/\omega(\omega_c/\omega + \omega/\omega_D)}{\gamma^2 + (\omega_c/\omega + \omega/\omega_D)^2}\right)\right] ,$$

(6.83)

$$\alpha = \frac{K^2\omega_c}{2c_0}\left(\frac{\gamma}{\gamma^2 + (\omega_c/\omega + \omega/\omega_D)^2}\right) ,$$

(6.84)

where

$$\omega_c = \frac{n_0 q\mu}{\epsilon} \qquad \text{(conductivity relaxation frequency),} \qquad (6.85)$$

$$\omega_D = \frac{c_0^2}{D_n} \qquad \text{(electron 'diffusion' frequency),} \qquad (6.86)$$

$$\gamma = 1 + \frac{\mu E_0}{c_0} = 1 - \frac{c_d}{c_0} \qquad \text{(drift parameter),} \qquad (6.87)$$

$$K^2 = \frac{d^2}{c\epsilon} \qquad \text{(square of the electromechanical} \qquad (6.88)$$
$$\text{coupling constant).}$$

Figure 6.11 shows c_s as a function of the drift parameter, γ. c_s has a minimum value when $\gamma = 0$. $\gamma\omega$ is the effective acoustic frequency observed from a frame of reference moving at the drift velocity c_d. When $\gamma = 0$, $c_d = c_0$, that is, the electrons drift at the acoustic velocity and can therefore achieve an equilibrium distribution. As a result the field of the bunched electrons almost cancels the longitudinal piezoelectric field of the acoustic wave. When $|\gamma|$ is sufficiently large, negligible bunching occurs.

Figure 6.12 shows the attenuation coefficient as a function of the drift parameter. For $\gamma > 0$, α is positive and the acoustic wave loses energy to the electrons. When $\gamma < 0$, α is negative, indicating that the acoustic wave gains energy from the electrons. This effect is known as **acoustic**

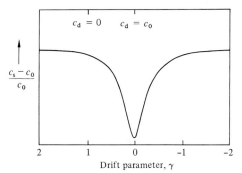

Figure 6.11. Acoustic velocity, c_s, as a function of the drift parameter, γ.

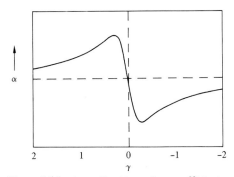

Figure 6.12. Acoustic attenuation coefficient, α, as a function of the drift parameter, γ. When $\gamma < 0$, α is negative and acoustic amplification takes place.

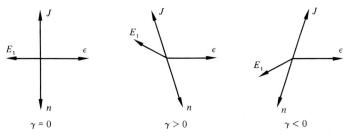

Figure 6.13. Phase relationships between acoustic strain, ϵ, electric field E_1, charge density, n, and current density, J. The case $\gamma < 0$ corresponds to amplification when the phase angle between J and E_1 is greater than $\frac{1}{2}\pi$ and smaller than π.

amplification and will be further discussed in section 6.7.2. When $\gamma = 0$, there is equilibrium between the acoustic wave and the electron distribution and the average force exerted on the acoustic wave by the electrons is zero. The acoustic attenuation is then also zero. The phase relations between the various quantities involved are shown in figure 6.13.

6.7.2 Acoustic amplification

As shown in the previous section, if a DC electric field is applied to a piezoelectric semiconductor so as to cause the interacting charge carriers to drift in the direction of wave propagation, then acoustic amplification (negative attenuation) is possible when the drift velocity exceeds the acoustic wave velocity. An experiment to investigate this effect was described by Hutson *et al.* (1961). Figure 6.14 shows the experimental arrangement used. 1 μs radio frequency pulses are applied to the top Y-cut quartz transducer to produce 15 MHz (or 45 MHz) shear waves which pass down through the CdS sample to a similar receiving transducer. The fused silica buffer rods are used to provide time delay and electrical insulation. The CdS crystal is oriented so that the shear wave propagates perpendicular to the CdS hexagonal axis with particle motion along the hexagonal axis so as to produce a piezoelectrically active wave. Conduction electrons were produced in the sample by illumination with yellow light from a high-pressure mercury arc. When the sample is in the dark it behaves essentially as an insulator and the corresponding shear wave signal level is used as a zero reference for attentuation measurement. A drift voltage pulse of approximately 5 μs duration could be applied during the time of transit of the signal through the sample by means of indium contacts on the sample surfaces. Figure 6.15 shows the observed effects of the electron drift on the ultrasonic attenuation. It may be noted that for drift fields greater than 700 V cm^{-1} the acoustic attenuation becomes negative and the output signal is then larger than the signal obtained under dark conditions. The crossover from positive to negative attenuation was found to correspond to a drift velocity of about 2×10^5 cm s^{-1}. Since the shear wave velocity is $1 \cdot 75 \times 10^5$ cm s^{-1}, the theoretical

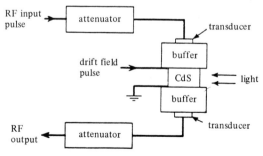

Figure 6.14. Experimental arrangement for observing acoustic amplification in CdS. (After Hutson *et al.*, 1961.)

prediction that acoustic amplification occurs when the drift velocity
becomes greater than the acoustic velocity is upheld. The agreement
between the data of figure 6.15 and the theory presented in the previous
section is described by Hutson *et al.* as semiquantitative. Later work has
shown that electron trapping at low carrier concentrations is a major
cause of the deviations.

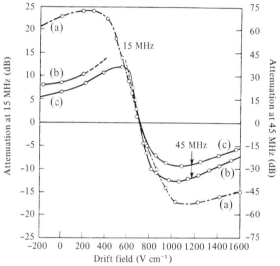

Figure 6.15. Effect of drift velocity on ultrasonic attenuation in CdS at: ω_c/ω of
(a) $1 \cdot 2$, (b) $0 \cdot 24$, and (c) $0 \cdot 21$. (After Hutson *et al.*, 1961.)

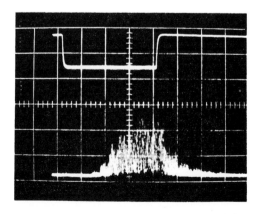

Figure 6.16. Acoustic oscillations in CdS. Top trace is the drift field pulse. Lower
trace is the signal from output transducer at 15 MHz. Horizontal scale: 20 μs per
large division. (After Hutson *et al.*, 1961.)

Hutson *et al.* made a further spectacular observation. If the drift field pulse is made longer than the round-trip transit time for shear waves in the sample, a spontaneous growth of acoustic waves was observed at the output transducer even in the absence of an input signal, as shown in figure 6.16. For instance, if a drift field of 1000 V cm^{-1} is applied, the wave in the sample will be amplified when it travels with the drift and attenuated when it travels against the drift. As may be observed from figure 6.15, the gain greatly exceeds the attenuation. If the duration of the drift field pulse is made sufficiently long, the amplitude of the acoustic oscillations reaches a steady-state saturation level.

References
Berry, B. S., Nowick, A. S., 1966, in *Physical Acoustics,* **IIIA**, Ed. W. P. Mason (Academic Press, New York).
Hutson, A. R., McFee, J. H., White, D. L., 1961, *Phys. Rev. Letters,* **7**, 237.
Hutson, A. R., White, D. L., 1962, *J. Appl. Phys.,* **33**, 40.
Kê, T. S., Tsien, C. T., 1956, *Scientia Sinica,* **5**, 625.
Kelvin, Lord, 1875, *Encyclopaedia Britannica.*
Lücke, K., 1956, *J. Appl. Phys.,* **27**, 1433.
McFee, J. H., 1966, in *Physical Acoustics,* **4A**, Ed. W. P. Mason (Academic Press, New York).
Mason, W. P., McSkimin, H. J., 1947, *J. Acoust. Soc. Am.,* **19**, 464.
Maxwell, J. C., 1867, *Philos. Trans. R. Soc. (London),* **157**, 49.
Nye, J. F., 1960, *Physical Properties of Crystals* (Oxford University Press, Oxford).
Randall, R. H., Rose, E. C., Zener, C., 1939, *Phys. Rev.,* **56**, 343.
Voigt, W., 1892, *Ann. d. Phys.,* **47**, 671.
White, D. L., 1962, *J. Appl. Phys.,* **33**, 2547.
Zener, C., 1948, *Elasticity and Anelasticity of Metals* (University of Chicago Press, Chicago).

REVIEW QUESTIONS

6.1 When a constant stress is applied to a viscoelastic solid, the strain becomes time-dependent. Discuss the reasons for this behaviour. (§6.3)

6.2 What are the main features of the anelastic (Zener) solid? How does energy loss arise when a periodic stress is applied? (§6.4)

6.3 In relation to the Zener solid, distinguish between the relaxed and unrelaxed elastic moduli. How is the energy loss related to these quantities? (§6.4)

6.4 How may Hooke's law for an isotropic solid be modified to include thermal strain? (§6.6)

6.5 What is the relationship between the isothermal and adiabatic elastic constants? (§6.6.1)

6.6 What method could you use to calculate the modulus defect for thermoelastic losses in a Zener solid? (§6.6.2)

6.7 How does the frequency of an acoustic wave determine the type of piezoelectric field set up in a material possessing both piezoelectricity and conductivity? (§6.7)

6.8 Show how to formulate a wave equation to describe the propagation of one-dimensional plane waves in a piezoelectric semiconductor. (§6.7.1)

6.9 Explain how the acoustic velocity and the acoustic attenuation coefficient depend on the value of the electronic drift velocity. (Figures 6.11 and 6.12)

6.10 How may the phenomenon of acoustic amplification be observed experimentally? (§6.7.2)

PROBLEMS

6.11 Show that the equation of motion for the propagation of a disturbance in the x_1 direction in an isotropic viscoelastic solid has the form

$$\rho \ddot{u}_1 = (\lambda + 2G)\frac{\partial^2 u_1}{\partial x_1^2} + (\chi + 2\eta)\frac{\partial^3 u_1}{\partial x_1^2 \partial t} \ ,$$

where G and λ are the Lamé elastic constants, χ and η are the compressional and shear viscosity coefficients respectively.

Hence derive an expression for the velocity of propagation and attenuation coefficient of a plane longitudinal wave in the x_1 direction in an isotropic viscoelastic solid.

6.12 Derive expressions for the velocity and attenuation of a transverse wave propagating in the x_1 direction in an isotropic viscoelastic solid.

6.13 Derive expressions for the velocity and attenuation of a longitudinal wave propagating in the x_1 direction in an isotropic Zener solid.

6.14 Set up the equations of motion for the thermoelastic behaviour of an isotropic solid. Hence find the wave equation for a longitudinal wave propagating in the x_1 direction and expressions for its velocity and attenuation.

Dislocation damping and interactions

7.1 Introduction

When elastic waves pass through a crystal containing defects, some of the wave energy will be used up in producing motion of the defects. The defects usually cause changes in the effective elastic constants of the crystal which in turn will cause changes in the velocity of propagation of the elastic waves.

One defect that reacts strongly to an applied elastic wave is the dislocation. Ths dislocations that occur in single crystals are often found to form networks in well defined slip planes. Dislocation networks in KCl that have been decorated by diffusion of Ag atoms are shown in figure 7.1.

Edge dislocations may be considered as analogous to stretched strings with strong nodal points where dislocations intersect or are pinned by impurities. In addition, there are weaker intermediate pinning points distributed more or less at random along each dislocation line. These are due to vacancies, interstitials, or impurities, and may vary in number when a crystal is deformed or irradiated. The dislocations present in a crystal act as very sensitive detectors of any changes in the population of point defects.

Figure 7.1. Dislocation networks in KCl. The dislocations have been decorated by diffusion of silver atoms. (After Amelinckx, 1964.)

7.2 Some dislocation properties

Before proceeding with the study of the effects produced by dislocations on the propagation of elastic waves in crystals, it is necessary to summarise some relevant properties of dislocations.

7.2.1 Slip

When a single crystal is subjected to an external stress, slip occurs, that is, there is a shearing motion between two parts of the crystal across a common lattice plane. This causes the appearance of slip lines or bands on the surface representing the relative displacement of crystal planes by an amount in the range of 20 to 500 atomic spacings. Detailed examination shows that there are steps of this magnitude in the crystal surface.

The planes along which slip occurs are usually those with small Miller indices, that is those with the highest atomic density. Examples of slip planes are the $\{111\}$ planes in fcc metals, and $\{110\}$, $\{112\}$, $\{123\}$ planes in bcc metals.

The slip directions are the directions of slip in the slip plane, and again these occur in the closest packed directions such as $\langle 110 \rangle$ in fcc metals, $\langle 111 \rangle$ in bcc metals, and $\langle 110 \rangle$ in NaCl-type crystals.

A slip system is a particular combination of slip plane and slip direction. More than one slip system may occur in most crystals. When slip does take place in a given crystal it may occur on only one system or on more than one (multiple slip). Slip systems for different types of crystals are discussed in detail in Hirth and Lothe (1968).

7.2.2 Resolved shear stress

Consider a tensile stress, σ_{11} applied along the x_1 axis of a single crystal in the form of a cylinder, as in figure 7.2. The stress component in the slip direction is $\sigma_{11} \cos\beta$. Since the area of the slip plane is $A/\cos\alpha$, where A is the area of the end face normal to σ_{11}, the shear stress resolved in the slip direction is

$$\sigma'_{12} = \sigma_{11} \cos\alpha \cos\beta .\qquad(7.1)$$

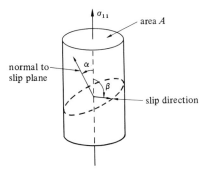

Figure 7.2. A tensile stress σ_{11} applied along the x_1 axis gives rise to a resolved shear stress $\sigma'_{12} = \sigma_{11} \cos\alpha \cos\beta$ in the slip direction.

The value of this stress required to initiate slip is called the **critical resolved shear stress**, σ_{cr}. The theoretical value of σ_{cr} for a perfect crystal has been computed to lie between $\frac{1}{5}G$ and $\frac{1}{30}G$, where G is the shear modulus. However, for most soft crystals σ_{cr} is in the range $10^{-5}G$ to $10^{-4}G$. For instance, for fcc and hcp metals $\sigma_{cr} \approx 7 \times 10^5$ N m^{-2}; for bcc metals $\sigma_{cr} \approx 7 \times 10^7$ N m^{-2}, and for NaCl-type crystals $\sigma_{cr} \approx 7 \times 10^5$ to 7×10^6 N m^{-2}. Some experiments on bulk copper and zinc have shown evidence of plastic deformation beginning at stresses as low as $10^{-9}G$. In figure 7.3 is shown a stress–strain diagram for a single crystal (fcc metal) with the critical shear stress indicated.

It is interesting to note that whiskers of normally soft materials, such as tin, yield at a value $\sigma_{cr} \approx \frac{1}{15}G$. A perfect two-dimensional bubble raft crystal model will produce plastic flow at $\sigma_{cr} \approx \frac{1}{30}G$. The low shear strength normally observed can be explained if it is assumed that the lattice contains imperfections which cause slip to occur at low values of the applied stress. The idea that slip propagates by the motion of dislocations was suggested independently in 1934 by Taylor, Orowan, and Polanyi.

Orientation relations between the direction of propagation of a stress wave, the slip plane, and the slip direction must be considered. For each slip system an **orientation factor**, Ω_j is introduced (Green and Hinton, 1966). Since there are usually a number of slip systems for a given crystal, the effect of all the slip systems that are operative must be considered.

The average orientation factor due to simultaneous dislocation motion on n slip systems may be defined as

$$\Omega = \frac{1}{n}\sum_{j=1}^{n}\Omega_j = \frac{1}{n}\sum_{j=1}^{n}\frac{\sigma_j^2}{2WG_j} \tag{7.2}$$

where n is the number of slip systems, σ_j is the resolved shear stress on the jth slip system, W is the energy density of the elastic wave per cycle, and G_j is the shear modulus on the jth slip system.

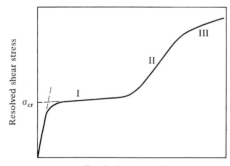

Figure 7.3. Resolved shear stress–strain curve for an fcc metal single crystal showing the region of easy glide (stage I of work hardening) and two further stages of work hardening (stages II and III).

7.2.3 Dislocation geometry

In general, the effects produced by dislocations may be discussed in terms of two 'pure' types, an edge dislocation and a screw dislocation. In the **edge dislocation**, $n+1$ atomic planes above the slip plane try to link with n atomic planes below the slip plane. Consequently one plane of atoms terminates on the slip plane, as shown in figure 7.4a. Such a dislocation therefore represents a form of atomic disorder and gives rise to an elastic stress field with a region of compression above the slip plane and a region of tension below the slip plane, as shown schematically in figure 7.4b.

To generate a **screw dislocation** a crystal block may be visualised being cut part way through with a knife and then sheared parallel to the edge of the cut by one atomic spacing, as shown in figure 7.5a. The successive atomic planes are then transformed into the surface of a helix. The stress field around a screw dislocation is mostly pure shear, as shown schematically in figure 7.5b.

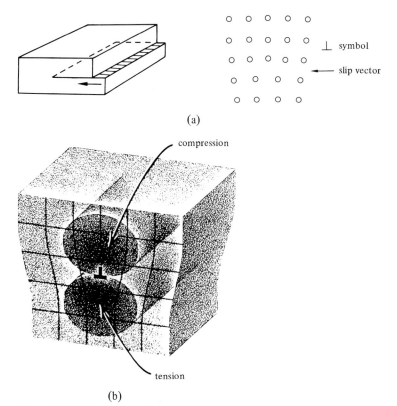

(a)

(b)

Figure 7.4. (a) Atomic configuration for an edge dislocation showing the extra half plane of atoms; (b) elastic stress field around an edge dislocation. (After Hayden *et al.*, 1965.)

In general, a dislocation may be curved so that in some locations it has the properties of an edge dislocation and in others it behaves as a screw dislocation. In order to define the type of dislocation more precisely the concept of the **Burgers vector, b,** is introduced.

Around any point along a dislocation line draw a closed circuit in the clockwise sense, moving along local lattice vectors in 'good' crystal (as distinct from 'bad' crystal near the dislocation line). Using the same sequence of lattice vectors, draw the corresponding circuit in a similarly oriented perfect crystal. The closing vector required is called the Burgers vector, **b,** of the dislocation, as shown in figure 7.6.

The sense of the dislocation line may be defined in terms of a **unit tangent vector, t,** whose direction is that of a right-handed screw. Then, an edge dislocation may be defined as one for which $b \cdot t = 0$ and for a

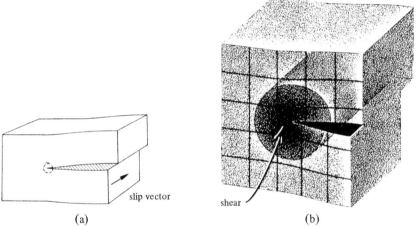

slip vector	shear
(a)	(b)

Figure 7.5. (a) Formation of a screw dislocation; (b) stress field around a screw dislocation. (After Hayden *et al.*, 1965.)

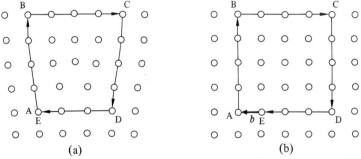

(a)   (b)

Figure 7.6. Burgers circuit for an edge dislocation: (a) closed circuit ABCDE enclosing an edge dislocation; (b) corresponding circuit in a perfect lattice showing the Burger's vector **b**. (After Sinclair, 1971.)

right-handed screw dislocation $b \cdot t = b$. If these extreme conditions are
not satisfied the dislocation line or segment is of mixed type, as illustrated
in figure 7.7.

Two important results follow from the introduction of the Burgers
vector:

(i) A given dislocation has everywhere the same Burgers vector, and
therefore cannot end within a crystal except at the surface, a crystal
boundary, at another dislocation, or more general imperfection.
Dislocations in the interior of a crystal therefore tend to form closed
loops or interconnecting networks.

(ii) The vector sum of the Burgers vectors of all dislocations meeting at a
node of a dislocation network is zero:

$$\sum_{i=1}^{n} b_i = 0 \ .$$

This theorem is illustrated in figure 7.8.

Sometimes b differs from a lattice vector. The dislocation line must
then lie along the boundary of a fault in the crystal, that is a surface
across which the atomic arrangement differs from that of the main part of
the crystal. The simplest surfaces of this type are called stacking faults.

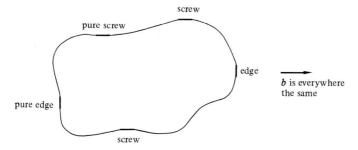

Figure 7.7. General dislocation line of mixed type.

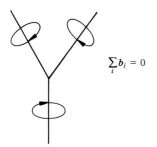

Figure 7.8. A dislocation node. The Burger's vectors are measured by means of
clockwise Burgers circuits as seen looking outwards from the node.

I

Many examples of dislocations may be found other than in crystals, for example in botany or geophysics. For instance, figure 7.9 shows a dislocation in a corncob.

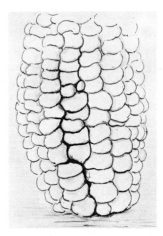

Figure 7.9. A dislocation in a corn cob (drawing by Christine Pollard).

7.2.4 Strain energy of dislocations

7.2.4.1 *Screw dislocation*

Consider a cylindrical crystal of length l with a screw dislocation of Burgers vector b along its axis, as in figure 7.10. The elastic shear strain, $\epsilon_{r\theta}$, in a thin annular section of radius r and thickness dr is

$$\epsilon_{r\theta} = \frac{b}{2\pi r} \; .$$

In an elastic continuum the corresponding shear stress would be

$$\sigma_{r\theta} = G\epsilon_{r\theta} = \frac{Gb}{2\pi r}$$

where G is the shear modulus in the slip plane. Although Hooke's law does not hold in the core region of a dislocation it may be applied at a reasonable distance away from the core.

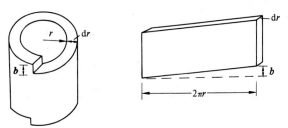

Figure 7.10. Screw dislocation configuration in a cylindrical crystal.

The energy per unit volume of the annulus is

$$\frac{dE}{dV} = \tfrac{1}{2}\sigma_{r\theta}\,\epsilon_{r\theta} = \frac{Gb^2}{8\pi^2 r^2}$$

$$dE = \frac{Gb^2 l\,dr}{4\pi r}$$

since $dV = 2\pi r l\,dr$.

The total elastic energy of the screw dislocation is found by integrating between some suitable lower limit, r_0, and the upper limit, R. (The integral would be infinite if either $r_0 = 0$ or $R = 0$.)

$$E = \int_{r_0}^{R} \frac{Gb^2 l\,dr}{4\pi}\,\frac{dr}{r} = \frac{Gb^2 l}{4\pi}\ln\frac{R}{r_0}\ . \tag{7.3}$$

A convenient and reasonable value for r_0 is $r_0 = b$, that is, one lattice constant. Since E is relatively insensitive to the value of R/r_0, it is convenient to take $\ln(R/r_0) = 4\pi$ so that $E \approx Gb^2 l$.

7.2.4.2 Edge dislocation

Application of standard elasticity theory (see for example Hirth and Lothe, 1968) shows that the energy of an edge dislocation is

$$E = \frac{Gb^2 l}{4\pi(1-\mu)}\ln\frac{R}{r_0} \approx \frac{Gb^2 l}{1-\mu}\ , \tag{7.4}$$

where μ is Poisson's ratio. If $\mu = 0\cdot33$, the energy of an edge dislocation is about $\tfrac{3}{2}$ that of a screw dislocation of the same length.

The energy of both edge and screw dislocations is proportional to b^2. Hence, the most stable dislocations will be those with minimum values of b, that is, those dislocations with b in close-packed directions.

7.2.4.3 Line tension

Since the strain energy is proportional to the length of the dislocation line, the line energy can be equated to a **line tension**, C, such that

$$C = \frac{\partial E}{\partial l} \approx Gb^2\ . \tag{7.5}$$

7.2.5 Force on a dislocation

Consider a crystal in the form of a cube of side l containing an edge dislocation of Burgers vector b. A shear stress σ_{12} acts on the upper and lower faces to produce slip of magnitude b, as shown in figure 7.11. The work done by the applied stress is $\sigma_{12}l^2 b$. If f is the force per unit length acting on the dislocation, the work done in moving the dislocation a distance l across the slip plane is fl^2. Assuming that f is caused by σ_{12}, we obtain $fl^2 = \sigma_{12}l^2 b$, or

$$f = \sigma_{12}b\ . \tag{7.6}$$

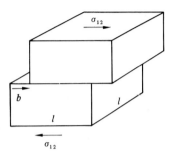

Figure 7.11. Crystal in the form of a cube containing an edge dislocation with Burger's vector **b**.

7.2.6 Dislocation pinning

As a dislocation moves along the slip plane it may encounter impurity atoms or point defects which can cause the motion of the dislocation line to be impeded. The stress causing the motion of the dislocation exerts a force normal to the line, so that if parts of the line become pinned the line between the pinning points will bow out, as shown in figure 7.12. The normal force on the dislocation is

$$F = \sigma_{12}bl .$$

This is balanced by the resolved line tension so that

$$\sigma_{12}bl = 2C\sin\theta ,$$

$$\sigma_{12} = \frac{2C\sin\theta}{bl} = \frac{2Gb\sin\theta}{l} . \tag{7.7}$$

Thus, increasing the stress will cause increased bowing with the maximum stress occurring when $\theta = 90°$ and

$$\sigma_{12}\Big|_{max} = \frac{2Gb}{l} .$$

If the dislocation does not meet any obstacles, $\sin\theta$ will have a value approaching zero, and the dislocation is able to move with a very small applied stress. Higher stresses are required if l becomes smaller because of increased pinning.

Sometimes a dislocation will avoid becoming pinned by changing to another slip plane. When an edge dislocation changes plane, the process

Figure 7.12. A dislocation line bowed out between two pinning points.

is called **climb**. If a screw dislocation changes plane the process is called **cross-slip**. These two processes involve the study of the influence of kinks and jogs on dislocation motion.

7.2.7 Kinks and jogs

When a dislocation glides in its slip plane it experiences a periodic force that arises from periodic changes in crystal potential energy; these, in turn, are due to the varying degrees of misfit of the dislocation to the basic lattice. Peierls (1940) and Nabarro (1947) have found expressions for the misfit energy and for the periodic displacement potential experienced by the dislocation. In figure 7.13 is shown a dislocation with segments lying in different Peierl's 'valleys' but still in the same glide plane. A **kink** is defined as a part of the dislocation that lies across a Peierl's energy 'hill'. Kinks may appear in pairs owing to thermal fluctuations.

A **jog** is a segment of a dislocation line that has moved in a direction normal to the glide plane. A simple jog involves a step of one atomic spacing. When the jog extends over more than one atomic spacing it is called a **superjog**. In figure 7.14 is shown a jog in an edge dislocation and the corresponding extra half-plane of atoms. A jog tends to impede the

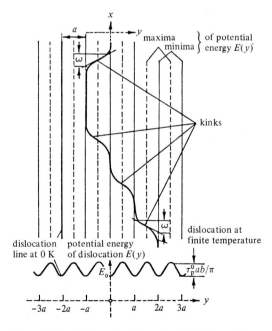

Figure 7.13. Dislocation line showing kinks. For part of the dislocation line to move from one position of minimum potential energy to another, a shear stress greater than or equal to the Peierls stress τ_p is required. (After Seeger *et al.*, 1957.)

motion of the dislocation so that a greater amount of work is required to move a jogged dislocation than a straight one. Jogs act both as sources and sinks for point defects.

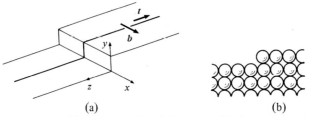

(a)   (b)

Figure 7.14. (a) Jog in an edge dislocation; (b) the corresponding extra half-plane. b is Burger's vector, t is the unit vector tangent to the dislocation line. (After Hirth and Lothe, 1968.)

7.2.8 Dislocations in ionic crystals

With ionic crystals an added factor that influences the atomic configurations is the need to preserve charge neutrality over the whole crystal. For an edge dislocation an extra pair of atomic planes are necessary. Two possible configurations are shown in figure 7.15.

In an ionic crystal a simple jog in an edge dislocation has a charge of $\pm\frac{1}{2}e$. This type of jog can not therefore be neutralised by a point defect since the latter carries an integral value of charge. A superjog of height an even number of atomic spacings will be neutral. In figure 7.16 is shown a charged jog and a neutral jog. An edge dislocation in equilibrium can therefore have a net charge, which it retains if it glides. It can also gain or lose charges if it interacts with point charges as it glides. Overall charge neutrality is maintained if the point of emergence of an edge dislocation at a surface bears a charge. The presence of charges on a dislocation also gives rise to a cloud of charged point defects in the nearby regions of the crystal. When a screw dislocation moves there is a charge

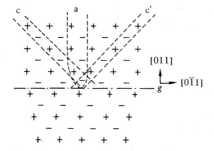

Figure 7.15. Edge dislocation in NaCl. The dislocation line is in the [100] direction. The atomic arrangement shown is in the (100) plane perpendicular to the line. The glide plane g is (011). The extra pair of atomic planes are shown by a. Alternatively, either one of the planes c and c' may be considered to be an 'extra half-plane'. (After van Bueren, 1961.)

motion back and forth along the dislocation line. This charge motion occurs by the motion of charged kinks. Kinks in edge dislocations are uncharged.

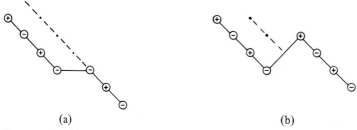

(a)   (b)

Figure 7.16. (a) Charged jog in dislocation shown in figure 7.15; the effective charge is $-\frac{1}{2}e$. (b) Neutral jog in the same dislocation.

7.3 Experimental evidence for dislocation damping

Some properties of solids, such as internal friction, are found to be very sensitive to structural imperfections. These imperfections may be point defects such as vacancies, interstitials, or impurity atoms; line defects of which dislocations are the main type; or surface-type defects such as stacking faults, grain boundaries or fractures. In contrast, other properties such as elastic constants, resistivity, and density are structure-insensitive in that they depend mainly on the basic crystal lattice and are only perturbed to a small extent by imperfections.

Internal friction measurements have been made over a frequency range of at least 14 powers of ten, ranging from very low frequencies of 10^{-4} Hz to 10^{10} Hz or more. The majority of investigations on single crystals and polycrystalline materials have been made in the range 10^4 to 10^8 Hz employing either the composite oscillator or pulse methods.

The first claim that the damping of ultrasonic waves in single crystals was due to the vibrations of dislocations was made by Read (1940). He observed that the damping in a single crystal of copper appeared to be composed of two parts; one of which was a function of the strain amplitude, and a residual part which remained at low strain amplitudes. Read's measurements are shown in figure 7.17. These results were confirmed by Gordon and Nowick (1956) with measurements on single crystals of NaCl, as shown in figure 7.18.

Hikata *et al.* (1956) measured the damping of longitudinal waves in polycrystalline aluminium while a tensile stress was applied. They plotted a stress–strain graph together with a damping-strain graph (figure 7.19) and found that even in the linear elastic region large changes in damping occurred. This result suggests an additional source of strain in the crystal that is too small to affect the elastic strain, but the effects of which are directly measurable through the damping.

If dislocations are a major source of damping in single crystals there should be a strong directional dependence between the particle motion of the applied wave and the slip direction. Alers (1955) deformed zinc single crystals, which have only one slip system, by pure shear parallel to the basal plane. During the deformation ultrasonic waves at $7 \cdot 8$ MHz

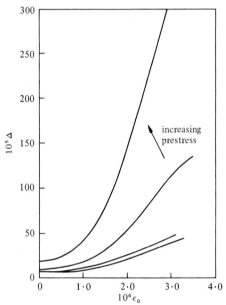

Figure 7.17. Decrement, Δ, as a function of strain amplitude, ϵ_0, and applied compressive stress for copper single crystals, measured at 30 kHz. (After T. A. Read, 1941.)

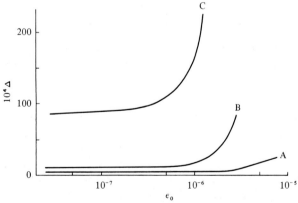

Figure 7.18. Internal friction of NaCl crystals subjected to various degrees of plastic deformation: (a) fully annealed; (b) compressed $0 \cdot 04\%$; (c) compressed $2 \cdot 3\%$. (After Gordon and Nowick, 1956.)

were propagated perpendicular to the basal plane. When the wave was a shear wave, Alers found that the ultrasonic attenuation increased with increasing deformation; for a longitudinal wave the attenuation remained practically unaltered. Since the shear wave, but not the longitudinal wave, has a strong shear component in the slip plane, it was concluded that the dislocations created during the deformation were responsible for the increase in the attenuation of the shear wave.

Further interesting measurements were made on zinc by Waterman (1958), as discussed in section 6.6.2.3. The attenuation of $7 \cdot 8$ MHz shear and longitudinal waves was determined as a function of frequency. The attenuation of the longitudinal wave agrees very well with theoretical values computed by Lücke (1956) for thermoelastic losses. The higher attenuation of the shear waves provides further evidence that the motion of dislocations in their slip planes involves absorption of ultrasonic energy.

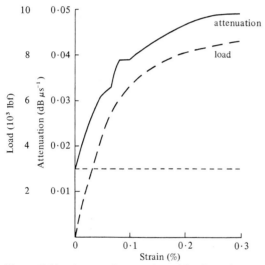

Figure 7.19. Attenuation–strain and load–strain curves for a polycrystalline aluminium specimen. The ultrasonic measurements were made at 5 MHz. (After Hikata *et al.*, 1956.)

7.4 The Granato–Lücke theory of dislocation damping

There have been many contributions to the development of the vibrating string model as a mechanism to explain dislocation damping. Mott (1952) assumed that a pure crystal contains a network of dislocations and he calculated the reduction in the elastic constants when the dislocations were set into vibration by an applied alternating stress. Koehler (1952) introduced the idea that a dislocation line segment pinned by point defects (assumed to be impurity atoms) might vibrate in the manner of a forced damped vibration of a stretched string under the action of an alternating stress field.

Granato and Lücke (1956a; 1956b) subsequently developed a
comprehensive theory of dislocation damping in a form that is amenable
to experimental testing. It is assumed that an undeformed single crystal
of high purity contains a dislocation network with two types of pinning
point: (a) strong nodal pinning where dislocations intersect, and
(b) weaker intermediate pinning caused by point defects or impurity
atoms. Two characteristic loop lengths may therefore be introduced: the
network loop length, L_N, determined by the nodal pinning; and a shorter
loop length, L_c, determined by the intermediate pinning. In practice a
distribution of values of L_c is found. The average value of L_c will be
denoted by L.

The effect of gradually increasing the applied alternating stress is shown
in figure 7.20. With no applied stress the dislocation lines are straight
with an equilibrium concentration of pinners. As the external stress is
increased the lines bow out with increasing amplitude of vibration, until
at a critical value of the resolved shear stress catastrophic breakaway
occurs from the intermediate pinning points, leaving the line vibrating
between the nodal points. At high enough stresses dislocation multiplication
may occur. Only the resolved shear stresses having the correct orientation
with respect to the slip system are effective in producing this motion.

In addition to the elastic strain, ϵ^{el}, there is a dislocation strain, ϵ^{dis}.
The dislocation stress–strain law is in general a function of frequency and
is nonlinear. The form of the stress–strain law corresponding to this
model is shown in figure 7.21.

Granato and Lücke postulate that the energy loss is made up of two
types:
(a) Since the measurement is a dynamic one, some damping mechanism
causes a phase difference between stress and strain. This type of loss is of

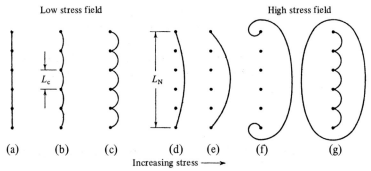

Low stress field   High stress field

(a)   (b)   (c)   (d)   (e)   (f)   (g)

Increasing stress ⟶

Figure 7.20. The successive drawings indicate schematically the bowing out of a
pinned dislocation line by an increasing applied stress. The length of loop determined
by defect pinning is denoted by L_c, and that determined by the network by L_N. As
the stress increases, the loops L_c bow out until breakaway occurs. For very large
stresses, the dislocations multiply according to the Frank–Read mechanism. (After
Granato and Lücke, 1956a.)

a resonance character being dependent on frequency and independent of amplitude.

(b) For large enough stress the loops L_c break away becoming loops of length L_N. They then collapse and are again pinned giving rise to a hysteresis loop. This type of loss is proportional to the area enclosed by the stress–strain graph. The loss is then dependent on amplitude and independent of frequency.

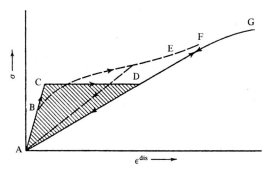

Figure 7.21. The solid line drawing shows the stress–strain law that results for the model shown in figure 7.20. The elastic strain has been subtracted so that only the dislocation strain is shown. The path ABCDEF is followed for increasing stress, while the path FA is followed for decreasing stress. The dashed line curve is that which would result if not all of the loops have the same length, but there is a distribution of lengths L_c. (After Granato and Lücke, 1956a.)

7.4.1 Low-amplitude theory

The theory developed by Granato and Lücke is a phenomenological one and therefore ideas based on continuum theory may be used. Temperature effects have been excluded, so that strictly the theory only applies at zero temperature. Lattice interactions have also been ignored, so that effects such as dislocation kinking have not been included.

Consider a vibrating dislocation line, with equal loop lengths $L = L_c$ as in figure 7.20b, lying along the x axis and having lateral displacement in the y direction. We shall assume that Newton's law of motion may be applied to the system so that

$$\frac{\partial^2 \sigma_{ik}}{\partial x_k^2} = \rho \frac{\partial^2 \epsilon_{ik}}{\partial t^2} \ , \tag{7.8}$$

where ρ is the mass per unit volume. The total strain, ϵ, is given by

$$\epsilon = \epsilon^{el} + \epsilon^{dis} \ , \tag{7.9}$$

where the elastic strain is found from Hooke's law

$$\sigma_{ik} = c_{iklm} \epsilon_{lm} \ . \tag{7.10}$$

Since only the resolved shear stress in the appropriate slip plane is effective in causing dislocation movement, the subscripts in the above equations may be dropped if the equations are written in terms of the resolved shear stress and strain. For instance, equation (7.10) will now read

$$\sigma = G\epsilon^{\text{el}} , \tag{7.11}$$

where G is the shear modulus of the (perfect) crystal lattice.

The dislocation strain will be represented by

$$\epsilon^{\text{dis}} = \Lambda b \overline{\xi} , \tag{7.12}$$

where Λ is called the **dislocation density** and is the total length of dislocation line per unit volume, b is the magnitude of the Burgers vector, and $\overline{\xi}$ is the average displacement of a dislocation line of length l given by

$$\overline{\xi} = \frac{1}{l} \int_0^l \xi(y) \, dy . \tag{7.13}$$

Equations (7.8) to (7.13) may be combined to give

$$\frac{\partial^2 \sigma}{\partial x^2} - \frac{\rho}{G} \frac{\partial^2 \sigma}{\partial t^2} = \frac{\Lambda \rho b}{l} \frac{\partial^2}{\partial t^2} \int_0^l \xi(y) \, dy . \tag{7.14}$$

Equation (7.14) is a general equation of motion written in terms of the dislocation strain.

Let us now assume that the pinned dislocation vibrates as a damped string according to the equation of motion given by Rayleigh (1894) and applied to dislocations by Koehler (1952):

$$A \frac{\partial^2 \xi}{\partial t^2} + B \frac{\partial \xi}{\partial t} - C \frac{\partial^2 \xi}{\partial y^2} = b\sigma(x,y,t) , \tag{7.15}$$

where $\xi = \xi(x, y, t)$ is the displacement of an element of the dislocation from its equilibrium position, the displacement being assumed to lie in the slip plane; A is the **effective mass** per unit length; B is the **damping force** per unit length; C is the **effective line tension** in a bowed out dislocation; and $b\sigma$ is the driving force per unit length of the dislocation exerted by the applied shear stress. The constants A and C have the values

$$A = \pi \rho b^2 ,$$

$$C = \frac{2Gb^2}{\pi(1 - \mu)} , \tag{7.16}$$

where μ is Poisson's ratio.

An expression for B has been calculated by Leibfried (1950) on the assumption that the primary cause of damping is scattering of lattice phonons by dislocations:

$$B = \frac{3kzT}{10c_t a^2} \; ; \tag{7.17}$$

here k is Boltzmann's constant, z is the number of atoms per unit cell, c_t is the transverse wave velocity, and a is the lattice spacing. According to this relationship B depends linearly on temperature. While Leibfried's formula has been applied successfully to metal single crystals, doubts have been raised concerning its applicability to ionic crystals where charge effects may become important.

The boundary conditions for this problem are

$$\xi(x, 0, t) = \xi(x, 1, t) = 0 . \tag{7.18}$$

Equations (7.14) and (7.15) must now be solved subject to the boundary conditions. Let us assume for the moment that the dislocations are normal to the direction of wave propagation. Solutions for which σ is periodic and independent of y are of most interest. The displacement of the dislocations can therefore be written in terms of a series of modes of the form $\sin(m\pi y/l)$. If a uniform periodic stress is applied to the dislocations, only the odd modes will be excited since these are the only ones that have a net strain. We are looking for a solution in terms of the wave velocity, c, and the attenuation coefficient, α.

Assume a trial solution of the form

$$\sigma = \sigma_0 \exp(-\alpha x) \exp \left[i\omega \left(t - \frac{x}{c} \right) \right] . \tag{7.19}$$

This leads to the displacement in terms of a Fourier series

$$\xi = \frac{4b\sigma_0}{A} \sum_{n=0}^{\infty} \frac{1}{2n+1} \sin \frac{(2n+1)\pi y}{l} \frac{\exp[i(\omega t - \delta_n)]}{[(\omega_n^2 - \omega^2)^2 + (\omega d)^2]^{\frac{1}{2}}} \tag{7.20}$$

where

$$d = \frac{B}{A} , \qquad \omega_n = (2n+1)\frac{\pi}{l}\left(\frac{C}{A}\right)^{\frac{1}{2}} , \qquad \delta_n = \tan^{-1}\frac{\omega d}{\omega_n^2 - \omega^2} .$$

For most purposes the first term of the series is sufficient, particularly in discussing attenuation measurements. However in dealing with high frequency velocity computations this simplification leads to serious errors.

In writing the solutions it is convenient now to redefine σ and ϵ in terms of an applied longitudinal stress and strain, which are the parameters measured in most experiments. To do this, the orientation factor, Ω, defined in equation (7.2), is introduced into the equations, as proposed by Granato and Lücke (1956a). It is also necessary to replace G by E, the Young's modulus of the specimen.

The required solutions may now be written

$$\frac{\Delta c}{c_0} = \frac{c_0 - c(\omega)}{c_0} = \left[\frac{\Delta_0}{2\pi} \omega_0^2 \Lambda L^2 \Omega \frac{\omega_0^2 - \omega^2}{(\omega_0^2 - \omega^2)^2 + (\omega d)^2}\right] , \tag{7.21}$$

$$\alpha(\omega) = \frac{\Delta_0}{2\pi c} \omega_0^2 \Lambda L^2 \Omega \frac{\omega^2 d}{(\omega_0^2 - \omega^2)^2 + (\omega d)^2} , \tag{7.22}$$

where

$$\Delta_0 = \frac{8Eb^2}{\pi^3 C} , \qquad c_0 = \left(\frac{E}{\rho}\right)^{1/2} , \qquad \omega_0 = \frac{\pi}{L}\left(\frac{C}{A}\right)^{1/2} ;$$

c_0 is the velocity of propagation in a perfect crystal. In practice this velocity is taken to be the value when all the dislocations are immobile, as occurs after long irradiation of a crystal or after prolonged annealing. Truell *et al.* (1969) have pointed out that c_0 can be found in principle by measuring the velocity at a frequency sufficiently high that the dislocations cannot follow the stress wave driving force. ω_0 is the value of ω for which α is a maximum.

The fractional modulus change, or modulus defect, $\Delta E/E$, may be introduced in place of the fractional velocity change, since $\Delta E/E = 2c/c_0$. In figure 7.22 is shown a plot of equation (7.21) and in figure 7.23 a plot of equation (7.22). The value for d corresponds to overdamped vibrations as occurs normally with dislocations. Also shown in figure 7.23 is a plot of the decrement versus frequency. The **decrement**, Δ, is defined as the ratio of the energy loss per cycle, ΔW, to twice the total vibrational energy, E_0, stored in the specimen, that is,

$$\Delta = \frac{\Delta W}{2E_0} .$$

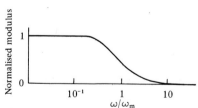

Figure 7.22. Change in normalised modulus near the frequency of maximum decrement, ω_m, for the case of overdamped vibrations, $d = B/A = 100\omega_m$.

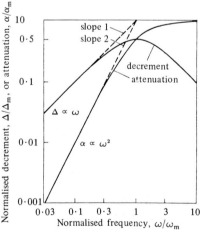

Figure 7.23. Normalised attenuation and decrement as a function of ω/ω_m for the overdamped case ($d = 100\omega_m$).

Since

$$\Delta = \alpha\lambda = \frac{2\pi c\alpha}{\omega} \ ,$$

equation (7.22) becomes

$$\Delta = \Delta_0 \omega_0^2 \Lambda L^2 \Omega \frac{\omega d}{(\omega_0^2 - \omega^2)^2 + (\omega d)^2} \ . \tag{7.23}$$

Equation (7.23) has a maximum given by

$$\Delta_m = \tfrac{1}{2}\Delta_0 \Lambda L^2 \Omega \tag{7.24}$$

at a frequency

$$\omega_m = \frac{\omega_0^2}{d} = \frac{\pi^2 C}{B L^2} \ . \tag{7.25}$$

For frequencies much greater than ω_m, α approaches a limiting value, α_∞, given by

$$\alpha_\infty = \frac{4\Omega E b^2 \Lambda}{\pi^2 B} \ . \tag{7.26}$$

This limiting attenuation is independent of loop length and line tension. If B is known, a measurement of α_∞ gives a measure of the dislocation density, Λ.

The theory so far has assumed that all the loop lengths are equal to L_c. In practice it is to be expected that there will be a distribution of loop lengths. If we assume that the values of L_c are randomly distributed along the dislocation line according to Koehler's (1952) distribution function

$$N(l)\,dl = \frac{\Lambda}{L^2}\exp\left(-\frac{l}{L}\right)dl \ , \tag{7.27}$$

where $N(l)\,dl$ is the number of loops which have lengths between l and $l+dl$, and L is now defined as the **average loop length**, then equations (7.24) and (7.25) become

$$\Delta_m = 2\cdot2\Delta_0\Lambda\Omega L^2 \ , \tag{7.28}$$

$$\omega_m = 0\cdot084\frac{\pi^2 C}{B L^2} \ . \tag{7.29}$$

The effects of the two distributions on the values of the decrement are shown in figure 7.24.

Measurement of Δ_m and ω_m allows the determination of $\Lambda L^2/C$ and $C/L^2 B$ respectively; their product yields Λ/B.

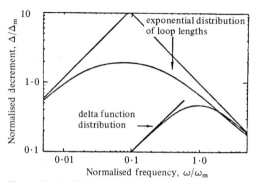

Figure 7.24. The maximum in the decrement as a function of frequency due to the damped resonance for an exponential distribution of loop lengths and a delta-function distribution of loop lengths. The low- and high-frequency asymptotes for both curves have slopes of plus 1 and minus 1, respectively. (After Stern and Granato, 1962.)

7.4.1.1 *Low-frequency approximation*

When $\omega \ll \omega_0$, equations (7.21) and (7.23) may be written

$$\frac{\Delta c}{c_0} = \frac{4Eb^2\Omega}{\pi^4}\frac{1}{C}\Lambda L^2 \tag{7.30}$$

$$\Delta_I = \frac{8Eb^2\Omega\omega}{\pi^5}\frac{B}{C^2}\Lambda L^4 \tag{7.31}$$

where Δ_I is the **amplitude-independent decrement**. Thus, if measurements are made on a given material at a fixed frequency, for which Λ, Ω, ω, B, and C are constant, then

$$\frac{\Delta c}{c_0} \propto L^2 ,$$

$$\Delta_I \propto L^4 .$$

Both the velocity and the decrement are therefore very sensitive to any changes in the average loop length of the dislocations that may occur by processes such as deformation or radiation. Under these limiting conditions it is theoretically possible to derive values of L and Λ from equations (7.30) and (7.31):

$$L = \text{constant} \times \left(\frac{\Delta}{\Delta c/c_0}\right)^{\frac{1}{2}} , \tag{7.32}$$

$$\Lambda = \text{constant} \times \frac{(\Delta c/c_0)^2}{\Delta} . \tag{7.33}$$

7.4.2 Experimental verification of low-amplitude theory

The velocity dispersion associated with the dislocation resonance peak has been observed by Granato *et al.* (1957) and by Truell and Granato (1963).

Granato *et al.* found that the velocity for megahertz pulses in an annealed crystal of NaCl increased by 0·5% over the frequency range 20–100 MHz. When the crystal was deformed by 0·06% the velocity showed a dispersion of 4% between the same frequencies with the effect centred at a lower frequency of approximately 35 MHz, as shown in figure 7.25. During plastic deformation fresh dislocations are produced and, provided the deformation is small, the overall effect is that of increasing the average loop length, L. According to equation (7.29) this will have the effect of decreasing the resonant frequency. At frequencies below the resonant frequency the dislocations can follow the motion of the applied stress leading to a reduction in the rigidity of the specimen. At frequencies above the resonant frequency the dislocations can no longer follow rapid changes of stress and the modulus tends to the value for a perfect crystal.

   Direct observation of the resonance curve for the absorption has been reported by Alers and Thompson (1961) for copper (99·999% pure crystals, well annealed) and by Stern and Granato (1962) for copper (99·999% pure crystals, annealed and also after cobalt gamma irradiation). The results of Alers and Thompson's experiment are plotted in figure 7.26. The decrement shown is that due to the dislocations alone, the background damping having been subtracted out. Merkulov and Yakovlev (1960) measured the attenuation as a function of frequency for NaCl deformed by 1%. Their results are shown in figure 7.27. The maximum of the resonance curve is seen to move towards higher frequencies during recovery from deformation. The average loop length will be expected to decrease during recovery, owing to pinning by defects created during the deformation process. The attenuation tends to the same value for all curves at high frequencies in agreement with the theory, which predicts that at very high frequencies the attenuation should be independent of loop length. In this investigation Merkulov and Yakovlev investigated thoroughly the dependence of the velocity and attenuation on pre-strain, frequency, orientation, ultrasonic amplitude, recovery time, and ageing, and found for all these experiments good agreement with the Granato–Lücke model.

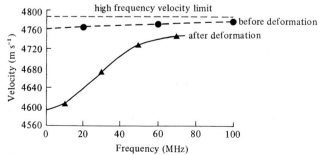

Figure 7.25. Velocity dispersion for compressional elastic waves propagating in the [100] direction in NaCl. Deformation increases the magnitude of the dispersion from about 0·5 to 4%, and moves it to lower frequencies. (After Granato *et al.*, 1957.)

Important confirmation of the validity of the Granato–Lücke theory at low frequencies came from experiments of Thompson and Holmes (1956) who measured the changes of modulus and damping in pure single crystals of copper under fast neutron irradiation (in-pile reactor neutrons). Assuming that the dislocation density does not change under the levels of irradiation employed and also assuming that the number of pinning points increased linearly with time, Thompson and Holmes showed that the modulus change depended on the second power of the average loop length, while the damping change depended on the fourth power. These results give striking confirmation of equations (7.30) and (7.31). Thompson and Holmes found that, even in specimens of nominally the same purity, the size of the measured effect varied considerably from specimen to specimen. In the case of copper the original state of the crystals could be restored by high temperature annealing. Further discussion of the effects of deformation and irradiation in relation to the vibrating string model will be discussed in later sections of this chapter.

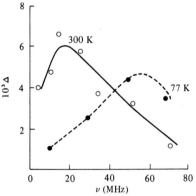

Figure 7.26. Variation of the logarithmic decrement, Δ, with frequency, ν, for longitudinal waves propagating in the [110] direction in a well annealed pure copper single crystal. (After Alers and Thompson, 1961.)

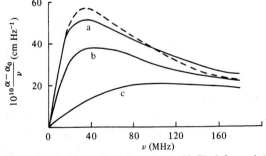

Figure 7.27. Normalised damping in NaCl, deformed 1%, as a function of frequency and recovery time after deformation after: (a) 20 min, (b) 2 h, (c) 60 h. The dashed curve is the theoretical one for equal loop lengths. (After Merkulov and Yakovlev, 1960.)

7.4.3 High-amplitude theory

Dislocation breakaway occurs when the force exerted on a dislocation segment exceeds that exerted on the segment by the pinning points. Since in practice there will be a distribution of pinning points along the dislocation lines leading to a distribution of loop lengths, breakaway will occur over a range of strain values. The condition for breakaway may be written

$$\frac{b\sigma(l_1 + l_2)}{2} = f_m \approx \frac{E_c}{b} , \tag{7.34}$$

where l_1 and l_2 are the lengths of two adjacent loops, b is the magnitude of the Burgers vector, f_m is the maximum force exerted by a pinning point, and E_c is the pinning energy. It appears therefore that the longest loops will breakaway first, leading to a catastrophic process.

In finding the amplitude dependence of the decrement, Granato and Lücke (1956a) assumed that:
(i) all the network lengths L_N are of the same size;
(ii) $L_N \gg L_c$;
(iii) at zero stress the loop lengths l are distributed randomly according to Koehler's distribution

$$N(l)\,dl = \frac{\Lambda}{L_c^2} \exp\left(-\frac{l}{L_c}\right) .$$

With these assumptions the following relationship is found for both the decrement and the modulus change:

$$\Delta_H = \frac{\Delta E}{E}\bigg|_H = \frac{c_1}{\epsilon_0} \exp\left(-\frac{c_2}{\epsilon_0}\right) , \tag{7.35}$$

where

$$c_1 = \frac{\Omega \Delta_0 \Lambda L_N^3}{\pi L_c} c_2 , \qquad c_2 = \frac{K\eta b}{L_c} , \qquad \Delta_0 = \frac{8Eb^2}{\pi^3 C} , \qquad K = \frac{G}{4RE} ,$$

ϵ_0 is the strain amplitude, R is the resolved shear stress factor (defined by Granato and Lücke) and η is **Cottrell's misfit parameter**. The subscript H denotes hysteresis loss. Δ_H is the decrement over and above Δ_I.

According to equation (7.35), $\ln(\Delta_H \epsilon_0)$ should be a linear function of $1/\epsilon_0$. The slope of the line, c_2, should be proportional to the pinning point density, while the intercept, c_1, should be proportional to the dislocation density.

7.4.4 Experimental verification of the high-amplitude theory

The data reported by Read (1941) and shown in figure 7.17 may be replotted as in figure 7.28, where $\ln(\Delta_H \epsilon_0)$ is plotted against $1/\epsilon_0$ (referred to as a Granato–Lücke plot). It is observed that good straight lines are obtained having the same slope which indicates no change in the pinning point density. The intercepts, however, increase indicating an increasing dislocation density, which is to be expected since there is an increase in the applied stress from line to line. In figure 7.29 is shown a plot of the dislocation densities derived from figure 7.28.

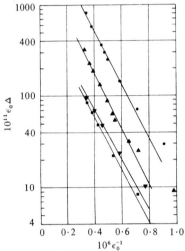

Figure 7.28. Measurements made by T. A. Read of the decrement, Δ, of $99 \cdot 998\%$ pure copper single crystal after applied compressive loads of 0, 40, 80, and 100 kPa are here plotted by the method of the present theory. The lowest curve is that for no applied stress, while the higher curves correspond to the successively larger loads. According to the theory, it would be expected in this case that the slopes of the successive curves would not change much, but that the intercepts, which are proportional to the dislocation density, would increase. (After Granato and Lücke, 1956b.)

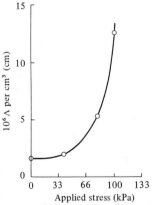

Figure 7.29. The dislocation densities, Λ, derived from figure 7.28, are here plotted as a function of the applied load. The derived dislocation density increases very rapidly before any measurable strain occurs. (After Granato and Lücke, 1956b.)

Fiore and Bauer (1964) determined the decrement both as a function of the strain amplitude and temperature for a copper crystal containing a small amount (0·0057 at.%) of germanium (figure 7.30); corresponding Granato–Lücke plots are shown in figure 7.31. The decreasing slope as the temperature increases is interpreted to mean that at higher temperatures the equilibrium density of pinning points is decreased. Granato and Lücke assume that the concentration, c, of impurity atoms on the dislocation line is larger than the overall concentration, c_0, of impurities in the lattice. At temperatures high enough for diffusion to take place the concentration is assumed to take an equilibrium value given by Cottrell's equation:

$$\frac{d}{L_c} = c = c_0 \exp\frac{Q}{kT} , \tag{7.36}$$

where d is the atomic spacing along the dislocation line and Q is the interaction energy between a dislocation and an impurity atom. From equation (7.36) a value of Q of 0·3 eV is found for germanium.

Numerous other investigations indicate reasonable agreement between experiment and theory for pure materials. A number of modifications and extensions of the basic theory are discussed in Granato and Lücke (1968). The effects of thermal motion on breakaway have been investigated by Teutonico et al. (1964). They show that the form of the solutions for dislocation breakaway remains the same but the stress level at which breakaway occurs is reduced because of the additional effect of thermal breakaway. In the zero-temperature theory the decrement, Δ_H, is proportional to $\exp(-\sigma_M/\sigma_0)$, but in the modified theory this factor becomes $\exp(-\sigma_{th}/\sigma_0)$; here σ_m is the mechanical breakaway stress, σ_{th} is the corresponding quantity including thermal breakaway, and σ_0 is the

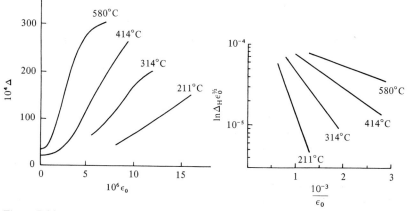

Figure 7.30. Specimen damping versus maximum strain amplitude, ϵ_0, for a copper-0·0057% germanium alloy. (After Fiore and Bauer, 1964.)

Figure 7.31. Data of figure 7.30 replotted. (After Fiore and Bauer, 1964.)

maximum stress amplitude during a cycle. As a consequence there should
be a change in the slopes of the lines in a Granato–Lücke plot of the data.

7.5 Deformation effects

When a gradually increasing stress is applied to a single crystal, fine slip
lines are first observed on one slip system, which then gradually broaden
into bands with the appearance of slip lines on other systems. In the easy
glide region (stage I) the dislocations travel long distances and may reach
the crystal surface. Easy glide ends when multiple glide occurs to such an
extent that the motion of the dislocations becomes restricted. As shown
by Young (1963), multiplication of dislocations begins even before the
yield stress is reached. The increase in dislocation density with increasing
stress has been studied by a number of workers. Kovacs (1967) has
studied work hardening in fcc metals and shows that in stage II the
dislocation density, Λ, is proportional to the square of the critical resolved
shear stress, while in stage III Λ is proportional to σ_{cr}^3.

As well as dislocations, vacancies, interstitials, and other defects are
produced by deformation. The concentration, c, of point defects after a
plastic strain ϵ is found to follow the relationship

$$c \approx A\epsilon^p \, ,$$

where p lies between 1 and 2 and A lies between 10^{19} and 10^{21} cm^{-3}.
For instance, in a typical metal after 10% deformation, the defect
concentration is of the order of 10^{18} to 10^{19} cm^{-3}.

Komnik (1968) has observed that, in alkali halide crystals, mechanical
polygonization occurs, that is vertical dislocation walls form during plastic
deformation. This effect begins during stage I of work hardening and
covers the greater part of the crystal during subsequent deformation. The
effect begins in 'soft' crystals at strains of about $0 \cdot 2\%$ and in 'hard'
crystals at strains of about 1% (only after occupation of the whole crystal
by slip trails, that is dipoles, dislocations, dislocation loops, etc., formed
by interaction of moving dislocations with one another and with various
imperfections in the original structure). The main difference between soft
and hard crystals is the impurity content.

Since mechanical deformation changes the dislocation density and
presumably the average loop length, it is to be expected that there will be
changes in the dynamic elastic moduli and decrement when stress waves
propagate through a deformed crystal. The early measurements of Read
(1940) reported in section 7.3 have already demonstrated this effect.

7.5.1 Time-dependent modulus and damping following deformation (Köster effect)

After deformation of a single crystal the modulus is generally found to
decrease and the damping to increase. This behaviour is in accordance
with the predictions from equations (7.30) and (7.31) if either or both Λ
and L increase during deformation. At the end of the deformation process

both the modulus and the damping recover towards their original values. This recovery process has been called the Köster effect.

In an investigation carried out by Gordon and Nowick (1956), recovery curves for different strain levels were recorded in deformed crystals of NaCl. In figure 7.32 are shown curves for the recovery of the resonant frequency of the crystal and two curves for the recovery of the damping, one corresponding to high-amplitude damping and the other to low-amplitude damping. Measurements of modulus and damping were made by an ultrasonic method at 85 kHz. The data shown are for a crystal compressed by 4·2%. Many other similar recovery curves have been reported for ionic and for metal crystals. Further details of the recovery process in NaCl have been described by Merkulov and Yakovlev (1960).

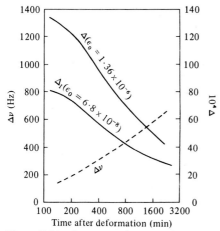

Figure 7.32. Recovery of internal friction and resonant frequency at room temperature, as a function of time after deformation, for an NaCl single crystal. The deformation was 4·2% compression and the resonant frequency 85 kHz. The internal friction is measured at two strain amplitudes. (After Gordon and Nowick, 1956.)

7.5.2 Theory for time-dependent deformation effects

Granato *et al.* (1958) have developed a theory which assumes that the observed changes in modulus and decrement with time result from dislocation pinning by point defects created during the deformation process. The high-amplitude decrement and modulus defect are given by

$$\Delta_{\text{H}} \approx \left(\frac{\Delta E}{E}\right)_{\text{H}} = A_1 \exp\left(-\frac{A_2}{L}\right) \tag{7.37}$$

and the low amplitude values by

$$\Delta_{\text{I}} = A_3 L^4 , \tag{7.38}$$

$$\left(\frac{\Delta E}{E}\right)_{\text{I}} = A_4 L^2 \tag{7.39}$$

where

$$A_1 = \frac{2 \cdot 5 \Omega \Lambda L_N^3}{\pi^2 L} \frac{K \eta d}{L \epsilon_0}, \qquad A_2 = \frac{K \eta d}{\epsilon_0}, \qquad A_3 = \frac{120 \Omega \Lambda B \omega}{\pi^3 C},$$

$$A_4 = \frac{6 \Omega \Lambda}{\pi^2}.$$

The total number of pinning points is assumed here to be the sum of the network pinners and pinning points due to point defects. If there are k different kinds of defects, then the average loop length L can be expressed by

$$\frac{1}{L} = \frac{1}{L_N} + \sum_{i=1}^{k} \frac{c_i}{d}, \tag{7.40}$$

where c_i is the concentration of the ith type of defect on the dislocations at time t after deformation.

Two types of defect will be involved in the study of recovery effects: c_{10} which represents the concentration of defects produced by the deformation, and c_{20} which represents the concentration of impurities. We shall assume that (i) dislocation interaction effects may be neglected (that is, the deformation is not too great), and (ii) the impurities are immobile at the temperature of the measurements whereas the defects represented by c_{10} are mobile. We shall further assume that the defects produced by the deformation diffuse towards the dislocations according to the Cottrell–Bilby relation

$$\eta_1(t) = \eta_{10} \alpha \Lambda \left(\frac{ADt}{kT} \right)^{2/3}, \tag{7.41}$$

where $\eta_1(t)$ is the number of defects which have diffused to the dislocations after a time t, η_{10} is the number of such defects in the lattice, α is a constant (equal to ~ 3), A is a parameter which measures the strength of the Cottrell attraction, and $D = D_0 \exp(-U/RT)$ is the diffusion coefficient.

In terms of concentrations equation (7.41) will read

$$c_1(t) = c_{10} \frac{4\alpha}{d^2} \left(\frac{ADt}{kT} \right)^{2/3}, \tag{7.42}$$

since $\eta_{10} = c_{10} N$, where N is the number of atoms per unit volume. The factor Na^3 equals 4 for fcc materials, and 8 for NaCl-type materials.

After a sufficient time the point defects will reach an equilibrium concentration on the dislocations, postulated to be given by Cottrell's equation

$$c_1 = c_{10} \exp \frac{Q}{kT}, \tag{7.43}$$

where Q is the interaction energy between a point defect and the dislocation. If there are sufficient other defects, the network pinning points may be neglected. Then equations (7.37) to (7.39) become more explicitly

$$\Delta_H = A_1 \exp\left[-A_2(c_{10}+c_{20})(1+\beta t^{2/3})\right] , \tag{7.44}$$

$$\Delta_I = \frac{A_3 d^4}{(c_{10}+c_{20})^4} \frac{1}{(1+\beta t^{2/3})^4} , \tag{7.45}$$

$$\left(\frac{\Delta E}{E}\right)_I = \frac{A_4 d^2}{(c_{10}+c_{20})^2} \frac{1}{(1+\beta t^{2/3})^2} , \tag{7.46}$$

$$\beta = \frac{c_{10}}{(c_{10}+c_{20})} \frac{4\alpha}{d^2} \left(\frac{AD}{kT}\right)^{2/3} . \tag{7.47}$$

The curves predicted by equations (7.45) and (7.46) are shown in figure 7.33. Experimental data points from a number of investigations have been superimposed by Granato et al. (1958) with good agreement. It is noted that both curves start from infinite slopes at $t = 0$ and that the decrement changes with time much faster than does the modulus.

The following predictions may be made from equations (7.44) to (7.47):
(i) A plot of $\ln\Delta_H$ against $t^{2/3}$ should give a straight line. The slope of this line should be proportional to the concentration of defects produced by the deformation, be very sensitive to temperature, and also be a function of the strain amplitude.
(ii) In equation (7.47) for the constant β, which determines the rate of the recovery process, it is predicted that pure materials (for which c_{20} is smaller) will recover faster than impure ones.

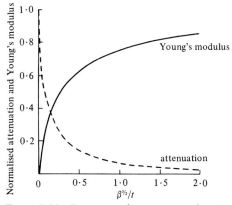

Figure 7.33. Decrement (or attenuation) and modulus changes with time following plastic deformation as predicted by the pinning theory. Both the vertical and horizontal scales have been normalized. In those cases for which data for both the decrement and the modulus are available, the same normalisation factor is used for both quantities. (After Granato et al., 1958.)

(iii) Recovery will be sensitive to temperature since D is a function of temperature.

(iv) Recovery will be faster for more heavily deformed materials (c_{10} will be larger).

(v) Plots of $\log\beta$ against $1/kT$ should be straight lines with slopes $\frac{2}{3}U$.

In principle, all unknowns may be determined experimentally by measuring Δ_H, Δ_I, and $(\Delta E/E)_I$ as functions of time and of temperature. However, it is very difficult to determine $(\Delta E/E)_I$ experimentally since the initial rate of change is so fast and it is difficult to find the end point for very long times.

7.5.2.1 Application of the theory to the data of Gordon and Nowick (1956)

The following analysis scheme is suggested by Granato et al. for the examination of deformation recovery data:

(1) Plot $\ln\Delta_H$ against $t^{2/3}$. Applying this to the data of Gordon and Nowick (shown in figure 7.32) we obtain the plot shown in figure 7.34. It is observed that there is excellent agreement between the experimental points and the predicted line. From the slope of the line we can calculate the term

$$A_2 c_{10} \frac{8\alpha}{d^2}\left(\frac{AD}{kT}\right)^{2/3}.$$

The constant A_2 can be found from the strain amplitude since $A_2 = K\eta d/\epsilon_0$.

(2) From modulus measurements find a value for β, where

$$\beta = \frac{c_{10}}{c_{10}+c_{20}}\frac{8\alpha}{d^2}\left(\frac{AD}{kT}\right)^{2/3}.$$

(3) By combining steps (1) and (2) find $(c_{10}+c_{20})$. This yields the value of the constant A_4, and thence the dislocation density Λ, since $A_4 = 6\Omega\Lambda/\pi^2$.

(4) Since β is now fully determined, check for the fourth power dependence of equation (7.45) by plotting $\lg\Delta_I$ against $\lg(1+\beta t^{2/3})$. Such a plot is shown in figure 7.35. The measured slope of the line is $-4\cdot3$, in good agreement with the theoretical value of $-4\cdot0$.

(5) Using the value of Δ_I at $t = 0$ find B from equation (7.38).

(6) Determine L_N from the intercept of the plot of $\ln\Delta_H$ against $t^{2/3}$.

All the values obtained from these steps depend on the value chosen for K in equation (7.37). In table 7.1 are shown values of all the constants determined by the above procedure for Gordon and Nowick's data on NaCl with two different values for K. Granato et al. (1958) include in their paper a detailed discussion of these and other results. They conclude that agreement between theory and experiment is good for deformations between $0\cdot4$ and $4\cdot0\%$.

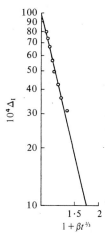

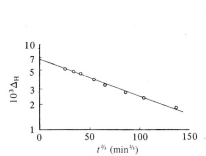

Figure 7.34. The logarithm of the amplitude dependent decrement, Δ_H, as a function of time after deformation, t, to the two-thirds power. The points shown are computed from the curves of Gordon and Nowick given in figure 7.32. (After Granato *et al.*, 1958.)

Figure 7.35. Plot of $\lg\Delta_I$ versus $\lg(1+\beta t^{2/3})$. The measured slope of the line is $-4\cdot3$ compared with the theoretical slope of $-4\cdot0$. (After Granato *et al.*, 1958.)

Table 7.1. Values of the constants found by applying the theory of Granato *et al.* (1958) to the data of Gordon and Nowick (1956) for two different values of K.

	$K = \frac{1}{50}$	$K = \frac{1}{5}$
A_2	$1\cdot5 \times 10^3$	$1\cdot5 \times 10^1$
$c_{10}\dfrac{4\alpha}{d^2}\left(\dfrac{AD}{kT}\right)^{2/3}$ $(s^{-2/3})$	$4\cdot5 \times 10^{-7}$	$4\cdot5 \times 10^{-8}$
$c_{10}+c_{20}$	$2\cdot7 \times 10^{-3}$	$2\cdot7 \times 10^{-1}$
A_1	$0\cdot33$	$0\cdot33$
$\dfrac{A_3 d^4}{(c_{10}+c_{20})^4}$	$1\cdot05 \times 10^{-2}$	$1\cdot05 \times 10^{-2}$
$\dfrac{A_4 d^2}{(c_{10}+c_{20})^2}$	$3\cdot06 \times 10^{-2}$	$3\cdot06 \times 10^{-2}$
L_{20} (cm)	$2\cdot1 \times 10^{-5}$	$2\cdot1 \times 10^{-4}$
Λ (cm^{-2})	$1\cdot1 \times 10^9$	$1\cdot1 \times 10^7$
L_N (cm)	$0\cdot69 \times 10^{-4}$	$6\cdot9 \times 10^{-4}$
ΛL_N^2	$5\cdot2$	$5\cdot2$
B	$4\cdot6 \times 10^{-2}$	$4\cdot6 \times 10^{-4}$
$c_1\exp\left(-\dfrac{2U}{3RT}\right)$	$.1\cdot2 \times 10^{-17}$	$1\cdot2 \times 10^{-17}$
U (kcal mol^{-1})		
for $c_1 = 10^{-5}$	25	27
$c_1 = 10^{-6}$	23	25
$c_1 = 10^{-7}$	21	23

Experiment E7.1   Time and amplitude dependence of Young's modulus in KBr

(Platkov, V. Y., 1969, *Sov. Phys.—Solid State,* 11, 343.)

This investigation deals with time-dependent changes in the Young's modulus of KBr crystals when driven by ultrasonic oscillations into the amplitude-dependent region. Measurements were made using a composite oscillator method at frequencies of $77 \cdot 7$ and 102 kHz. In a previous paper (Platkov and Startsev, 1967) measurements were made of both the Young's modulus and decrement as a function of strain amplitude in crystals of KBr, KCl, and RbI (figure E7.1). Both the modulus and decrement become dependent on the strain amplitude when a certain critical value is exceeded. The modulus decreases with increasing strain and the decrement increases. In the amplitude-dependent region both the modulus and decrement depend on the time and this leads to a hysteresis effect as shown in figure E7.2.

Platkov (1969) presents further measurements of the variations in the modulus as a function both of strain amplitude and of time, as shown in figure E7.3. The Young's modulus defect, $\Delta E/E$, is defined as

$$\frac{\Delta E}{E} = \frac{E - E_0}{E_0}$$

where E is the instantaneous value of Young's modulus and E_0 is its value at amplitudes less than critical.

According to the Granato–Lücke high-amplitude theory (section 7.4.3), $\ln(\Delta_H \epsilon)$ should be a linear function of $1/\epsilon$. Granato–Lücke plots for the data in figure E7.3 are shown in figure E7.4 for four different values of the time. In a vibration experiment, $\epsilon^{1/2}$ is used in the ordinate as the average value of the strain in the specimen. It is observed that these plots consist of two straight line portions. The intersection of the two portions corresponds to the critical amplitude, ϵ_c'', for the start

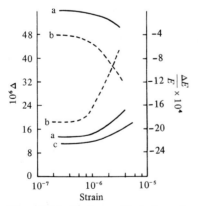

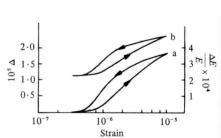

Figure E7.1. Strain amplitude dependence of internal friction and Young's modulus in KBr: (a) before deformation, (b) after deformation, and (c) after annealing. (After Platkov and Startsev, 1967.)

Figure E7.2. Hysteresis of the internal friction and Young's modulus defect in KBr single crystals: (a) decrement, Δ, and (b) modulus effect, $\Delta E/E$. (After Platkov and Startsev, 1967.)

of dislocation breakaway. The time dependence of Young's modulus is observed when the oscillation amplitude exceeds a critical value, ϵ'_c, which is found to correspond to a value of $\Delta E/E$ of 2×10^{-5}. The ratio of ϵ''_c/ϵ'_c is found to have a value of 3. The value of ϵ'_c depends on temperature.

When the crystal is driven with a vibration amplitude $\epsilon > \epsilon'_c$ and then the drive voltage is removed, Young's modulus returns to its original value over a period of time, as shown in figure E7.5. When the recovery data are plotted in semilogarithmic coordinates, as in figure E7.6, a relaxation time, τ, may be derived from the slope of the line. Values between 2 and 6 min were determined for the specimens examined. No dependence of τ was found on the amplitude of previous excitation or on the type of impurity present in the crystal. The range of values for τ agrees with the values

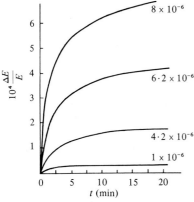

Figure E7.3. Time dependence of Young's modulus as a function of strain (vibration) amplitude over the range 1×10^{-6} to 8×10^{-6}.

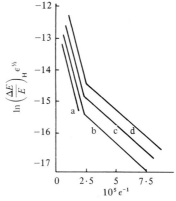

Figure E7.4. Granato–Lücke plots for four different excitation periods: (a) 30 s, (b) 1 min, (c) 3 min, and (d) 20 min.

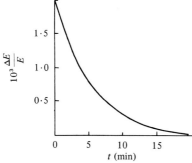

Figure E7.5. Recovery of Young's modulus following oscillation with strain amplitude $\epsilon > \epsilon'_c$.

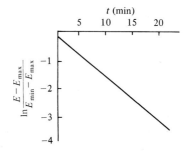

Figure E7.6. Recovery of Young's modulus in semilogarithmic coordinates. $E_{\min}$ is the value of the modulus at the start of excitation; $E_{\max}$ at the end of recovery; E is the instantaneous value of the modulus.

found in the previous paper (Platkov and Startsev, 1967) derived from the growth curves for $\Delta E/E$. The approximate relation for the growth curves was found to be

$$\frac{E_0 - E}{E_0 - E_1} = 1 - \exp\left(-\frac{t}{\tau}\right),$$

where E_0 is the value of the modulus in excitation at an amplitude less than critical, E is the instantaneous value of the modulus, and E_1 is the equilibrium value of the modulus.

In discussing these results, the author concludes that two different mechanisms act in two different amplitude ranges. The time dependence of Young's modulus starts at very small deformation amplitudes; at higher amplitudes the phenomenon of dislocation breakaway occurs. The time dependence of Young's modulus could only be attributed to a change in the distribution function of pinning centres along the dislocation lines. The theory of this effect (Alefeld, 1965) leads to the condition for the time dependence of Young's modulus:

$$\epsilon > \left(\frac{b}{l}\right)^{3/2} \frac{(kTCb)^{1/2}}{3Gb^3} = \epsilon_c' , \tag{E7.1}$$

where b is Burgers vector, C is the tension along the dislocation line, l is the average distance between pinning centres, and G is the shear modulus. With appropriate values for the constants: $kT = 0 \cdot 025$ eV, $\epsilon_c' = 5 \times 10^{-7}$, $b = 4 \cdot 65$ Å, the value for l is $0 \cdot 4$ μm.

Using this value of l, the author obtained a value of 2×10^{-10} cm^2 s^{-1} for the **coefficient of pipe diffusion**, D^P, for mobile pinning centres along a dislocation line. Yamafuji and Bauer (1965) give the following formula for D^P

$$D^P = \frac{2l^2}{\tau} , \tag{E7.2}$$

where τ is the time characteristic of the return of the distribution function to its original value. The coefficient of pipe diffusion for pinning centres along the dislocation line exceeds the value for bulk diffusion by several orders of magnitude.

The theory of Yamafuji and Bauer also allows a value to be found for F_b, the free binding energy between pinning centres and dislocations, from the ratio

$$\frac{\sigma_c''}{\sigma_c'} \approx \left(\frac{2F_b}{kT}\right)^{1/2} , \tag{E7.3}$$

where σ_c' is the critical stress for the start of time dependence, and σ_c'' is the critical stress for dislocation breakaway. The value of F_b found was $\sim 0 \cdot 1$ eV.

7.5.3 Bordoni relaxation peaks

Bordoni (1949; 1954) observed in several fcc metals, such as Pb, Al, Ag, Cu, relaxation peaks which appeared after deformation both with single crystals and polycrystalline specimens. The peaks are not present in fully annealed crystals. The temperature at which maximum decrement occurs depends, according to equation (6.44), on the frequency of measurement, indicating the presence of a relaxation phenomenon. Values of U and ω_0 are in the ranges $0 \cdot 05$ to $0 \cdot 2$ eV and $10^8 - 10^{12}$ s^{-1} respectively. The observed relaxation peaks indicate the presence of more than one basic relaxation mechanism. Some typical peaks observed in copper are shown in figure 7.36.

Seeger *et al.* (1957) proposed a theory for the formation of these peaks in terms of the kink theory for dislocations. If the frequency of the applied stress is of the same order of magnitude as ν_f, the frequency at which pairs of kinks are generated by thermal motion, then appreciable energy loss will occur from the applied stress wave. If the stress wave frequency is either much greater than or much less than ν_f, relatively little energy will be lost. Calculations based on this hypothesis lead to an activation energy that depends on the dislocations present but not on the pinning points. The theory of Seeger *et al.* does not account for all the known properties of these relaxation peaks and subsequently a number of alternative explanations have appeared.

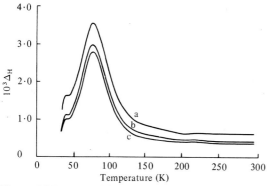

Figure 7.36. Internal friction of a pure copper single crystal deformed 10% by rolling. Strain amplitudes: (a) $1 \cdot 9 \times 10^{-5}$; (b) $2 \cdot 3 \times 10^{-6}$; (c) $1 \cdot 5 \times 10^{-7}$. (After Paré, 1961.)

7.5.4 Acoustic emission

When a solid is deformed, strain energy may be released in a number of forms including exo-electron emission, thermal emission, and acoustic emission. In ionic solids there is in addition electric dipole emission. Radiated stress waves have been observed (Kaiser, 1953; Schofield, 1958; 1963) over a strain range from the onset of plastic flow up to fracture processes. The acoustic energy is emitted in the form of pulses having durations in the microsecond range. The spectrum of the pulses is correspondingly broad, acoustic emission having been detected over a range from audio frequencies up to several megahertz. Audible emission is a familiar phenomenon in the sounds made by metals and timber under stress. The rate of emitted pulses appears to be closely connected with suddenly occurring events such as dislocation breakaway, sudden movement of piled-up dislocations, and crack formation. Other processes which occur less violently produce less emission. The role of dislocation processes in producing acoustic emission has been investigated by James and Carpenter (1971).

7.5.4.1 *Measurement techniques*

For the detection of acoustic emission either a piezoelectric crystal plate having a resonant frequency in the high kilohertz range or a nonresonant detector may be attached to the surface of the specimen. Because of the low level of the signals, it is necessary to use a low-noise amplifier before processing or displaying the output. A serious practical problem is to compensate for the characteristics of each of the elements involved in the measurement chain. Since the basic acoustic emission signal is a stress pulse of very short duration, it will act as an impulse source and trigger off any resonant element in the system, thus producing resonances in the specimen, transducer, specimen mount, amplifier, filters, recorder, etc.

7.5.4.2 *Nature of acoustic emission signals*

At low levels of plastic deformation the individual stress pulses cannot be resolved and therefore appear as a continuous signal of low amplitude. The recorded output is, however, related to the amount of deformation, as shown in figure 7.37. When larger amounts of strain energy are involved, for example in crack growth, the acoustic emission signal appears in the form of individual bursts at irregular time intervals. In figure 7.38 is shown an example of these bursts together with part of a time-expanded trace.

In a discussion of the nature of the primary signal Stephens and Pollock (1971) conclude that the source of acoustic emission is a pulselike function rather than an oscillatory function of stress. Thus, the observed oscillations, such as the ones shown in figure 7.38, are mainly structural and instrumental in origin.

James and Carpenter (1971) made a careful examination of possible relationships involving emission rate and dislocation density for single crystals of irradiated LiF, annealed LiF, NaCl, and Zn during deformation by compression at constant strain rate. They found reasonable correlation

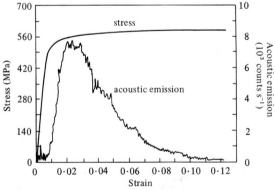

Figure 7.37. Acoustic emission and stress plotted against strain for a 7075-T6 aluminium tensile specimen. (After Dunegan and Harris, 1969.)

between acoustic emission pulse rate, ζ, and the rate of change of mobile dislocation density:

$$\zeta = 10^{-4}\frac{d\Lambda_m}{dt},$$

where Λ_m is the mobile dislocation density (total length of dislocation line per unit volume). James and Carpenter concluded that the pulses were generated by dislocation breakaway from pinning points, and estimated that several thousand centimetres of dislocation line length were involved in the generation of an individual acoustic pulse. They suggest that as a consequence, 10^5–10^6 segments of dislocation line participate in a short interval of time in a process of stimulated dislocation breakaway. The stress wave from a number of simultaneous breakaways propagates along the slip plane and triggers the breakaway of other segments.

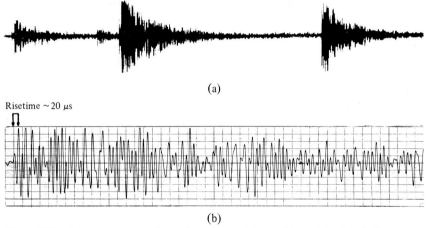

(a)

Risetime ~ 20 μs

(b)

Figure 7.38. (a) Acoustic emission pulses recorded in a mild steel (embrittled) specimen when strained into the plastic region just past the yield point. Instrumentation bandwidth 10 kHz to 300 kHz. Trace time-expanded by a factor 8192. x scale 0·5 mm s^{-1}; y scale 25 mV mm^{-1} ($\times$ 80 dB). (b) Part of a pulse shown above time-expanded by a factor 32768. (By courtesy of Dr. B. Woodward, Australian Atomic Energy Commission Research Establishment.)

7.5.4.3 Types of analysis

Depending on the nature of the problem various methods of analysis have been developed. Counting the burst rate is often sufficient; the pulse amplitudes are of significance, and spectral analysis has been attempted in order to identify the origin of the emission. The various significant parameters and the type of information they reveal have been discussed by Stephens and Pollock (1971) and are summarised in table 7.2.

K

Table 7.2. Parameters relating to acoustic emission signals.

Emission parameter	Type of information carried
waveform	fine structure of source event
frequency spectrum	nature of source event; integrity of specimen
amplitude	energy of source event
amplitude distribution	type of damage occurring
rate	rate of damage occurring
distribution in time	type of damage occurring; integrity of specimen
relative arrival times at several transducers	source location

7.5.4.4 *Applications of acoustic emission*

Acoustic emission has been observed in single crystals of metals and ionic compounds, polycrystalline materials, concrete, rocks, ice, graphite, adhesives, and various bonded structures. Apart from fundamental investigations into the origins of the primary pulses and the sources of the observed bursts at high stress levels, numerous studies have been made with regard to the nondestructive testing of structures and materials. With the aid of suitable transducers attached to structures such as girders, reactor pressure vessels, rocket motor casings, etc., the behaviour of the structure under load may be examined. A great deal of work is in progress with the aim of identifying characteristic emission from particular sources such as slip, crack formation, structural failure. The physical location of sources may be estimated from the measurement of the transit times of the pulses made with more than one transducer.

Investigation into the behaviour of materials involves a number of topics such as dislocation activity, work-hardening, crack growth, and fatigue. Twinning and phase changes during deformation give rise to characteristic acoustic emission. In making acoustic emission measurements it is important to eliminate as far as possible all other sources of noise. The environment under which measurements are made can be important in cases where crack formation is being investigated. As observed originally by Kaiser (1953), if a specimen is being examined up to a given stress level, then on subsequent examination no acoustic emission will be observed until this level is exceeded. In practice, therefore, an acoustic emission test should be conducted during the first stress cycle.

Experiment E7.2 Acoustic emission and dislocation kinetics
(James, D. R., Carpenter, S. H., 1971, *J. Appl. Phys.*, **42**, 4685.)

Emitted stress waves from single crystals of irradiated LiF, and annealed LiF, NaCl, and Zn were recorded during constant strain-rate compressive deformation and then analysed. A silent deformation machine was constructed using Teflon on all bearing surfaces to eliminate metal-to-metal contact. The acoustic emission signal was measured from the side of the sample with a PZT-5 piezoelectric accelerometer.

The signals were amplified, filtered, and recorded on an instrumentation recorder. The total system gain was of the order of 10^4 over a bandwidth of 200 Hz to 200 kHz. The acoustic pulse rate was measured by playing the tape back into an oscilloscope and visually counting the pulses which were above the noise level.

The experimental results for annealed pure NaCl are shown in figure E7.7. The crystals were annealed at 600°C in a vacuum for 24 h and then allowed to cool slowly in the furnace. The strain rate was $1 \cdot 595 \times 10^{-5}$ s^{-1}. The figure shows some activity almost immediately after applying the load, a maximum just after yield, and a decline in the rate in the easy glide region. In analysing the results from all the specimens tested the following possible relations between pulse rate, ζ, and dislocation kinetics were examined:

(a) $\zeta = \alpha \dfrac{\mathrm{d}\Lambda_{total}}{\mathrm{d}t}$,

(b) $\zeta = \delta \Delta_m$,

(c) $\zeta = \beta \dfrac{\mathrm{d}\Lambda_m}{\mathrm{d}t}$,

where α, β, δ are proportionality constants, and Λ_m is the mobile dislocation density.

In investigating relation (a) an interrupted stress–strain test was performed on as-cleaved optically-clear LiF, and etch pit counting was performed in regions which involved more than 1000 pits per photograph. In figure E7.8 is shown the resulting total dislocation density as a function of the total strain. It is observed that the graph shows an ever increasing function after yield. Since the tests were performed at a constant strain rate, figure E7.8 also represents dislocation density versus time if the abscissa is divided by the total strain rate. All the acoustic emission rate data presented show the onset of activity before yield, a distinct maximum after deviation from linearity, and a significant decline in the pulse rate after the maximum. This behaviour could not have been caused by multiplication of dislocations alone, because the behaviour was near zero before yield and was ever increasing after yield.

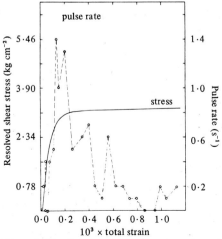

Figure E7.7. Pulse rate as a function of total strain for annealed NaCl.

In investigating relation (b), since the mobile dislocation density is not a quantity amenable to easy measurement, an attempt was made to relate this quantity to the velocities of the dislocations. One would expect the velocity to be zero in the elastic range. At yield the velocity should rapidly rise from zero to accommodate the increasing plastic strain rate. After yield one might expect constant velocity, or, more realistically, a decrease in velocity as the lattice becomes populated with pinned dislocation forests. Theoretical examination of the velocities implied by relation (b) gave unreasonable values and therefore this hypothesis was disallowed.

Assuming relation (c), an expression was computed for the relative velocity in terms of pulse rate and strain, and the computed values plotted as in figure E7.9. The calculated velocity was zero until deviation from linearity of the stress–strain curve, the velocity then increased rapidly to a maximum in the early plastic region and tended to decline in the far plastic region. The assumed relationship leads to a very realistic velocity behaviour.

Estimation of magnitudes involved leads to the equation

$$\zeta \approx 10^{-4}\frac{d\Lambda_m}{dt} \ . \tag{E7.4}$$

The smallest measurable stress was assessed as $\sim 5 \times 10^{-3}$ g cm^{-2}.

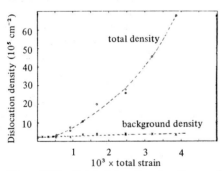

Figure E7.8. Total dislocation density as a function of total strain for LiF.

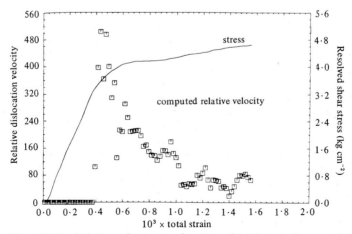

Figure E7.9. Relative velocity as a function of total strain for annealed NaCl.

Since the mobile dislocation density is involved in equation (E7.4), three possible mechanisms which could change this quantity are:
(i) movement of a previously stationary or pinned dislocation,
(ii) the operation of a multiplication source which produces fresh dislocation line length,
(iii) any interaction of a moving dislocation with something which impedes or halts its motion.
The authors consider that the major contribution to the generation of acoustic emission pulses at the onset of yield appears to be dislocation breakaway from pinning points.

It is proposed that acoustic emission in the macroscopic elastic range is caused by dislocation breakaway from minor (intermediate) pinning points, whereas breakaway from the major (nodal) pinning points at yield produces the maximum observed there. An estimate was made of the length of dislocation line participating in the generation of an acoustic pulse. It was concluded that the length associated with an individual pulse was of the order of several thousand centimetres. Therefore the change in Λ_m cannot be due to a single event because of the great length estimated. A large number of individual segment breakaways would be involved in producing an observed pulse. If an average segment length is taken to be 100 μm, the number of segments involved in the generation of an acoustic pulse would be of the order of 10^5 cm^{-3}. It seems probable therefore that an avalanche effect or stimulated breakaway occurs. The stress wave generated by a few simultaneous events could act as a trigger for further breakaway events. If the elastic wave travels the entire length of the slip plane it would take approximately 5 μs. A number of factors could tend to broaden the stress pulse and lead to the observed rise time of 20–30 μs (in the case of irradiated LiF). The authors comment that the concept of stimulated breakaway may be a reasonable explanation for observed discontinuous plastic flow, jerky dislocation motion, slip line broadening, and other features of crystal plasticity.

Experiment E7.3. Acoustic emission and crack propagation
(Evans, A. G., Linzer, M., Russell, L. R., 1974, *Mater. Sci. and Eng.*, 15, 253.)

By measuring acoustic emission during the fracture of polycrystalline alumina, it was shown that acoustic emission is obtained during macrocrack growth, which can be used for failure indication. Acoustic emission is also obtained as a result of the formation of nonpropagating grain-size microcracks at the surface.

Simultaneous measurements of acoustic emission rate, dN/dt, crack growth rate, da/dt, and stress intensity factor, K, were made on double torsion specimens with configuration shown in figure E7.10. K characterises the stress field ahead of the crack tip. Measurements were performed at room temperature in water and at elevated temperatures in air. The acoustic emission at high temperatures was transmitted to the transducer through an alumina waveguide. Figure E7.11 shows the crack growth rate and acoustic emission rate as a function of the stress intensity factor at room temperature in water. The crack growth rate data can be fitted with the approximate relation:

$$\frac{da}{dt} = \delta K^n ,$$

where δ and n are constants. A similar relation can be fitted to the acoustic emission rate

$$\frac{dN}{dt} = \beta K^{n'},$$

where β and n' are constants. At room temperature in water n' is significantly larger than n, so that to a first approximation

$$\frac{dN}{da} \propto K^2.$$

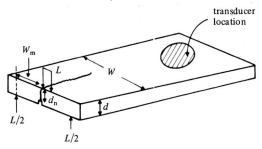

Figure E7.10. Double torsion specimen used for studying slow crack growth. W_m, W, d, and d_n are specimen dimensions. The stress intensity factor is given by

$$K = L W_m \left[\frac{3(1+\mu)}{W d^3 d_n \xi} \right]^{\frac{1}{2}},$$

where μ is Poisson's ratio and $\xi \approx 1 - \frac{5}{4}(d/W)$ is a correction factor for thick specimens.

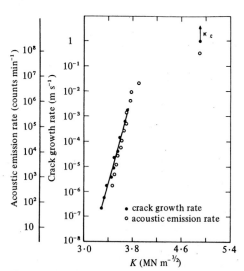

Figure E7.11. The crack growth rate and acoustic emission rate plotted against the stress intensity factor, K, for Lucalox tested at room temperature in water. Data obtained using double torsion specimen.

Strength tests were performed on beams in four-point flexure. Tests were conducted at constant displacement rate or at constant stress in air or in water. The time dependence of the acoustic emission rate at constant stress is shown in figure E7.12. The data can be fitted quite well with straight lines on log plots, at least until fracture is imminent when a large increase in the acoustic emission rate is evident. The magnitude of the stress has a marked effect on the initial emission rate; doubling the stress increases the emission rate by more than an order of magnitude, although the time dependence is relatively unaffected by stress. The increase in the acoustic emission rate prior to fracture occurred over a relatively short time (about 20 s) at the higher stress shown, whereas at the lower stress detectable emissions were observed well before fracture (about 2 h). This observation has considerable importance for the practical application of acoustic emission to failure prediction.

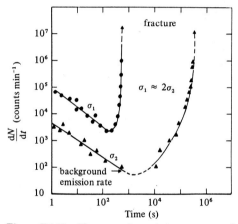

Figure E7.12. The acoustic emission rate as a function of time at constant stress, σ. $\sigma_1 \approx 0 \cdot 95 \, \sigma_f$, where σ_f is the fast fracture stress; $\sigma_2 \approx 0 \cdot 5 \, \sigma_f$.

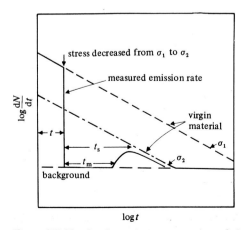

Figure E7.13. A schematic representation of the acoustic emission rate at constant stress, after a stress reduction from σ_1 to σ_2.

The stress dependence of the acoustic emission rate was also found to be strongly related to prior stress history. As shown in figure E7.13, if the stress is decreased from σ_1 to σ_2 at time t, no emission is observed until a certain minimum time, t_m, has elapsed. The emission rate then increases until, after a time t_s, it coincides approximately with the rate observed for the virgin material. The converse effect is shown in figure E7.14.

Mechanical polishing was found to reduce the acoustic emission rate considerably. This suggests that the formation of stable grain-size microcracks at the surface is one source of acoustic emission.

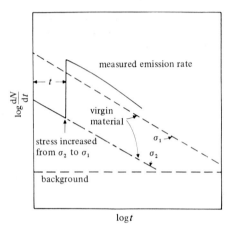

Figure E7.14. A schematic representation of the acoustic emission rate at constant stress, after a stress increase from σ_2 to σ_1.

Origin of acoustic emission

Macrocracks. The acoustic emission from a large precrack at room temperature originates from the increments in the primary crack and from subsidiary microcracking in the process zone around the crack tip. There is no emission from the other potential sources, dislocation motion and twinning, because these do not accompany crack propagation at room temperature.

Bulk stressing. The acoustic emission that occurs prior to that from macrocrack propagation could be due to three types of event: dislocation motion, twinning, or microcracking. The formation of grain-size microcracks prior to fracture is readily detected by optical microscopy, but dislocation motion and twinning are more difficult to detect and consequently have not been observed.

Expressions are developed for the total number of acoustic emissions at constant stress and under constant displacement rate (stress rate) conditions. The agreement between the theoretical expressions and the experimental data is then discussed. The remarkably good correlation between the prediction of the microcracking model for acoustic emission and the experimental observation lends credence to the assumption that this process is the primary contribution to the acoustic emission during bulk stressing, prior to macrocrack growth. The effect of mechanical polishing on the acoustic emission rate is then readily explained by the removal of machining-induced microcracks at the surface which are the precursors for the grain-size microcracks that form when a bulk stress is applied.

7.6 Radiation effects

Irradiation of a crystal often provides a flexible means of introducing point defects into the crystal in a controllable manner. A number of macroscopic properties will be affected by a change in the population of point defects within the crystal. The basic physical effects produced by atomic radiations have been described by Thompson and Paré (1968) while the effects produced by electromagnetic radiation have been described by Gordon (1965).

When a beam of atomic particles strikes a solid, the primary effect is to produce interstitial atoms and vacancies. A minimum **displacement energy**, E_d, of about 25 eV is required to displace an atom into an interstitial site. Such displaced atoms are referred to as primary knock-ons; if they in turn have sufficient energy, they will produce secondary or tertiary knock-ons. For instance, with high energy neutrons the energy of the primary knock-ons may be as high as $10^4 - 10^5$ eV which is sufficient to produce further secondary displacements. Associated with these displacements are electron excitation and ionisation processes. When the particle energy becomes less than E_d, the energy is dissipated in thermal vibrations of the lattice. In turn, thermal motion may cause some recombination of interstitials and vacancies.

Thompson and Paré, using a hard-sphere model, calculated the total number of displaced atoms, $n(E_p)$, and thus of interstitial–vacancy pairs, produced by a primary knock-on of energy E_p:

$$n(E_p) \approx 2^N = \frac{E_p}{2E_d} \; ,$$

where N is the number of generations of displaced atoms. For fast neutrons obtained from a reactor, $E_p \approx 25$ keV. If these are incident on a copper specimen with $E_d = 25$ eV, then $n(E_p) = 500$. Such a displacement spike or cascade may extend over a range of 10^5 atom sites. The associated thermal spike may well be sufficient to cause marked local heating.

The lattice may modify these effects through the phenomena of focusing and channeling. Focusing has the overall effect that vacancies produced by the displacement spike remain near the path of the primary knock-on while the interstitials may end up some distance away. A channeled atom ends up as an interstitial at a large distance from the origin of the primary knock-on. If the channeled atom intercepts a dislocation it may initiate a new displacement spike.

Particles that have been used in studies of material properties include fast neutrons derived from a reactor or accelerator, thermal neutrons, charged particles, fast electrons, and electromagnetic radiation ranging from the ultraviolet to gamma rays. In the latter case the photons cause displacements by Compton scattering of electrons. The efficiency in

producing displacements is much less than for fast neutrons, but the rate
of production of defects is often sufficient for studies at low levels such
as internal friction.

7.6.1 Effects of radiation on dislocations

Dislocations are particularly sensitive to the advent of additional point
defects which lead to an increase in pinning, and hence to changes in the
elastic constants and damping. For instance, in copper single crystals both
gamma and neutron radiation can cause an increase in the elastic constants
of up to several per cent and a decrease in the internal friction by factors
of up to 30. The particular defects responsible and the mechanism of
pinning depend on the type of crystal.

During an irradiation experiment the average loop length, L, will vary
with time owing to the production of new pinning points. If L_0 is the
average loop length before irradiation, then the number of pinning points
(per unit volume) will be

$$n_0 = \frac{\Lambda}{L_0} \ .$$

If $L(t)$ is the average loop length at time t, then the number of pinning
points at time t will be

$$n(t) = \frac{\Lambda}{L(t)} \ .$$

But $n(t) = n_0 + n_r$, where n_r is the number of pinning points produced
by the radiation. Hence,

$$\frac{\Lambda}{L(t)} = \frac{\Lambda}{L_0} + n_r \ , \tag{7.48}$$

or

$$L(t) = \frac{L_0}{1 + n_d} \ , \tag{7.49}$$

where $n_d = n_r/n_0$ is the **pinning point density** (the number of pinning
points added to each original length L_0 in time t).

If the radiation level is constant with time, it is reasonable to assume
that the number of pinning points produced by the radiation, n_r, is a
linear function of time, that is

$$n_r = Pt$$

where P is the total number of pinners arriving per second at the
dislocations (per unit volume). It is also convenient to write

$$n_d = \gamma t \tag{7.50}$$

where $\gamma \equiv PL_0/\Lambda$ is the fractional increase in pinners per second (that is,
the number of pinners per second per original loop).

Since we are dealing with a process in which the changes are diminishing with time, due to the pinning up of the dislocations, it is convenient to use normalised quantities which represent the fraction remaining, at any given time t, of the total possible change. We therefore introduce a **normalised decrement**, $Z(t)$, and a **normalised modulus**, $Y(t)$, defined as follows:

$$Z(t) = \frac{(\Delta_I)_t}{(\Delta_I)_0} = \frac{\Delta(t) - \Delta_\infty}{\Delta_0 - \Delta_\infty} , \tag{7.51}$$

$$Y(t) = \frac{(\Delta E/E)_t}{(\Delta E/E)_0} = \frac{E_\infty - E(t)}{E_\infty - E_0} \approx \frac{v_\infty - v(t)}{v_\infty - v_0} . \tag{7.52}$$

The decrement Δ_I is the low-amplitude dislocation decrement and is found by subtracting from the total decrement the background decrement due to non-dislocation causes. The latter value is taken to be equal to the measured decrement when all available dislocations have been fully pinned after long irradiation. Infinity subscripts refer to values after long irradiation and zero subscripts to pre-irradiation values (at $t = 0$).

Expressions for $Y(t)$ and $Z(t)$ in terms of the average loop length at time t may be found by combining equations (7.51) and (7.52) with equations (7.30) and (7.31):

$$Y(t) = \left[\frac{L(t)}{L_0}\right]^2 , \tag{7.53}$$

$$Z(t) = \frac{B(t)}{B_0}\left[\frac{L(t)}{L_0}\right]^4 . \tag{7.54}$$

Thus, the fractional change in loop length and the pinning point density may be found from measurement of $Y(t)$ alone:

$$\frac{L(t)}{L_0} = Y(t)^{1/2} , \tag{7.55}$$

$$n_d = [Y(t)]^{-1/2} - 1 . \tag{7.56}$$

Equations (7.53) and (7.54) provide the means for a strict test of the applicability of the Granato–Lücke theory to a given experiment. Provided $B(t)/B_0$ remains constant during an irradiation experiment, then a graph of $Y^2(t)$ versus $Z(t)$ should give a straight line. Such a result implies that the same basic mechanism is responsible both for the changes in velocity and the changes in the damping. Since we have used the low-frequency approximation, these conclusions apply to the frequency range below about 100 kHz.

7.6.2 Some experimental results for copper and ionic crystals

In figure 7.39 are shown the variations in Young's modulus and damping for a copper crystal when exposed to fast neutrons in a reactor. The neutrons produce an increase in the population of point defects within the crystal and some of these interact with dislocations to form new pinning points. As a consequence, the elastic modulus of the specimen increases and the damping diminishes with time. This behaviour is in agreement with equations (7.30) and (7.31), assuming that L decreases with time due to the pinning process.

Similar results have been obtained for a number of other metals and ionic crystals. In figure 7.40 are shown results for a NaCl crystal when irradiated with high energy neutrons derived from a Van de Graaff accelerator using the reaction $^9Be(dn)^{10}B$. In order to verify that the vibrating string model is applicable to such experiments a plot may be

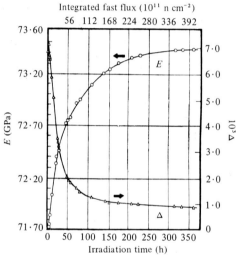

Figure 7.39. Young's modulus E and logarithmic decrement Δ as functions of neutron irradiation time for a copper crystal. The flux is $3 \cdot 1 \times 10^6$ n cm^{-2} s^{-1} and the maximum strain amplitude is 5×10^{-8}. (After Thompson and Holmes, 1956.)

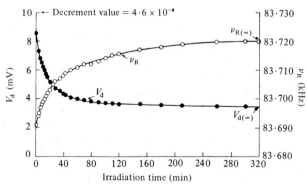

Figure 7.40. Variation in drive voltage V_d and resonant frequency ν_R with neutron irradiation for a NaCl crystal. Dose rate = 5×10^7 n s^{-1}. (After French and Pollard, 1970.)

made of Z against Y^2, or of $\log Z$ against $\log Y$. According to equations (7.53) and (7.54), the latter plot should have a slope of 2. Such a graph is drawn in figure 7.41 for the data shown in figure 7.40 together with data obtained with a different dose rate. The close agreement indicates that in these experiments the ratio $B(t)/B_0$ is essentially constant and hence other relevant theoretical parameters may be computed from the data. For example, in figure 7.42 are shown curves for the pinning point density as a function of irradiation time computed from equation (7.56) for four different dose rates. The initial part of each curve is essentially a straight line. The subsequent departure from linearity suggests the possibility of clustering of defects. Such an explanation has been discussed by Thompson and Paré (1968) in relation to experiments on copper. The sensitivity of the ultrasonic method is such that it is easy to detect the arrival of one new pinning defect per 100 dislocation loops.

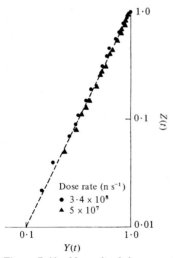

Figure 7.41. Normalised decrement $Z(t)$ against normalised modulus $Y(t)$—broken curve; line predicted by Granato-Lücke theory. (After French and Pollard, 1970.)

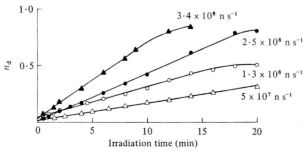

Figure 7.42. Number of pinning points added per dislocation segment n_d plotted against irradiation time for a series of neutron dose rates. (After French and Pollard, 1970.)

Experiment E7.4 Mechanism for dislocation pinning in the alkali halides
(Bauer, C. L., Gordon, R. B., 1962, *J. Appl. Phys.*, 33, 672.)

Early experiments by Frankl (1953) had shown the connection between x-irradiation and bleaching effects on the internal friction of NaCl both at room temperature and liquid nitrogen temperature. Frankl postulated that F-centres were involved in the pinning process. In the experiments of Bauer and Gordon the alkali halide specimens form part of a composite resonator operating at about 90 kHz. Prior to irradiation the specimens were deformed by compression along the long axis. The resonator assembly was placed in a low-temperature cryostat with provisions for cooling with helium exchange gas and with windows to allow the entrance of both 90 kV x-rays from a tungsten target and either white or monochromatic light.

The first experiment consisted of determination of the rate of change of modulus with radiation time at different temperatures. NaCl crystals were subjected to x-irradiation at 78, 196, 273 and 298 K while the modulus and damping were recorded continuously. As shown by the superimposed curves of figure E7.15, there is no appreciable temperature dependence of the rate of change of modulus with x-irradiation. This result implies that dislocation pinning does not involve a thermally activated process and, therefore, that the pinning defects are created in the immediate vicinity of the dislocations. These results are in marked contrast to those for metals such as copper in which the dislocation pinning rate diminishes with decreasing temperature (see Thompson and Paré, 1968, for a full discussion).

A series of experiments were then conducted on the effects of illumination, both white light and monochromatic, on dislocation pinning. It was found that white light illumination at room temperature caused a small increase in modulus, corresponding to additional pinning. Experiments with monochromatic light showed that F light (465 nm) produces additional pinning at both high and low temperatures (78 K). On the other hand, monochromatic light of wavelength 630 nm was found to produce unpinning at low temperatures. This light has been called by the authors 'u light'. The frequency dependence of u light that causes unpinning in NaCl at 78 K is shown in figure E7.16; the temperature dependence of u light is shown in figure E7.17. The experiments indicate that pinning produced by x-irradiation consists of (a) reversible pinning which occurs at low temperatures, and (b) a stable type of pinning which grows with x-irradiation time and at a faster rate with increasing temperature.

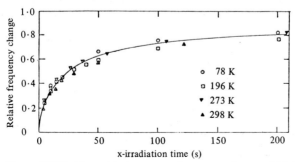

Figure E7.15. Normalised resonant frequency change of NaCl specimens plotted as a function of x-irradiation time at various temperatures.

The fact that unpinning only occurs at low temperatures indicates that this effect does not involve the diffusion of structural defects. It seems that both x-irradiation and illumination with F light are capable of producing the same type of results. Unpinning in a number of other alkali halides was also studied. Since unpinning at low temperatures involves the measurement of the change of elastic modulus, this effect cannot be measured by optical techniques alone.

It follows from these experiments that any mechanism put forward to explain dislocation pinning must satisfy the following conditions (Gordon, 1965):

1. The defects responsible for dislocation pinning must be created in the immediate vicinity of the dislocations.

2. The defects must interact with the dislocations strongly enough to produce distinct pinning points.

3. The pinning defect must have a well-defined optical absorption band.

4. It must be simple enough to decompose when excited by light of wavelength corresponding to its absorption band.

5. It must be such that it can occur in any alkali halide with the rock salt structure (except, possibly, LiF).

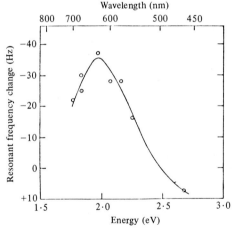

Figure E7.16. Spectral dependence of the resonant frequency change observed upon illumination of an NaCl specimen previously x-irradiated at 78 K. 25 nm bandwidth.

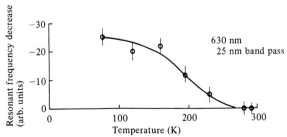

Figure E7.17. Temperature dependence of the effect of u light on the resonant frequency of an NaCl crystal.

6. Warming must result in the conversion of the temporary pinning point to a permanent one.

7. It must be possible to create the pinning defects either by x-irradiation of transparent crystals or F-light illumination of coloured crystals.

In one model which fits all the experimental facts the terminating row of ions in the extra half-plane of an edge dislocation in an NaCl type lattice can be schematically represented by a row of alternating Na$^+$ and Cl$^-$ ions as in figure E7.18a. Suppose that during irradiation a Cl$^-$ ion located near the dislocation and below its slip plane loses an electron, for example by capture of a hole. Suppose further that the Cl atom so formed moves one or two atomic distances so as to form a jog on the dislocation and a Cl$^-$ vacancy, as in figure E7.18b. The vacancy and the Cl atom could then capture an electron each, forming an F centre and a Cl$^-$ ion attached to the extra half-plane of the dislocation, as in figure E7.18c. It is hypothesised that the defect complex formed in this way is the dislocation pinning point produced by x-irradiation at low temperatures.

To reverse the sequence of events and thereby unpin the dislocation, all that is required is that the F centre be ionised (provided it has not diffused away from the dislocation). As the F centre lies in a region of dilation, illumination other than by F light, that is u light, will be required to ionise it. The resultant configuration is shown in figure E7.18d. There is now an electrostatic attraction between the Cl$^-$ ion and the Cl$^-$ vacancy; hence the two will tend to recombine, resulting in a return to the initial configuration of the dislocation line (figure E7.18a). Warm up will convert temporary pinning to permanent pinning because the F centre can diffuse beyond recombination range, while illumination with u light dissolves the pinning points.

Consideration of the distribution of ions around the core of an edge dislocation in NaCl, as computed by Huntington et al. (1955) and shown in figure E7.19, indicates that a negative ion located in position ① could act as a pinning point. In order for a dislocation to move along the slip plane, this halogen ion must move very close to another negative ion. The electrostatic repulsion, coupled with elastic forces, could immobilise the dislocation at this point. Further considerations show that an F centre located in a highly strained region of the crystal near a dislocation will have its absorption band shifted towards the red end of the spectrum. It is estimated that the F band of NaCl would be shifted to the observed u band if the F centre were located at a distance of approximately $0 \cdot 7$ Å away from the dislocation core.

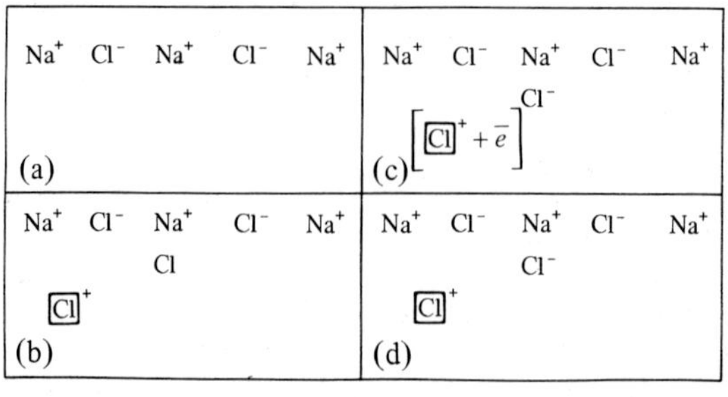

Figure E7.18. Schematic representation of the events that may occur during the creation of a dislocation pinning point, as discussed in the text.

In summary, then, the defect produced by low temperature x-irradiation which pins dislocations in plastically-deformed alkali halide single crystals was identified as a jog on a dislocation formed by a negative ion positioned directly below the extra half-plane, in close association with an F centre just one ionic jump distance away.

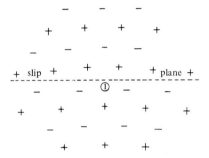

Figure E7.19. Distribution of ions around the core of an edge dislocation in NaCl. A negative ion in position ① could act as a pinning point.

Experiment E7.5   Effects of fast neutrons on internal friction and Young's modulus of NaCl
(French, I. E., Pollard, H. F., 1970, *J. Phys. C.*, 3, 1866.)

An ultrasonic regenerative measuring system was used which involved a three-component resonator consisting of two matched quartz bars and the NaCl specimen. The composite resonant frequency was approximately 85 kHz. Signals proportional to the elastic modulus and internal friction were recorded continuously. The strain amplitude was kept constant at 5×10^{-8}. Neutron irradiation was carried out with a 3 MeV van de Graaff accelerator. Deuterons of energy 2·62 MeV were allowed to strike a beryllium target, high energy neutrons resulting from the reaction $Be^9(dn)B^{10}$.

Some of the experimental results have already been discussed in section 7.6.2 where it was shown that there was apparent agreement with the Granato–Lücke equations and that the rate of dislocation pinning was directly proportional to the neutron dose rate. In addition to these continuous irradiation experiments, further experiments were carried out in which the radiation was interrupted periodically during runs covering a temperature range from 10–29°C. Some typical experimental results are shown in figure E7.20 in which the main features are (i) immediately the radiation is interrupted the resonant frequency of the specimen becomes constant or nearly so, and (ii) the decrement continues to decrease for a period of some minutes after interruption of the radiation. In figure E7.21 are shown plots of the normalised decrement, Z, at two different temperatures. The time taken for Z to become constant after interruption is seen to depend markedly on temperature. At lower temperatures Z changes over much longer periods. The rate at which Z decreases prior to interruption of the radiation appears to be independent of temperature in the range tested.

The time constant for the part of the Z curve with the radiation off was measured and values for four different temperatures are shown in table E7.1. The fact that the rate of dislocation pinning is directly proportional to neutron dose rate, and in addition that the pinning process ceases as soon as the radiation is interrupted, suggests that pinning is caused by defects produced quite close to the dislocations. There is no evidence for any diffusion process involved in the basic dislocation pinning. These conclusions are in agreement with those reached by Bauer and Gordon (1962).

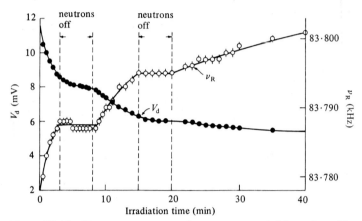

Figure E7.20. Variation in resonant frequency ν_R and drive voltage V_d with time during an interrupted irradiation.

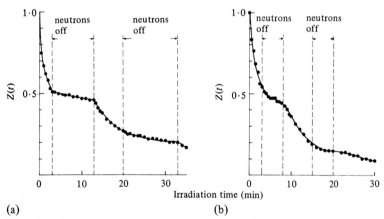

Figure E7.21. Variation in normalised decrement $Z(t)$ during an interrupted irradiation (a) at $10°C$, (b) at $22°C$.

Table E7.1. Variation of time constant for $Z(t)$ decay with temperature.

Experiment number	Neutron dose-rate (neutrons s^{-1})	Temperature (K)	Time constant for decay of $Z(t)$ (s)
19	$2 \cdot 6 \times 10^8$	302	60
16	$2 \cdot 6 \times 10^8$	295	85
17	$2 \cdot 6 \times 10^8$	290	156
18	$2 \cdot 6 \times 10^8$	283	350

The fact that the decrement kept changing during interruptions to the radiation may be interpreted as a change in the damping parameter B according to equation (7.54). The decrease in the ratio $B(t)/B_0$ for the experiments was found to lie between 10 and 25%. The time over which B changed was found to increase with decreasing temperature. This suggests that the change in B involves some process that is diffusion-controlled so that the value of B at some time t after interruption is given by an equation of the form

$$B(t) = B_i \exp\left(-\frac{t}{\tau_B}\right) \tag{E7.5}$$

where B_i is the damping coefficient immediately before interruption and τ_B is the relaxation time. If the process involved is diffusion-controlled, a plot of $\ln \tau_B$ against $1/kT$ should produce a straight line from which the appropriate activation energy can be determined, as shown in figure E7.22. The value of the activation energy found was 0.7 ± 0.1 eV which is in agreement with the value for positive ion vacancies in NaCl.

The conclusion was therefore reached that the diffusion process, and hence the resulting interaction with the vibrating dislocations, involves positive ion vacancies. In explanation of the results, the authors proposed a two-region model for an ionic crystal. Region 1 is a very small volume immediately around the dislocation core and only defects which are created in this region can act as pinning points. Region 2 is a larger volume around the dislocation; the defects in this region affect dislocation motion such that the damping parameter B is a function of the number of effective defects in this region. During interruptions to irradiation no new defects are formed so that the number of effective defects, and hence B, will decrease due to processes such as clustering and recombination. The time interval over which these changes in the number of effective defects occur must be dependent on temperature since the processes involved are controlled by diffusion. Although no particular mechanism is

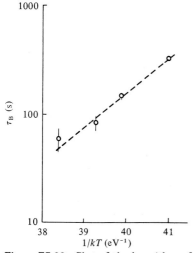

Figure E7.22. Plot of the logarithm of relaxation time for the damping process, τ_B, against $1/kT$.

proposed for the changes in B, it is suggested that a likely mechanism would be electrical interaction between positive ion vacancies and charged dislocations. This explanation would be in agreement with the relationships discussed in section 7.7 between charges on the dislocation core and defects present in the compensating charge cloud region.

7.7 Electrical effects in ionic crystals

Dislocations in ionic crystals have a resultant charge, as discussed in section 7.2.8. In order to preserve overall charge neutrality in the crystal the dislocations are surrounded by a compensating charge cloud. Dislocation charges have an important effect on the mechanical and electrical properties of ionic crystals.

When an ionic crystal is deformed, either statically or by means of a stress wave, electrical charges appear at the surface. The first investigation of the latter effect was that of Amelinckx *et al.* (1959). They bent a crystal of NaCl in a four-point bending jig, and measured induced surface electric charge by means of a gold electrode evaporated on the central part of the crystal and an electrometer. The specimen was caused to vibrate in the frequency range 10–100 Hz by means of a loudspeaker coil system. The amplitude of vibration was small so that no appreciable plastic deformation took place. Under the action of the applied alternating stress the dislocations vibrate relative to their charge clouds, thus giving rise to a system of vibrating dipoles which induce charges on the surface electrode. Opposite faces of the crystal should acquire the same charge in the same phase of vibration. Amelinckx *et al.* observed an alternating electrical signal at the same frequency as the applied stress wave, and found a phase difference between the applied signal and the surface electric signal.

Whitworth (1964) investigated the production of charges on dislocations in single crystals of NaCl that had been plastically deformed either by uniaxial compression or by bending. The advantage of deforming a crystal by bending is that the process introduces an excess of edge dislocations of one mechanical sign. Whitworth showed that the process of dislocation multiplication does not give rise to charged dislocations directly. The dislocations acquire their charge as they move through the crystal lattice. He assumed that when a dislocation passed a vacancy on its slip plane, the vacancy became attached to the extra half-plane of the dislocation and moved with it. The sign of the dislocation charge was determined by indentation tests. Whitworth found that the indentor always acquired a positive potential with respect to more distant parts of the crystal. This result is consistent with the edge dislocations having a nett negative charge. The value of the charge was found to be of the order of one electronic charge per 60 atomic planes.

In a further investigation Whitworth (1966) studied the effects produced by a cyclic tensile and compressive stress applied along the axis of pure single crystals of NaCl that had been plastically bent. It was found that under the action of the applied stress the crystals exhibited piezoelectric properties. Whitworth comments that, when a dislocation is first formed by deformation, it is expected to acquire its charge by sweeping up relatively immobile positive ion vacancies as it moves. Such a charged dislocation will not be in thermal equilibrium. When left undisturbed for long periods, mobile positive ion vacancies may diffuse to form a charge cloud around the dislocation and possibly to change its charge. The charge on the dislocations was determined experimentally after the crystal had been resting for at least a day and found to be in the range $-(2$ to $4) \times 10^{-11}$ C m^{-1}. This value was found to be approximately doubled when further cyclic stressing was conducted at high amplitudes. A charge of -2×10^{-11} C m^{-1} corresponds to one electronic charge per 14·5 positive ion sites on the dislocation core.

Robinson (1972a) presented a theory for the relationship between the mechanical damping, compliance defect, piezoelectric moduli, and electric susceptibility due to charged dislocations. Using a composite oscillator system at 40 kHz he carried out an experiment on a bent single crystal of KCl, and found the strain amplitude dependence of the relevant piezoelectric modulus to be similar to that of the compliance defect and mechanical damping. From these results he determined the variation of dislocation charge with strain amplitude and found it to range between 1 and 3% of the maximum allowed on a dislocation. For the charge cloud radius he obtained the value of 20 Å. This figure is a measure of the displacement at which the dislocations carrying the average charge and with a mean loop length become amplitude dependent.

Further contributions to the theory of charged dislocations include those of Eshelby et al. (1958), Brown (1961), Robinson and Birnbaum (1966), Whitworth (1968) and Robinson (1972b). Whitworth (1968) examined the dislocation structure in detail and found that the charge on a dislocation in an ionic crystal is determined by an excess of jogs or vacancies. Divalent impurities and associated vacancies greatly affect this charge.

Brown (1961) has shown that, when dislocations in an ionic crystal are caused to vibrate with respect to their charge clouds, the force per unit length of dislocation is given by

$$F = -\frac{q^2 \kappa^2}{2\epsilon_r \omega}[\tan^{-1}\omega\tau - \tfrac{1}{2}i\ln(1 + \omega^2\tau^2)]v_0\exp(i\omega t), \qquad (7.57)$$

where q is the charge per unit length of dislocation, κ^{-1} is the effective charge cloud radius, ϵ_r is the static relative permittivity (dielectric constant), τ is the relaxation time of the charge cloud, and v is the dislocation velocity. Equation (7.57) may be written in complex form

$$\hat{F} = -\hat{B}\hat{v} = -(B_1 + iB_2)\hat{v} = -\frac{B_0}{\omega}[\tan^{-1}\omega\tau - \tfrac{1}{2}i\ln(1 + \omega^2\tau^2)]v_0\exp(i\omega t), \qquad (7.58)$$

where $\hat{B}$ is the complex damping parameter and B_0 is the frequency-independent part of $\hat{B}$ defined by

$$B_0 = \frac{q^2 \kappa^2}{2\epsilon_r} \ . \tag{7.59}$$

Brown also shows that

$$\kappa^2 = \frac{4\pi e^2}{\epsilon_r kT} \sum n_i^\infty \tag{7.60}$$

where e is the charge on an electron, k is Boltzmann's constant, T is the absolute temperature, and n_i^∞ is the density of the charged point defect of type i well away from the dislocation. Hence,

$$B_0 = \frac{2\pi q^2 e^2 \sum n_i^\infty}{\epsilon_r^2 kT} \ . \tag{7.61}$$

In figure 7.43 is shown a phase diagram relating force, dipole moment, dislocation displacement, and velocity. When $\omega = 0$, we have $\phi = \pi$ and $\psi = 0$. Thus, for slow moving (nonoscillating) dislocations, the force opposes the velocity and the dipole moment is in phase with the velocity. As ω increases, $\phi \to \frac{1}{2}\pi$. Thus, for dislocations oscillating within a practically stationary charge cloud ($\omega\tau \gg 1$), the force opposes the displacement and the dipole moment is in phase with the displacement.

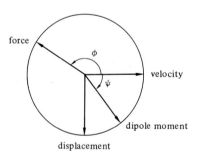

Figure 7.43. Phase-vector diagram relating dislocation velocity and displacement with the force and the dipole moment. (After Brown, 1961.)

7.7.1 Robinson–Birnbaum theory of dislocation damping

In the Granato–Lücke theory the damping parameter B is assumed to be constant. No particular mechanism is postulated but the authors suggest that phonon damping arising from the interaction between moving dislocation segments and lattice phonons is the most likely mechanism (Leibfried's equation 7.17). Changes in damping occur mainly through changes in the average loop length, as do changes in the modulus. Granato and Lücke postulate that the same basic mechanism is responsible for both changes. While these assumptions have proved successful in dealing with metals, modifications to the theoretical approach are required for ionic crystals.

Robinson and Birnbaum (1966) take the main cause of damping in ionic crystals to be electrical interaction between charged dislocations and their compensating charge clouds. When the dislocations vibrate, the charge clouds attempt to follow this motion. The resulting electrical interaction is found to depend on frequency. In the kilohertz range and at room temperature, where $\omega \approx 10^5$ rad s^{-1} and $\tau \approx 10^2$ s, we have $\omega\tau \gg 1$, and the dislocations vibrate with respect to practically immobile charge clouds. Robinson and Birnbaum also predict a maximum in the damping at iso-electric temperatures where the charge on the dislocations becomes zero. The theory allows for an increase in the parameter B during irradiation.

The theory of Robinson and Birnbaum starts from Koehler's equation of motion (7.15), but introduces the specific expression for B computed by Brown (1961). Thus the damping becomes dependent on the dislocation charge. The equation of motion is now solved by a Fourier transform method. The following expressions for the modulus defect and decrement in the amplitude-independent region are found on the assumption that (i) the dislocation segments are vibrating in their fundamental mode, (ii) all segments are of equal length (δ-function distribution)

$$\left(\frac{\Delta E}{E}\right)_{\mathrm{I}} = \left(\frac{16\Omega Eb^2\Lambda}{\pi A}\right)\frac{\omega_0^2 - \omega^2 - B_2/A}{(\omega_0^2 - \omega^2 - B_2/A)^2 + (B_1/A)^2} , \qquad (7.62)$$

$$\Delta_{\mathrm{I}} = \left(\frac{16\Omega Eb^2\Lambda}{\pi A}\right)\frac{\omega_0^2 - \omega^2 - B_2/A}{(\omega_0^2 - \omega^2 - B_2/A)^2 + (B_1/A)^2} , \qquad (7.63)$$

where $B_1 = B_0\tan^{-1}\omega\tau$ and $B_2 = \frac{1}{2}B_0\ln(1 + \omega^2\tau^2)$.

As shown by Bielig (1971), the low-frequency and low-temperature approximations to these equations are

$$\left(\frac{\Delta E}{E}\right)_{\mathrm{I}} = \frac{16\Omega Eb^2\Lambda L^2}{\pi^3 C[1 - (B_0L^2/\pi^2 C)\ln\omega\tau]} , \qquad (7.64)$$

$$\Delta_{\mathrm{I}} = \frac{8\Omega Eb^2 B_0\Lambda L^4}{\pi^4 C^2[1 - (B_0L^2/\pi^2 C)\ln\omega\tau]^2} . \qquad (7.65)$$

As before, a normalised modulus and decrement may be defined for time-dependent experiments and expressions similar to equations (7.53) and (7.54) derived.

The main practical differences between the Granato–Lücke and Robinson–Birnbaum theories lie in the shapes of the modulus and damping versus frequency curves. Figure 7.44 shows the damping curves predicted by the two theories. Since the Robinson–Birnbaum theory allows explicitly for the dislocation charge, any change in this quantity, such as occurs in a radiation experiment, could result in a variation in the damping parameter B.

The variations in B and q that occur during neutron irradiation were investigated by Anderson and Pollard (1976) on bent pure NaCl single crystals. They used a composite oscillator system at 84 kHz for determining changes in modulus, decrement, and the dislocation voltage developed at the surface of the crystal. In figure 7.45 are shown the experimental decrement values, a theoretical curve assuming that B depends only on loop length changes, and the variation in B as a function of radiation time. It has been assumed that the deviations between the experimental and theoretical curves are a measure of the interaction between the charged core regions of the dislocations and their surrounding charge cloud regions. In figure 7.46 are shown the experimental values of the dislocation voltage, a theoretical curve based on loop length changes only, and the computed curve for the variations of dislocation charge as a function of radiation time.

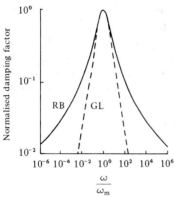

Figure 7.44. Normalised plots of damping versus frequency for the Granato–Lücke theory (GL) and the Robinson–Birnbaum theory (RB). (After Robinson, 1972b.)

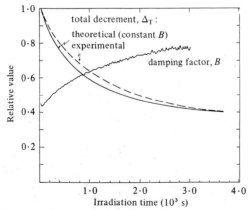

Figure 7.45. Experimental decrement values, theoretical curve (assuming that B depends only on loop length changes), and the variation in B as a function of neutron irradiation time for an NaCl crystal at 10°C (Anderson and Pollard, 1976).

At greater vibration amplitudes, the modulus defect and decrement become dependent on the amplitude. According to Granato and Lücke the dependence on amplitude is due to the breakaway of dislocations from intermediate pinning points. According to Robinson and Birnbaum the effect is caused by dislocations breaking through their compensating charge clouds.

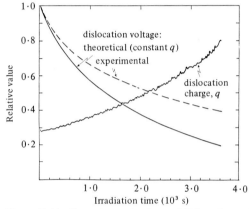

Figure 7.46. Experimental values of dislocation voltage, theoretical curve assuming that only the loop length is changing, and the variation in dislocation charge as a function of neutron irradiation time for an NaCl crystal at 10°C (Anderson and Pollard, 1976).

Experiment E7.6 Photoplastic effect in alkali halides
(Cabrera, J. M., Agullo-Lopez, F., 1974, *J. Appl. Phys.*, **45**, 1013.)

This paper reports an extensive investigation into the photoplastic effect first reported by Nadeau (1964). The photoplastic effect consists of a reversible change in the flow stress when light is shone on a coloured alkali halide crystal during plastic straining. Since the effect deals with the interaction between charged dislocations and charged defects it has an important bearing on the plastic behaviour of dielectric materials and on the understanding of the charge structure of dislocations in ionic compounds. The effect has been studied on coloured crystals of NaCl, KCl, KBr, and KI as a function of strain level, F-centre concentration, strain rate, light intensity, and light wavelength.

The alkali halides used in the experiments were coloured either additively or by γ irradiation or by hydrogen doping followed by γ irradiation. The coloured samples ($3 \times 4 \times 12$ mm^3) were strained in compression in an Instron testing machine. At a certain point of the stress–strain curve they were subjected to light from an arc or quartz-iodine lamp either directly or through a high-intensity monochromator. Figure E7.23 shows the photoplastic effect at room temperature in KBr, KI, NaCl, and KCl crystals which contain colour centres (mostly F-centres). On illuminating the crystal during deformation an abrupt rise in flow stress is observed which has practically an elastic slope. When the illumination is continued, a sharp yield drop is usually found and afterwards a region of steady flow sets in. On turning the light off the flow stress decreases rapidly to the initial level (or lower) and the pre-illumination flow behaviour is recovered. The effect shown in the figure is essentially the same regardless of the method used to create the colour centres.

Experiments show that the magnitude of the photoplastic effect, $\Delta\sigma$, decreases monotonically with strain rate. A curve of the type

$$\Delta\sigma \propto (1 + \text{const} \times v)^{-1}$$

where v is the cross-head speed, has been fitted to the experimental data.

The effect also depends on the intensity of the exciting light. For low intensities a smoother growth of $\Delta\sigma$ with strain is observed than that shown in figure E7.23. Figure E7.24 provides a comparison of the photoplastic effect at low and high intensities. The experimental curves obtained for the photoplastic effect versus light intensity could be fitted with a theoretical relationship of the type

$$\Delta\sigma \propto \left(1 + \frac{\text{const}}{I}\right)^{-1},$$

where I is the light intensity.

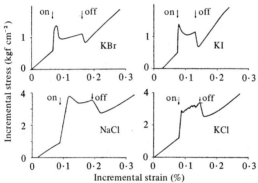

Figure E7.23. Structure of the room-temperature photoplastic effect for various alkali halides. The start and completion of the illumination are marked on each curve.

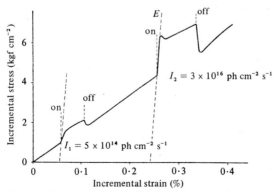

Figure E7.24. Comparison between the structure of the photoplastic effect at low and high illumination intensities.

Experiments were conducted to test the effect of F-centre concentration on the photoplastic effect. It was found that the effect increases monotonically with F-centre concentration for the crystals tested. Experiments to test the dependence on the wavelength of the light showed close agreement with the F-centre absorption band, as shown in figure E7.25 for samples of NaCl, KCl, and KBr with low F-centre concentrations. When high F-centre concentrations are used, a dip occurs in the curve near the maximum of the F band owing to excessive optical density.

The dependence of the photoplastic effect on light intensity for NaCl at 77 K is shown in figure E7.26. Experimental data for this type of experiment were shown to agree with a curve of the type

$$\Delta\sigma \propto I\left(1 + \frac{\text{const}}{I}\right)^{-1}.$$

As shown in the insert to figure E7.26, the photoplastic effect occurs less rapidly at 77 K than at room temperature. The spectral dependence also closely follows the curve of the F-centre absorption band at 77 K. It is therefore clear that F-centres appear to be directly responsible for the effect at room as well as at low temperatures.

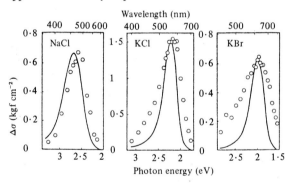

Figure E7.25. Dependence of the photoplastic effect on light wavelength for low F-centre concentrations: $n_F = 9 \times 10^{15}$ cm^{-3} (NaCl), $n_F = 1 \cdot 1 \times 10^{16}$ cm^{-3} (KCl), $n_F = 1 \cdot 0 \times 10^{16}$ (KBr). Circles represent experimental measurements of the photoplastic effect. Solid lines are F-centre absorption bands.

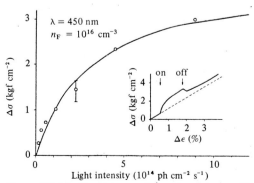

Figure E7.26. Dependence of the photoplastic effect on light intensity for NaCl at 77 K. The inset displays the structure of the effect at this low temperature.

Theoretical considerations

A theoretical model is proposed which takes into account the light-induced formation of an electron cloud around charged moving dislocations. In ionic crystals, moving dislocations created during plastic flow are electrically charged. Many investigators have concluded that dislocations in pure alkali halides at room temperature carry a negative electrical charge. There is some evidence that in irradiated crystals with relatively high defect concentrations the charge on moving dislocations is positive owing to the capture of negative ion vacancies during glide. It is therefore assumed that moving dislocations in coloured crystals become positively charged.

It is assumed that, when light is shone on a straining crystal, electrons are photoionsed from the F-centres. These electrons are then attracted to the dislocation core because of the strong electric fields generated by the charged dislocations (axially symmetric fields of $\sim 10^5$-10^6 V cm^{-1} are to be expected at ~ 1000 Å from the dislocation line). An electron cloud, determined by Boltzmann's law, should then be formed around the dislocation. The free electrons will be finally trapped at some impurities or lattice defects and form a cloud of trapped charge following the same distribution as that of the free electrons. The potential barrier produced by the trapped electron cloud opposes dislocation motion and induces hardening. If the light intensity is high enough, the mobile dislocations will be stopped until the applied stress overcomes the potential barrier and plastic flow can proceed.

Mathematical analysis of the model shows that the increase in the macroscopic flow stress is given by

$$\Delta\sigma = \frac{2^{1/2}}{\pi}\frac{\rho_+^2}{b^2}fF(a) , \qquad (E7.6)$$

where ρ_+ is the linear density of positive charge at the dislocation core, b is the magnitude of the Burger's vector, $f = \rho_-/\rho_+$ is the trapping factor, and $F(a)$ is a factor determined by the shape of the Boltzmann cloud surrounding the dislocation line.

When expressions involving the light intensity are included, the final equation for the photoplastic effect is

$$\Delta\sigma = C_1\left(1 + \frac{C_2 v}{I}\right)^{-1} \qquad (E7.7)$$

where $C_1\,[=(2^{1/2}/\pi)(\rho_+/b^2)F(a)]$ and C_2 are parameters, and v is the dislocation velocity.

Comparison with experiment

Application of equations (E7.6) and (E7.7) to the experimental data produces satisfactory agreement. The experiments show that light induces hardening in all cases, that is, the photoplastic effect is positive for all temperatures whatever the colour centre concentration. Negative effects obtained by Nadeau were probably due to the presence of divalent positive ion impurities, such as Cd^{2+} or Zn^{2+}.

The authors also comment on experiments on light-induced changes in internal friction in x-irradiated NaCl as reported by Bauer and Gordon (1962) and described in Experiment E7.4. The latter attributed the increase in resonant frequency with light excitation to additional pinning of the dislocations caused by the light. The suggestion is made in this paper that the effect may also be due to the back stress generated by the light-induced electron cloud surrounding the dislocations. The saturation observed with increasing illumination time is similar in the two sets of experiments.

References

Alefeld, G., 1965, *Phil. Mag.,* 11, 809.
Alers, G. A., 1955, *Phys. Rev.,* 97, 863.
Alers, G. A., Thompson, D. O., 1961, *J. Appl. Phys.,* 32, 283.
Amelinckx, S., 1964, *Solid State Phys.,* Supp.6, p.58 (Academic Press, New York).
Amelinckx, S., Vennik, J., Remaut, G., 1959, *J. Phys. Chem. Solids,* 11, 170.
Anderson, A. R., Pollard, H. F., 1976, *J. Phys. C.: Solid State Phys.,* 9, 247.
Bauer, C. L., Gordon, R. B., 1962, *J. Appl. Phys.,* 33, 672.
Bielig, G. A., 1971, *J. Appl. Phys.,* 42, 4758.
Bordoni, P. G., 1949, *Ricerca Sci.,* 19, 856.
Bordoni, P. G., 1954, *J. Acoust. Soc. Am.,* 26, 495.
Brown, L. M., 1961, *Phys. Stat. Solidi,* 1, 585.
Dunegan, H., Harris, D., 1969, *Ultrasonics,* 7, 160.
Eshelby, J. D., Newey, C. W. A., Pratt, P. L., Lidiard, A. B., 1958, *Phil. Mag.,* 3, 75.
Fiore, N. F., Bauer, C. L., 1964, *J. Appl. Phys.,* 35, 2242.
Frankl, D. R., 1953, *Phys. Rev.,* 92, 573.
French, I. E., Pollard, H. F., 1970, *J. Phys. C.: Solid State Phys.,* 3, 1866.
Gordon, R. B., 1965, in *Physical Acoustics,* 3B, Ed. W. P. Mason (Academic Press, New York).
Gordon, R. B., Nowick, A. S., 1956, *Acta Metall.,* 4, 514.
Granato, A. V., Lücke, K., 1956a, *J. Appl. Phys.,* 27, 583.
Granato, A. V., Lücke, K., 1956b, *J. Appl. Phys.,* 27, 789.
Granato, A. V., Hikata, A., Lücke, K., 1958, *Acta Met.,* 6, 470.
Granato, A. V., Lücke, K., 1968, in *Physical Acoustics,* 4A, Ed. W. P. Mason (Academic Press, New York).
Granato, A. V., De Klerk, J., Truell, R., 1957, *Phys. Rev.,* 108, 895.
Green, R. E., Hinton, T., 1966, *Trans. Met. Soc. AIME,* 236, 435.
Hayden, H. W., Moffatt, W. G., Wulff, J., 1965, *The Structure and Properties of Materials,* Ed. J. Wulff, Volume 3 *Mechanical Behaviour* (John Wiley, New York).
Hirth, J. P., Lothe, J., 1968, *Theory of Dislocations* (McGraw-Hill, New York), chapter 3.
Hikata, A., Truell, R., Granato, A. V., Chick, B., Lücke, K., 1956, *J. Appl. Phys.,* 27, 396.
Huntington, H. B., Dickey, J. E., Thomson, R., 1955, *Phys. Rev.,* 100, 1117.
James, D. R., Carpenter, S. H., 1971, *J. Appl. Phys.,* 42, 4685.
Kaiser, J., 1953, *Arch. Eisenhüttenw.,* 24, 43.
Koehler, J. S., 1952, in *Imperfections in Nearly Perfect Crystals,* Ed. W. Shockley (John Wiley, New York).
Komnik, S. N., 1968, *Sov. Phys.—Solid State,* 10, 118.
Kovaks, I., 1967, *Acta Met.,* 15, 1731.
Leibfried, G., 1950, *Z. Physik,* 127, 344.
Lücke, K., 1956, *J. Appl. Phys.,* 27, 1433.
Merkulov, L. G., Yakovlev, L. A., 1960, *Sov. Phys.—Acoustics,* 6, 239.
Mott, N. F., 1952, *Phil. Mag.,* 43 (7), 1151.
Nabarro, F. R. N., 1947, *Proc. Phys. Soc.,* 59, 256.
Nabarro, F. R. N., 1967, *The Theory of Crystal Dislocations* (Oxford University Press, Oxford).
Nadeau, J. S., 1964, *J. Appl. Phys.,* 35, 669.
Orowan, E., 1934, *Z. Phys.,* 89, 605, 634.
Paré, V. K., 1961, *J. Appl. Phys.,* 32, 332.
Peierls, R. E., 1940, *Proc. Phys. Soc.,* 52, 23.
Platkov, V. Y., Startsev, V. I., 1967, *Sov. Phys.—Solid State,* 8, 1587.
Polanyi, M., 1934, *Z. Phys.,* 89, 660.

Rayleigh, Lord, 1894, *Theory of Sound* (reprinted by Dover Publications, New York), chapter 6.

Read, T. A., 1940, *Phys. Rev.,* **58**, 371.

Read, T. A., 1941, *Trans. AIME,* **143**, 30.

Robinson, W. H., 1972a, *Phil. Mag.,* **25**, 355.

Robinson, W. H., 1972b, *J. Mat. Sci.,* **7**, 115.

Robinson, W. H., Birnbaum, H. K., 1966, *J. Appl. Phys.,* **37**, 3754.

Schofield, B. H., 1958, WADC Tech. Report 58-194, ASTIA Document AD 155674.

Schofield, B. H., 1963, in *Proceedings of the Fourth Symposium on Physics and Non-destructive Testing* (Southwest Research Institute, San Antonio, Texas), p.63.

Seeger, A., Donth, H., Pfaff, F., 1957, *Discussions Faraday Soc.,* **23**, 19.

Sinclair, J. E., 1971, Ph. D. Thesis, University of New South Wales.

Stephens, R. W. B., Pollock, A. A., 1971, *J. Acoust. Soc. Am.,* **50**, 904.

Stern, R. M., Granato, A. V., 1962, *Acta Metall.,* **10**, 358.

Taylor, G. I., 1934, *Proc. R. Soc.,* **A145**, 362.

Teutonico, L. J., Granato, A. V., Lücke, K., 1964, *J. Appl. Phys.,* **35**, 220.

Thompson, D. O., Holmes, D. K., 1956, *J. Appl. Phys.,* **27**, 713.

Thompson, D. O., Paré, V. K., 1968, in *Physical Acoustics,* **3A**, Ed. W. P. Mason (Academic Press, New York).

Truell, R., Granato, A. V., 1963, *J. Phys. Soc. (Japan),* **18**, Supp.1, 95.

Truell, R., Elbaum, C., Chick, B. B., 1969, *Ultrasonic Methods in Solid State Physics* (Academic Press, New York), chapter 3.

Van Bueren, H., 1961, *Imperfections in Crystals* (North-Holland, Amsterdam).

Whitworth, R. W., 1964, *Phil. Mag.,* **10**, 801.

Whitworth, R. W., 1966, *Phil. Mag.,* **15**, 305.

Whitworth, R. W., 1968, *Phil. Mag.,* **17**, 1207.

Waterman, P. C., 1958, *J. Appl. Phys.,* **29**, 1190.

Yamafuji, K., Bauer, C. L., 1965, *J. Appl. Phys.,* **36**, 3288.

Young, F. W., 1963, *J. Phys. Soc. (Japan),* **18**, Supp.1, 1.

REVIEW QUESTIONS

7.1   How are the terms dislocation line tension, dislocation pinning, and jog defined? (§7.2)

7.2   What experimental evidence is there for dislocations being a cause of ultrasonic damping in crystals? (§7.3)

7.3   What are the assumptions made in formulating the Granato–Lücke model for low-amplitude dislocation damping? (§7.4.1)

7.4   Outline the Granato–Lücke theory for high-amplitude dislocation damping. (§7.4.3)

7.5   Discuss the experimental verification of the high-amplitude Granato–Lücke theory. (§7.4.4)

7.6   Show how the Granato–Lücke high-amplitude theory may be modified to account for time-dependent effects introduced by deformation. (§7.5.2)

7.7   What is meant by a Bordoni relaxation peak? (§7.5.3)

7.8   What is understood by the term acoustic emission? (§7.5.4)

7.9   What effect does a beam of radiation have on a crystal containing free dislocations? (§7.6 and E7.5)

7.10 Discuss the Bauer–Gordon mechanism for dislocation pinning in the alkali halides. (E7.4)

7.11 What evidence is there for dislocations in ionic crystals having a resultant charge per unit length? (§7.7)

7.12 What is the photoplastic effect? On what parameters does it depend? (E7.6)

PROBLEMS

7.13 From the solutions of the Granato–Lücke equations (for equal loop lengths) show that the maximum value of the decrement is given by

$$\Delta_m = \tfrac{1}{2}\Delta_0 \Lambda L^2 \Omega \ ,$$

and that Δ_m occurs at a frequency ω_m given by

$$\omega_m = \frac{\pi^2 C}{BL^2} \ .$$

7.14 Show that a value for the damping parameter, B, in the Granato–Lücke theory, may be found from a determination of α_∞, the value of α when $\omega \gg \omega_m$.

7.15 In the low-frequency approximation of the Granato–Lücke theory, when $\omega \ll \omega_0$, show that

$$\Delta_l \propto L^4 \quad \text{and} \quad \frac{\Delta c}{c_0} \propto L^2 \ .$$

7.16 If a random distribution of loop lengths is assumed (Koehler's distribution function), investigate how this affects the values of the decrement.

7.17 From the expression for high-amplitude decrement [equation (7.35)], show how to find expressions for the dislocation density and the pinning point density.

8

Acoustic visualisation methods

8.1 Introduction

There has always been great interest in the possibility of making sound waves visible. A visual pattern enables the whole field to be comprehended quickly, whereas in normal acoustic measurement procedures the record is obtained over a period of time. There is scientific interest in the problem of rendering complex sound fields visible and the possibility of direct observation of microscopic defects in solids. In engineering there is the need to observe structural flaws and defects in opaque materials, while in medical fields the ability to observe internal organs and tissues without the potential for damage caused by x-rays is very inviting. The recent development of acoustic holographic methods gives the possibility of observing three-dimensional images in real-time which would have many applications in the fields of physics, engineering, geology, medicine, metallurgy, oceanography, etc.

Many of the early methods that were explored involved either mechanical or electronic scanning of the charges developed on a piezoelectric plate caused by an incident ultrasonic beam originating from the object under examination. Sokolov (1941) described an electronic scanning instrument in which an electron beam removed charges from the surface of a quartz piezoelectric plate. A later version of this instrument (called by the author an 'ultrasonic microscope') used secondary electron emission. An incident electron beam was used to scan the quartz receiver and the secondary electrons emitted were collected and fed to a cathode-ray oscilloscope on which an image of the object under examination was formed.

A second method that received much early attention and has subsequently been extensively developed is that involving the deformation of a liquid surface. If an ultrasonic beam is incident on a liquid–air interface from within the liquid, the surface develops a relief pattern which is related to the acoustic pressure distribution. A number of devices have been developed in which this relief pattern may be observed either by reflected or transmitted light.

Another, older, device which is again receiving attention is the Pohlman cell. Pohlman (1939) showed that a suspension of minute platelets of aluminium in xylene, which normally are randomly distributed and therefore have a diffuse appearance, when acted upon by a sound wave behaves like a collection of Rayleigh discs. Each platelet rotates so that it tends to become perpendicular to the direction of the acoustic wave. Specular reflection from these oriented platelets can give rise to an optical image. Other methods investigated have depended on chemical and thermal effects.

Possibly the first demonstration of an acoustic hologram was reported by Kock and Harvey (1951) in an account of a number of ingenious scanning methods for producing visible images of sound fields. They showed that addition of a reference wave allowed phase information to be recorded optically. In this experiment no reconstruction of the original object was attempted. Although reconstruction is a relatively straightforward process in optical holography, there is no acoustical counterpart of the photographic plate, lack of which causes serious practical problems. Some progress has been made with image reconstruction by computer.

8.2 Bragg diffraction

Whilst studying the patterns of Laue spots produced when a beam of monochromatic x-rays is diffracted by a single crystal, Bragg (1913) derived a simple equation for the angular positions of the spots. Instead of treating the complicated diffraction effects produced by the individual atoms in the crystal, Bragg found that the positions of the observed spots, on a photographic film, could be predicted by a simple interference equation. As shown in figure 8.1, owing to the periodicity of the crystal lattice the diffracted x-rays appear to be reflected from successive planes of atoms. The path difference for rays 'reflected' from adjacent crystal planes is $2d \sin \theta$, where d is the spacing between planes and θ is the angle of incidence measured from the plane. Whenever this path difference is an integral multiple of the wavelength λ, constructive interference of the reflected radiation occurs. The Bragg condition for reinforcement (Bragg's law) is

$$2d \sin \theta = n\lambda . \tag{8.1}$$

A similar phenomenon to the above occurs with the interaction between a light beam and a sound wave. As we have already observed in section 5.5, a travelling sound wave can act as a phase diffraction grating to a transverse beam of light, giving rise to a series of diffraction patterns of different orders on either side of a central zero-order beam. At high ultrasonic frequencies, only the zero-order and first-order diffracted light

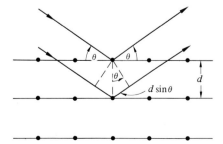

Figure 8.1. Imaginary atomic reflecting planes for the derivation of Bragg's law.

L

beams are observed. If the interaction between the light and the ultrasonic
wave is regarded as a collision between a photon and a phonon, then
conservation of momentum gives the condition

$$k_{inc} + q = k_{scat}.$$ (8.2)

where k_{inc} and k_{scat} are the wave vectors of the incident and scattered
phonons, and q is the wave vector of the phonon. Equation (8.2) may be
shown in diagram form, as in figure 8.2, where it has been assumed that
there is negligible change in the magnitude of the light vector. From
figure 8.2 it follows that if $|k_{inc}| = |k_{scat}| = k$, then

$$2k \sin\theta = q .$$

That is,

$$2\lambda^* \sin\theta = \lambda ,$$ (8.3)

where λ is the light wavelength and λ^* the sound wavelength. Note the
similarity between equations (8.1) and (8.3).

With a travelling ultrasonic wave the phase grating is moving with the
velocity of sound giving rise to a Doppler shift in the frequency of the
scattered light. Conservation of energy implies that

$$\hbar \omega_{scat} = \hbar \omega_{inc} + \hbar \omega^* .$$

That is,

$$\omega_{scat} = \omega_{inc} \pm \omega^* ,$$ (8.4)

where ω_{inc} and ω_{scat} are the frequencies of the incident and scattered
light and ω^* is the frequency of the ultrasonic wave. However, the
Doppler shift is normally very small since the ratio of light to sound
velocities is of the order of 10^5. Imaging methods based on Bragg
diffraction have been used to observe sound fields in liquids up to
frequencies of 50 MHz. At higher frequencies the absorption in liquids
becomes excessive. Solids may be used up to frequencies of the order of
1 GHz. It is desirable to use as high a frequency as is practicable since
the image resolution is directly proportional to the sound frequency.

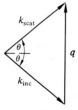

Figure 8.2. Condition governing photon-phonon collision. k_{inc} and k_{scat} are vectors
representing the incident and scattered light respectively, q is a vector representing the
phonon.

A typical Bragg-diffraction imaging system is shown in figure 8.3. The laser beam is expanded and then converted into a line source by a cylindrical lens. The light beam passes into the acoustic cell which contains a suitable liquid, such as water. Some of the light is scattered by the ultrasonic waves, the required image being received by the vidicon tube. The lateral magnification of the image is given by the ratio of the light wavelength to the sound wavelength (usually of the order of 1/100) but the vertical magnification (out of the plane of the paper) is unity. Hence, additional lenses are required to correct the resulting distortion.

With such a system it is possible to observe any parts of an object that absorb or scatter the sound waves, such as faults or holes in a metal plate, or the internal structure of biological objects (Landry et al., 1969). With solid specimens in the form of a plate, the imaging property of this method allows observation of any inhomogeneities in the material, while measurements may be made, as a function of sound frequency, of both the compressional and shear wave velocities in the material (Powers et al., 1972). Also, the attenuation of the sound waves can be determined from the intensity of the diffracted light beam at various points along the sound path.

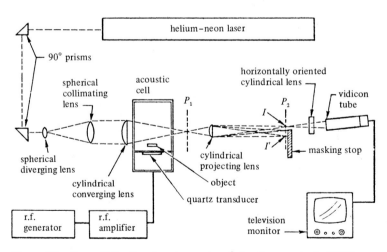

Figure 8.3. Bragg-diffraction imaging system (after Powers et al., 1972).

Experiment E8.1 Visualisation of high frequency sound waves
(Torguet, R., Carles, C., Rouvaen, J. M., Bridoux, E., Moriamez, M., 1974, *J. Appl. Phys.,* **45**, 1574.)

A Bragg-diffraction system is described which employs heterodyne detection and is sensitive to both the amplitude and phase of the sound waves. A fast two-dimensional display may be obtained over an area of 1 to 3 cm^2 of the sound field. The experimental arrangement is shown in figure E8.1. From the laser two beams are

obtained by means of a beam splitter. The beams impinge on a lead molybdate
flying-spot scanner and are steered by a time-varying angle θ. The beams are then
made to converge at a point in the sound field which is located in the focal plane of a
high-quality optical lens. The convergent beams make a constant angle $2\theta_B$, for any
θ, where

$$2\theta_B = \frac{d}{f} = \frac{\lambda}{\lambda_*} \, ,$$

f is the focal length of the lens, d is the distance between the two parallel light beams
impinging on the lens, λ is the light wavelength (6328 Å), and λ_* is the sound
wavelength. Optical heterodyning is then feasible if the scattered light is mixed with
part of the undeviated light (provided both beams are collinear).

As shown in the theoretical analysis below, the output of the photodetector is an
electrical current at the sound frequency. The desired phase information concerning
the sound wave is obtained by further heterodyning the detector output with part of
the driving voltage applied to the transducer (lithium niobate).

Two techniques may be used to display the entire width of the sound beam: (a) the
specimen may be moved mechanically in a vertical direction, slowly with respect to the
sweep speed of the laser flying-spot scanner; a potentiometer, coupled to the vertical
motion, provides a direct voltage which is applied to the vertical plates of a storage
oscilloscope; (b) the light beams may be scanned vertically with a slowly rotating
mirror inserted between the lens and the specimen.

With both methods the heterodyned signal is amplified and applied to the
oscilloscope for Z modulation. The horizontal plates of the oscilloscope are triggered
by the flying-spot scanner. In figure E8.2 are shown wavefront scans taken at 200 MHz
in a lead molybdate crystal.

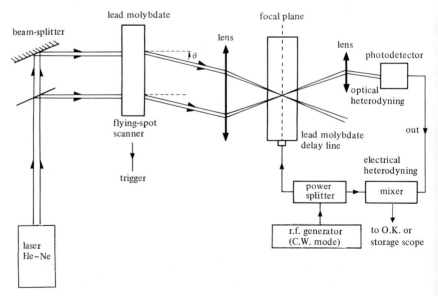

Figure E8.1. Experimental arrangement for visualisation of high-frequency sound by
Bragg diffraction.

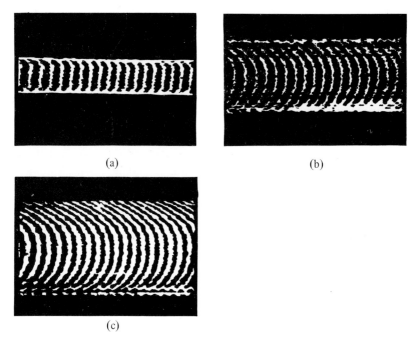

(a)   (b)

(c)

Figure E8.2. Wavefront scans at 200 MHz in a lead molybdate crystal specimen (a) near the lithium niobate transducer, (b) in the middle of the crystal specimen, (c) near the opposite face of the crystal specimen.

Theoretical analysis

The electric field of the reference light wave may be written

$$e = E \exp[i(\omega t - k \cdot r)] \ . \tag{E8.1}$$

The electric field of the diffracted light beam is

$$e_1 = h\Delta \exp\{i[(\omega + \omega^*)t - (k + k^*)r]\} \tag{E8.2}$$

where ω and k refer to the light wave, ω^* and k^* to the sound wave, h is a factor proportional to the square root of the acoustic power and to the effective photoelastic constant, Δ is the relative change in the dielectric constant of the crystal modulated by the sound wave, and r is the spatial coordinate denoting the point in the acoustic path at which the probing is done.

When the two beams are mixed at the photodetector, the output photocurrent is

$$i = \eta E^2[1 + h^2\Delta^2 + 2h\Delta \cos(\omega^* t - k^* \cdot r)] \tag{E8.3}$$

where η is a conversion efficiency factor for the photodetector.

The first two DC terms are filtered out and the AC part applied to a heterodyne receiver with input impedance $R = 50 \ \Omega$. The reference signal is a small part of the transducer driving voltage, $v = V \exp(i\omega^* t)$. The signal at the mixer is then proportional to

$$[V^2 + (2RE^2 h\Delta)^2 + 4VRE^2 h\Delta \cos(k^* \cdot r)]^{1/2} \ .$$

The first two terms act as a constant bias for the mixer diode. The rectified output is then proportional to $E^2 h\Delta \cos(k^* \cdot r)$ and consequently carries information concerning the amplitude and phase of the sound wave at the probed point in the sound field.

8.3 Brillouin scattering

A laser light beam, incident on a transparent liquid or solid, may be scattered by the thermal sound waves (phonons). The phonon spectrum ranges from low frequencies, of the order of 1000 Hz, up to frequencies of the order of 10^{13} Hz. (The lowest frequency and hence the longest wavelength corresponds approximately with twice the greatest linear dimension of the body.) The technique is suitable for measurements at frequencies above 1 GHz and does not require the use of an ultrasonic transducer. As shown in figure 8.4a, the light waves may be considered to be 'reflected' from the thermal waves as in Bragg diffraction. A similar equation then holds as the condition for constructive reinforcement,

$$2\lambda^* \sin\theta = \lambda , \tag{8.5}$$

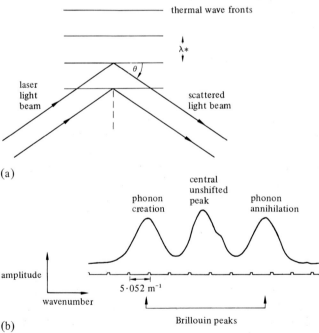

(a)

(b)

Figure 8.4. (a) Brillouin scattering of light by thermal sound waves (phonons). $\lambda*$ is the sound wavelength. (b) Spectrum of light (6328 Å) scattered from water. The central peak at the laser frequency is due to scattering from tiny particles suspended in the water. The line width is governed by the slit width of the spectrograph. The phonon frequency was determined to be 4·33 GHz. The sound velocity was calculated to be 1457 m s⁻¹ (after Benedek et al., 1964).

where θ is half the angle between the incident and scattered beams and λ is the light wavelength in the material. Thus, despite the enormous number of thermal waves present in the material, only waves satisfying the Bragg condition interact with the light beam. The interacting thermal waves must have the appropriate wavelength λ^*, and have a direction of propagation parallel to the bisector of the angle between the incident and scattered light beams. The smallest sound wavelength is observed under back-scattering conditions ($\theta = \tfrac{1}{2}\pi$) when $\lambda^* = \tfrac{1}{2}\lambda$.

Since the thermal waves (phonons) move with the velocity of sound for the material, there will be a Doppler frequency shift in the scattered light, $\Delta\nu = \nu' - \nu$, where ν' is the frequency of the scattered light:

$$\Delta\nu = \pm 2\nu \frac{c^*}{c} \sin\theta , \qquad (8.6)$$

where ν is the incident light frequency, c is the velocity of light in the material, $c = c_0/n$, where c_0 is the velocity of light in a vacuum, n is the refractive index of the material, c^* is the velocity of sound in the material.

Combining equations (8.5) and (8.6) we obtain

$$\Delta\nu = \pm \nu \frac{c^*}{c} \frac{\lambda}{\lambda^*} = \pm \nu^* . \qquad (8.7)$$

Two spectrum lines may therefore be expected, as originally predicted by Brillouin (1922). The Doppler frequency shift is simply equal to the phonon frequency ν^*.

The scattered light may be analysed by a high-resolution interferometer, such as a Fabry–Perot interferometer. A typical spectrum is shown in figure 8.4b. A strong central line of frequency ν is observed, caused by Rayleigh-type scattering from imperfections, on either side of which appear the two Brillouin components. Measurement of the frequency separation between the two Brillouin lines provides a means of determining the sound velocity from equation (8.6). The attenuation coefficient of the thermal wave may be found by measuring the bandwidth of the Brillouin lines.

8.4 Optical holography

In its original form, holography may be regarded as a technique of lens-less photography using coherent light. Waves scattered by an object are recorded on a photographic plate together with a reference wave. An image suitable for viewing may be formed by illuminating the processed plate with the reference wave alone. The image so formed retains all the information in the original beam from the object, that is, it is three-dimensional and may be in full colour. Holographic images may also be formed using sound waves, although a serious practical problem is the absence of the acoustical equivalent of the photographic plate.

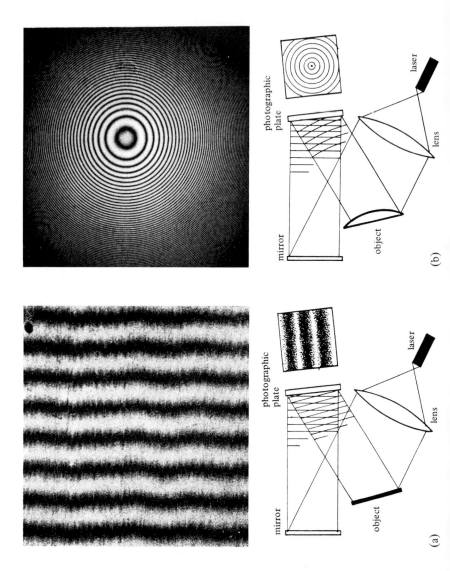

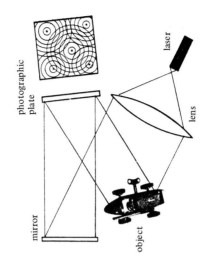

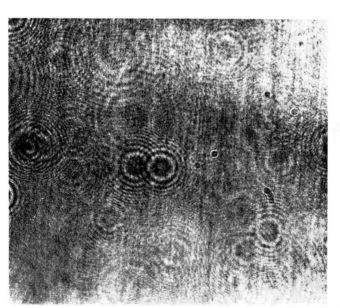

(c)

Figure 8.5. Holograms formed by (a) reflection from a plane mirror, (b) reflection from a spherical mirror, (c) scattering from an irregular object. (After Stroke, 1965.)

The principles of optical holography were enunciated by Gabor (1948, 1949). He introduced the term hologram, meaning complete picture, since both amplitude and phase are recorded. In optics, because of the high vibration frequencies involved, it is not possible to measure the phase of a light wave directly. In recording light, most detectors, such as a photoelectric cell or photographic plate, respond only to the light intensity and not to the phase. The only way that phase information can be recorded is to use coherent light and an interference technique. Gabor showed that all the information in a light beam could be recorded on a photographic plate as an interference intensity pattern, provided a coherent reference wave was superimposed on the light scattered by the object. The photographic image will then consist of a set of interference fringes, as shown schematically in figure 8.5. If the object is a plane mirror, as in figure 8.5a, then plane waves reflected by the mirror will interfere with plane waves from the reference beam to produce a series of straight interference fringes on the photographic plate. If the object is a spherical mirror, as in figure 8.5b, the interference pattern between spherical and plane waves is a series of concentric circles having the same form as a zone plate. From a real object, the scattered light will contain plane, spherical and other elements which will give rise to a complicated pattern, as in figure 8.5c.

Each point on the object forms its own zone plate by interference of the scattered light with the reference wave. When each zone plate is illuminated with laser light an image is reconstructed exactly at the point in space from which the original light came. Each point of a three-dimensional scene may therefore be reconstructed in sharp focus irrespective of its distance from the photographic plate.

Another important property of a hologram is that each portion contains information pertaining to the entire object irradiated. Thus, a small portion of a hologram may be used to represent the object but with loss of resolution determined by the aperture appropriate to the portion used. Holography is a technique for 'freezing' the incident light for reconstruction at any future time.

Gabor produced his original holograms using a filtered mercury source. However, such a source has a coherence length of only about $0 \cdot 1$ mm, that is, interference effects with mercury light may only be observed over this distance because of the short coherent wave-trains emitted. With the advent of the laser, coherence lengths of several metres became available. In addition, a laser produces a greater light intensity. Useful introductions to optical holography may be found in the review papers by Gabor and Stroke (1969) and by Gabor (1969). The former paper contains a number of coloured illustrations.

8.4.1 Simple theory of optical holography

The following simple theory of optical holography is given by Gabor (1969). Denote the complex amplitude of the reference wave by $\hat{A}(x,y,z) = |A|\exp i\phi_A$ and that of the object wave by $\hat{B}(x,y,z) = |B|\exp i\phi_B$. The combined amplitude is then $(\hat{A}+\hat{B})$. The time factor $\exp(i\omega t)$ is omitted as it cannot be observed optically.

The photographic plate records the combined intensity I, where

$$I = |A+B|^2 = (\hat{A}+\hat{B})(\hat{A}*+\hat{B}*) . \tag{8.8}$$

$\hat{A}*$ and $\hat{B}*$ are complex conjugates. Equation (8.8) may be written

$$I = (\hat{A}\hat{A}*+\hat{B}\hat{B}*)+(\hat{A}\hat{B}*+\hat{A}*\hat{B}) . \tag{8.9}$$

The first term in equation (8.9) is the sum of the intensities of waves A and B ($I_A = \hat{A}\hat{A}* = |A|^2$; $I_B = \hat{B}\hat{B}* = |B|^2$); the second term contains the interference effects. Since

$$\hat{A}\hat{B}* = |A||B|\exp[i(\phi_A-\phi_B)] = |A||B|\exp[-i(\phi_B-\phi_A)]$$

and

$$\hat{A}*\hat{B} = |A||B|\exp[i(\phi_B-\phi_A)] ,$$

equation (8.9) may be written

$$I = I_A+I_B+2|A||B|\cos(\phi_B-\phi_A) . \tag{8.10}$$

The final term shows clearly that the photographic plate now contains both amplitude and phase information.

To recover the object wave, the plate is now processed and illuminated with the reference wave A. For simplicity assume that the amplitude transmission coefficient of the plate, T, is proportional to I. The intensity transmission coefficient is then proportional to I^2, that is, the processing must be done with a gamma of -2 (this condition can be circumvented in practice). The transmitted amplitude behind the plate is given by

$$T\hat{A} \propto I\hat{A} = \underset{\substack{\text{illuminating}\\\text{wave}}}{\hat{A}(\hat{A}\hat{A}*+\hat{B}\hat{B}*)}+\underset{\substack{\text{twin}\\\text{wave}}}{\hat{A}^2\hat{B}*}+\underset{\substack{\text{reconstructed}\\\text{wave}}}{(\hat{A}\hat{A}*)\hat{B}} . \tag{8.11}$$

If the intensity $\hat{A}\hat{A}*$ of the reference wave A is uniform over the plate, then the last term of equation (8.11) represents a reconstruction of the object wave B immediately behind the photographic plate. By Huygens's principle, this condition should hold everywhere. Hence, on viewing the reconstructed wave field, the object appears to be in its original position. Further, all the properties of the original wave field are present, so that if the observer changes position, the object perspective will change. In other words, a full three-dimensional image may be viewed.

The pattern of lines and curves on the plate behaves like a diffraction grating to the reference wave. The reconstructed wave forms a virtual image of the object, as shown in figure 8.6. The twin, or conjugate, wave gives rise to a real image which is usually discarded.

There is now a variety of methods for producing holograms and numerous fields in which the technique has been applied. Applications include vibration analysis, pattern recognition, nondestructive testing, holographic microscopy, information storage, and the range of acoustical holography techniques.

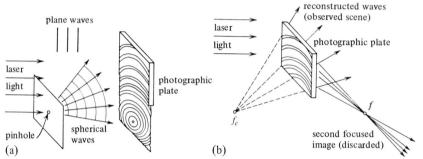

Figure 8.6. (a) Formation of a hologram from a point source; (b) reconstruction process giving a virtual image at the location of the original source and a real image that is discarded (after Kock, 1971).

8.5 Acoustical holography

In experiments described by Kock and Harvey (1951) acoustic holograms were produced by scanning the sound field in front of a sound source. The scanner consisted of a small black microphone to which was attached a neon lamp. The brightness of the light emitted by the lamp was proportional to the sound intensity. The microphone–lamp combination could be moved in vertical circular arcs and at the same time moved slowly away from the sound source. A camera, set for time exposure and placed at right angles to the direction of motion of the scanner, then recorded a light pattern which represented the sound field under examination. If the sound source is supplied with a signal from an oscillator and at the same time part of the oscillator output is added to the microphone signal, an acoustical fringe pattern is recorded, as shown in figure 8.7. At positions where the scanner is an integral number of wavelengths away from the source, the two signals will be in phase and the lamp will have maximum brightness. At the half-wavelength points the signals will be out of phase and the lamp brightness will be a minimum. It is noteworthy that in this acoustical analog experiment the reference signal is electronic. Acoustic holograms can be produced both with an acoustic reference signal and an electronic reference signal. The electronic signal plays the same role as the plane wave reference in the optical case and hence the resultant hologram is the equivalent of the optical zone plate produced by the interference of plane and spherical waves. Further details of these acoustic scanning methods may be found in Kock (1971).

The interest at the time of these experiments was in mapping sound fields and producing a visual display. A somewhat different approach is needed if the hologram is to be reconstructed in order to provide an image of the acoustical object.

Figure 8.7. Scanned sound field from a telephone receiver. A 4000 Hz signal was both fed to the telephone and used as a reference signal. The photograph then shows the actual wavefronts from the source (after Kock, 1971).

8.5.1 Liquid surface holograms
Mueller and Sheridon (1966) reported the production of an acoustical hologram on a water surface and its optical reconstruction. As shown in figure 8.8, the object under examination is irradiated by a diverging ultrasonic beam originating from a quartz crystal vibrating at 7 MHz. The waves scattered by the object and a coherent reference beam interfere at the surface of the liquid to form a stationary pattern. The hologram so formed may be photographed and the photograph reconstructed optically, or optical reconstruction may be achieved in real-time by illuminating the surface with laser light.

8.5.1.1 Theoretical analysis of liquid surface hologram
Following the analysis given by Brenden (1972), consider two ultrasonic beams incident on the surface at equal but opposite angles θ. The surface is taken to be in the x–y plane with the origin of coordinates at the surface. The waves leaving the two transducers may be written in terms of the acoustic pressure (omitting the common time factor)

$$p_A(x, y, z) = P_A \exp[i(k_y y + k_z z)] \quad \text{(reference wave)} \qquad (8.12)$$

$$p_B(x, y, z) = P_B \exp[i(-k_y y + k_z z)] \quad \text{(object wave)} \qquad (8.13)$$

where

$$k_y = \frac{2\pi}{\lambda*}\sin\theta \; ; \qquad k_z = \frac{2\pi}{\lambda*}\cos\theta \, ,$$

and $\lambda*$ is the acoustic wavelength.

At the surface ($z = 0$) the expression for p_B needs to be modified. Diffraction around the object causes x and y variations in the pressure amplitude P_B and a phase variation, $\phi(x, y)$, in the irradiated area. Thus, the appropriate equations at the surface are

$$p_A = P_A \exp(ik_y y) \quad \text{(reference wave)} \tag{8.14}$$

$$p_B = P_B(x, y)\exp[-ik_y y - i\phi(x, y)] \quad \text{(object wave)} . \tag{8.15}$$

Interference of these two waves at the surface produces an intensity distribution given by

$$I(x, y) = \frac{|p_A + p_B|^2}{2\rho c} . \tag{8.16}$$

At the surface, the change in momentum of the waves on reflection produces a pressure, the radiation pressure Π, which tends to elevate the surface. This radiation pressure is given by

$$\Pi(x, y) = \frac{2I}{c} . \tag{8.17}$$

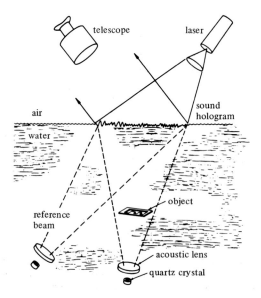

Figure 8.8. Method for producing a liquid surface hologram. Reconstruction may be in real-time, as shown, by illuminating the surface with laser light (after Mueller and Sheridon, 1966; Kock, 1971).

Pressures due to gravity and surface tension oppose the radiation pressure, so that for a unit area the force equation is

$$\Pi(x, y) = \rho g \zeta - \gamma \left(\frac{\partial^2 \zeta}{\partial y^2} + \frac{\partial^2 \zeta}{\partial x^2} \right), \tag{8.18}$$

where γ is the surface tension coefficient, ζ is the displacement in the z direction. When the radiation pressure is written in terms of the wave pressures, a solution of equation (8.18) is found to be

$$\zeta(x, y) = 2A \cos[2k_y y + \phi(x, y)] + B \tag{8.19}$$

where

$$A(x, y) = \frac{P_A P_B(x, y)}{\rho c^2 (\rho g + 4\gamma k_y^2)}, \tag{8.20}$$

$$B(x, y) = \frac{P_A^2 + [P_B(x, y)]^2}{\rho^2 c^2 g}. \tag{8.21}$$

The solution is valid only if

$$\frac{\partial^2 A}{\partial y^2}, \quad \frac{\partial^2 A}{\partial x^2} \ll A \text{ (or } B),$$

$$\frac{\partial^2 B}{\partial y^2}, \quad \frac{\partial^2 B}{\partial x^2} \ll B \text{ (or } A),$$

$$\frac{\partial^2 \phi}{\partial y^2} + \frac{\partial^2 \phi}{\partial x^2} \ll 4\gamma k_y^2.$$

Equation (8.19) represents the surface hologram. The surface has a configuration shown in simplified form in figure 8.9.

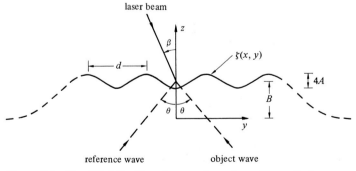

Figure 8.9. Simplified liquid surface configuration. The radiation pressure causes an overall bulge B and a ripple of amplitude $2A$. Normally $A \ll d$ (the ripple spacing). Superimposed on the ripples is a small oscillation at the ultrasound frequency.

8.5.1.2 Interaction of light with surface hologram

Consider a beam of coherent light of amplitude C incident on the surface at an angle β as shown in figure 8.9 and represented by

$$U_{\text{inc}}(y, z) = C \exp[i(\kappa_y y - \kappa_z z)] , \tag{8.22}$$

where

$$\kappa_y = \frac{2\pi}{\lambda} \sin\beta , \qquad \kappa_z = \frac{2\pi}{\lambda} \cos\beta ,$$

λ is the light wavelength.

After reflection the amplitude of the wave is reduced to D and there is a phase shift due to the change of path length caused by the surface ripples. For small values of the elevation $\zeta(x, y)$, the change of path length is approximately $2\zeta(x, y)$. Hence, the reflected wave may be represented by

$$U_{\text{refl}}(x, y, z) = D \exp[i\{\kappa_y y + \kappa_z[z + 2\zeta(x, y)]\}] , \tag{8.23}$$

or, with the aid of equation (8.19),

$$U_{\text{refl}}(x, y, z) = D \exp(2i\kappa_z B) \exp\{4i\kappa_z A \cos[2\kappa_y y + \phi(x, y)]\}$$

$$\times \exp[i(\kappa_y y + \kappa_z z)] . \tag{8.24}$$

Use is now made of the identity

$$\exp(i\sigma \cos\alpha) \equiv \sum_{n=-\infty}^{\infty} i^n J_n(\sigma) \exp(-in\alpha) \tag{8.25}$$

where $J_n(\sigma)$ is the nth order Bessel function. Let

$$\sigma = 4\kappa_z A(x, y)$$

and

$$\alpha = 2\kappa_y y + \phi(x, y) .$$

Equation (8.24) becomes

$$U_{\text{refl}}(x, y) = D \exp\{i\kappa_z[z + 2B(x, y)]\} \sum_{n=-\infty}^{\infty} i^n J_n[4\kappa_z A(x, y)]$$

$$\times \exp\{i[(\kappa_y - 2nk_y)y - n\phi(x, y)]\} . \tag{8.26}$$

Each value of n corresponds to a diffracted order of light. When $4\kappa_z A(x, y)$ is sufficiently small, J_0 approaches $1 \cdot 0$, J_{+1} approaches $2\kappa_z A(x, y)$, J_{-1} approaches $2\kappa_z A(x, y)$, and J_n for n not equal to -1, 0, or 1 approaches 0. Hence, when $4\kappa_z A(x, y)$ is small,

$$U_{\text{refl}}(x, y) = D \exp\{i\kappa_z[z + 2B(x, y)]\}[\exp(i\kappa_y y)$$

$$+ 2i\kappa_z A(x, y) \exp\{i[(\kappa_y - 2k_y)y - \phi(x, y)]\}$$

$$+ 2i\kappa_z A(x, y) \exp\{i[(\kappa_y + 2k_y)y + \phi(x, y)]\}]$$

$$= U_0(x, y) + iU_{+1}(x, y) + iU_{-1}(x, y) . \tag{8.27}$$

Comparison of the U_{+1} term with equation (8.15) shows that (a) the reflected light beam contains the same phase information as the ultrasonic object wave; and (b) since $A(x, y)$ is proportional to $P_B(x, y)$ through equation (8.20), the amplitude information is also preserved. The first-order light beam is therefore a scaled replica of the ultrasonic object wave and may be used to form a virtual image of the object. The U_{-1} term similarly gives rise to a real image.

Thus, an acoustical field may be transformed into an optical field by the liquid surface acting as a real-time hologram. The above analysis has assumed continuous waves. In practice the ultrasonic beam is often in the form of pulses of approximately 80 μs duration repeated every 2 or 3 ms. In some systems two steps are introduced: the liquid surface is first of all photographed and then laser reconstruction is carried out at a later time, the disadvantage of this procedure being the loss of the real-time ability.

A problem with any system involving sound–light conversion is the change of scale caused by the quite different wavelengths. It is advisable to use as high a frequency as possible since resolution depends on the acoustical wavelength. At frequencies in the megahertz range the ratio $\lambda/\lambda*$ is of the order of 10^{-3}. The reconstructed optical image is therefore very small. If attempts are made to enlarge the image, further problems arise. The optical radial magnification is proportional to the square of the lateral magnification and to $\lambda*/\lambda$. So an enlarged image will be greatly distorted in the radial direction.

8.5.2 Scanning methods

The basic method involves scanning a plane in which a hologram has been formed, and converting the resultant signal into light so that a photographic record may be made. The intensity of the optical image may be represented by an equation similar to equation (8.10). Electrical filtering may be used to eliminate the first two terms so that only the image-forming term remains. As shown by Brenden (1972), holograms may be reconstructed either by receiver scanning or by source scanning. Also, the source and receiver may be scanned simultaneously and with different velocity resulting in an increase in resolution.

A number of different types of scanning have been used including mechanical, electronic, laser beam, and multiple beam. Most scanning systems are two-stage systems; a hologram is first formed on photographic film and then reconstructed with laser light. Some success has been achieved with reconstruction via a computer.

In mechanical scanning, a transducer is moved systematically through the sound field and the signal used either to modulate a light source, as in Kock's original system, or to control the moving spot of a cathode-ray oscilloscope. In both cases an optical image can be built up over a period of time. The reference beam is usually an electronic signal of the same frequency as the source signal and simulates an on-axis plane wave.

If a reference signal of slightly different frequency is used, an off-axis reference wave is simulated. Since the sampling points in scanning are detected at different times, the frequency difference between source and reference waves is interpreted as a phase difference as occurs with an off-axis reference beam. Electrical signal conditioning may be used to improve the image; for instance filtering can be used to eliminate the conjugate image. Good mechanical and thermal stability is required because of the long scanning times.

In electronic scanning, the hologram is in the form of a charge pattern on a piezoelectric plate. An electron beam scans the plate producing secondary emission which can be recorded directly on film or used to activate a cathode-ray tube. This system provides increased sensitivity and speed of scanning. There is, however, the need to limit the angle of acceptance to a few degrees from normal incidence and the need to operate the transducer at a resonant frequency.

Multiple-beam scanning has been used for underwater holography. The combined transmitter/receiver consists of a square array of transducers, up to 100 elements on a side, and positioned half a wavelength apart. A coherent signal is transmitted by the array with a similar signal serving as reference signal to each element of the array. Waves reflected or scattered by an object are received by the array and form a holographic pattern. Each element of the array provides a sample of this pattern. One form of reconstruction, described by Kock (1973), uses a thick DKDP (deuterated potassium dihydrogen phosphate) crystal, a scanning electron gun, and associated equipment. The incoming holographic information modulates the beam of the electron gun which in turn forms a positive charge pattern as it scans over the surface of the crystal. The electric field within the crystal thus varies according to the hologram and modulates the refractive index of the crystal. Coherent light transmitted through the crystal then reconstructs the object. The hologram can be readily erased by flooding the crystal with electrons from the gun. The real-time system is capable of following a moving object and consequently a movie could be made from the reconstructions.

Experiment E8.2 Acoustic holograms via the Pohlman cell
(Lafferty, A. J., Stephens, R. W. B., 1971, *Opt. Laser Technol.*, **4**, 232.)

In the search for the acoustical equivalent of the optical photographic plate a possible contender is the Pohlman cell. In this cell a near-colloidal suspension of aluminium particles in xylene is held between two plates, one of which is transparent to sound and the other is an optical window. With no incident sound, the aluminium particles are randomly oriented and present a uniform appearance through the optical window. With the sound field present, the platelets of aluminium behave as Rayleigh discs and tend to reorient normal to the direction of incidence of the sound. Reflected light then takes on the intensity distribution of the incident sound field.

The couple acting on a disc in the sound field is given by

$$L = -\tfrac{2}{3}\rho_0 d^3 |v|^2 f(\rho_0/\rho_1) \sin 2\beta \,, \tag{E8.4}$$

where ρ_0 and ρ_1 are the densities of the xylene and aluminium, v is the particle velocity, d is the particle size, and β is the angle between the normal to the face of the particle and the direction of the incident sound.

The time taken for the formation of an image can be long and increases as the acoustic power decreases. At a power of 10^{-3} W cm^{-2} this time is approximately $0\cdot5$ s. The resolution of the cell is a function only of its aperture and the sound wavelength. The ultimate resolution limit depends on the size of the aluminium particles.

The system used for recording the acoustic hologram is shown in figure E8.3. Since continuous waves are used, the apparatus is placed in an anechoic tank, the sides of which are lined with dimpled rubber matting and the ends covered by a specially designed aluminium-loaded butyl rubber. The Pohlman cell stands on a small table inside the tank and is illuminated through a window cut in the side of the tank. The cell may be illuminated either by a mercury source or by a white light projector. The object and reference transducers are a pair of PZT ceramics operating at 1 MHz and mounted on carriages which allow for lateral and rotational movement. The brass Pohlman cell has a circular aperture 10 cm in diameter and its front face has a thin Mylar membrane stretched across it.

In the system described, the hologram was recorded on fast film and reduced by about one-tenth in size because of the differences between the light and sound wavelengths. The recorded hologram was reconstructed with a He–Ne laser (6328 Å) with output power of about 15 mW. Real-time reconstruction should be possible with the aid of a Pohlman cell.

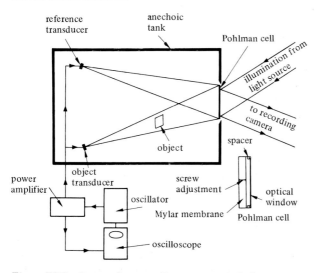

Figure E8.3. System for recording an acoustic hologram using a Pohlman cell.

Experiment E8.3 Vibration analysis by hologram interferometry
(Agren, C-H., Stetson, K. A., 1972, *J. Acoust. Soc. Am.,* **51**, 1971.)

Methods used for vibration analysis by holography depend on the interference of nonsimultaneous light waves. Suppose that holograms are taken of two waves, B and C, in succession using the same reference wave A. Then the recorded intensity, as in equation (8.9), will include the terms $\hat{A}^*\hat{B}$ and $\hat{A}^*\hat{C}$ which are additive. On illuminating the hologram with the reference wave A, the reconstructed wave, as in equation (8.11), will include the term $(\hat{A}\hat{A}^*)(\hat{B}+\hat{C})$. Hence, although the waves B and C were recorded at different times, they will be reconstructed together. An interference fringe pattern may then be formed between the two waves which will reveal any differences in displacement that may have occurred between the two objects.

Two methods have been developed for the study of a vibrating object. In the **time-average method** (Powell and Stetson, 1965) the time average of the waves scattered by a vibrating object is recorded over several seconds. The hologram then consists of the superposition of a number of different holograms corresponding to different phases of the vibrating cycle. In the reconstruction of such a hologram fringes appear that approximate to interference between two images of the object one at each extreme of the object's motion.

In the second method, sometimes called **real-time interferometry** and developed by Haines and Hildebrand (1966), a hologram is first made from the static object. This hologram is now reconstructed in such a way that the reconstructed image coincides exactly with the object. A new hologram is then recorded using the first hologram as reference wave. If the object vibrates or is deformed the holograms will superimpose forming an interference pattern. Thus the second reconstructed image of the object will have superimposed on it interference fringes due to the deformation of the object. Deformation amplitudes may be measured to half a wavelength of the laser light.

Agren and Stetson used both of these methods to study the vibrational properties of a treble viol. The classical treble viol has a weak and uneven tone and has not been the subject of such intensive development as occurred with the violin. The treble viol is a convenient size for study by holography and this paper is the first reported application of this technique to the study of string instruments. The purpose of the investigation was to measure the behaviour of the instrument and to use the results to design an improved version. Because of its construction it was considered satisfactory to study only the vibrations of a clamped top plate. The treble viol is a six-string bowed instrument, played vertically while held on or between the player's knees.

The viol top plate was clamped in a heavy wooden jig by means of clips along the edges and was excited magnetically through the bass bar. The vibrating plate was illuminated with light from a laser, part of the output of which acted as a reference beam. Scattered light from the plate and the reference beam were recorded together on a photographic plate with an exposure time of a few seconds. The procedure adopted was to use real-time interferometry to locate the various response peak frequencies of the vibrating viol plate. Time-average holograms were then made at each of these frequencies; figure E8.4 shows seven such holograms. Since the wooden viol plate tends to absorb moisture, it was found that the real-time fringes followed any deformation of the plate due to this cause.

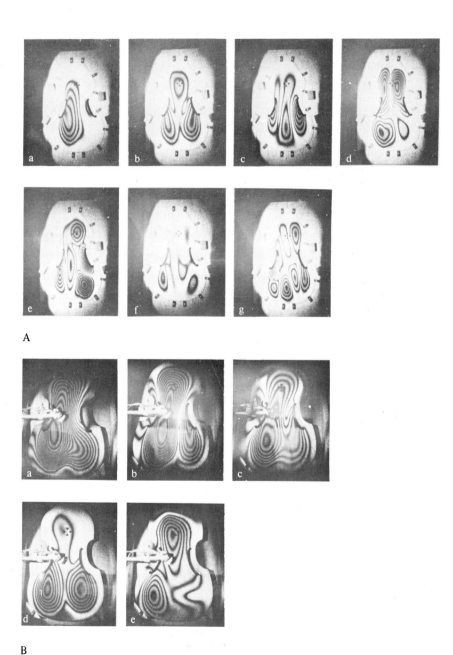

A

B

Figure E8.4. A. Seven hologram reconstructions of the classical treble viol plate vibrating at its frequencies of peak response (a) 484 Hz, (b) 628 Hz, (c) 760 Hz, (d) 878 Hz, (e) 960 Hz, (f) 1118 Hz, (g) 1408 Hz. B. Five reconstructions for response peaks of the assembled magnum viol top plate (a) 251 Hz, (b) 360 Hz, (c) 408 Hz, (d) 476 Hz, (e) 509 Hz. (By courtesy of K. A. Stetson.)

As a result of these measurements it was found that the whole series of vibration peaks occurred at too high frequencies to be suitable for the intended range of the instrument. This result confirmed suspicions that the traditional design of the treble viol was not an optimum one. As a result of this investigation a new magnum treble viol was designed. Alterations were made to the body size and shape and to the volume of the resonant air cavity. Musical tests made on the new instrument are reported to show "very little tendency to squeak compared to other viols, no observable wolf tones, no noticeable intermodulation with double stops, and very weak interstring coupling". Figure E8.4 shows five holograms of the top plate of the magnum viol.

The holographic investigation also revealed that caution was required in the application of normal-mode theory to a structure such as a viol plate. The interference fringes observed represent contours of constant amplitude but not necessarily constant phase. Further investigation showed that the holographic patterns consisted of response peaks composed of combinations of normal-mode vibrations. If the normal modes could be isolated they would characterise the plate itself, independent of the manner of excitation. Modes appear to combine by phasor addition. Thus, when modes combine in phase they give the illusion of real modes, while quadrature phase combinations may give discontinuous nodal lines or even point nodes. In order to decompose the patterns observed it would be necessary to know the phase of the combining modes and their relative amplitudes.

8.6 Biomedical applications

8.6.1 Propagation of waves in biological tissues

Since the velocity of a pressure wave depends primarily on the density and elastic properties of the medium, any variations in these parameters will cause an incident wave to be scattered in a variety of different directions. If the region exhibiting the variation is small compared with the wavelength of the incident radiation, scattering is of the Rayleigh type, that is the scattering tends to be isotropic and the intensity of the scattered waves is inversely proportional to the fourth power of the wavelength. At the other extreme, if the wavelength of the incident radiation is much less than the dimensions of the scattering region, specular reflection may occur. For incident waves having a wavelength of the same magnitude as the dimensions of the scatterer, the scattered waves form very complex patterns. In figure 8.10 is shown schematically the relationship between the diameter of the scatterer and the wavelength.

Scattering theory is usually discussed in terms of the scattering effects produced by cylinders and spheres (for instance, see Morse and Ingard, 1968). The conditions applicable to these bodies can be a useful guide in dealing with arbitrary shapes in practice. Figure 8.11a shows the angular distribution of scattered energy from a sphere of diameter d for the case when $\pi d/\lambda = 1$. For smaller ratios of $\pi d/\lambda$ (long wavelengths) a similar pattern is obtained, but with less overall scattered energy. Very long waves hardly 'see' a small object. When $\pi d/\lambda > 1$ (short wavelengths) there is a gradual transition to specular reflection conditions. Figure 8.11b shows the scattering pattern when $\pi d/\lambda = 5$. For larger ratios the back reflection pattern becomes smoother and the forward pattern long and narrow. In the short-wavelength limit, both for cylinders and spheres, half the incident energy is reflected specularly while half goes into the forward shadow-forming beam (the shadow is the result of interference between this beam and the incident radiation). Figure 8.10 includes a scale giving values of d for a typical ultrasonic wavelength of $1 \cdot 5$ mm for a number of values of $\pi d/\lambda$. Such a scale gives a rough guide to the type of scattering likely to be encountered in biological materials (assuming a velocity of 1500 ms^{-1} and a frequency of 1 MHz).

Under conditions when specular reflection occurs, the pressure reflection coefficient may be computed in terms of impedance ratios, as in section 3.3. Such reflections may be detected between biological tissue interfaces. Values of velocity and acoustic impedance for a number of materials of interest are shown in table 8.1. Also shown in this table are approximate values of the attenuation coefficient for a plane wave at 1 MHz. For most of the materials shown the attenuation, α, is linearly dependent on the frequency, ν. Exceptions are water for which $\alpha \propto \nu^2$, and blood for which $\alpha \propto \nu^{1 \cdot 3}$.

ν (MHz)	λ (mm)	Scattering zone (scattering theory applies)		Mixed zone		Geometrical zone (specular theory applies)	
		$\dfrac{\pi d}{\lambda}$					
		0·01		0·1	1	10	100
		d (mm)					
1	1·5	0·005		0·05	0·5	5	50
5	0·3	0·001		0·01	0·1	1	10
10	0·15	0·0005		0·005	0·05	0·5	5

Figure 8.10. Diagram showing the relationship between diameter of scatterer d, and wavelength, λ. Values of d have been calculated assuming $c = 1500$ m s^{-1}.

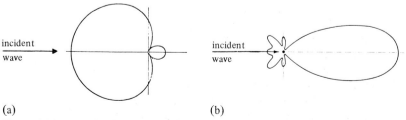

(a)   (b)

Figure 8.11. Angular distribution of scattered energy from a sphere of diameter d for the case (a) $\pi d/\lambda = 1$, (b) $\pi d/\lambda = 5$. In each case a plane wave is incident from the left.

The absorption mechanism for water is mainly that due to viscosity. In the case of blood, the attenuation around 1–3 MHz appears to be determined mainly by its total protein content. Biological soft tissues absorb wave energy mainly by molecular and thermal relaxation processes. Further discussion of absorption mechanisms and propagation in biological materials is given in Wells (1969).

Table 8.1. Acoustic properties of some biological materials (after Kossoff, 1971).

Material	Velocity $(m\ s^{-1})$	Density $(10^3\ kg\ m^{-3})$	Acoustic impedance $(10^6\ kg\ s^{-1}\ m^{-2})$	Attenuation at 1 MHz $(dB\ cm^{-1})$
Water (fresh, 39°C)	1530	1·00	1·53	0·05
Blood	1534	1·04	1·59	0·06
Fat	1440	0·97	1·40	0·5
Brain	1510	1·03	1·55	0·8
Liver	1590	1·06	1·68	1·3
Muscle	1590	1·03	1·64	1·6[a]
Bone	3360	2·0	6·62	4

[a] Along the fibres

8.6.2 Ultrasonic scanning methods

For many years scanning techniques based on pulse echo methods have been used for the examination of internal organs as well as in industrial non-destructive testing. Transducers employed are similar to those used in other pulse–echo work and include quartz, $BaTiO_3$ and PZT. For narrow-band applications the transducers may be used unbacked, when maximum power is obtained in the forward direction. When a broad-band response is required the transducer is provided with an absorbing backing, as discussed in Appendix 5.1. Greater efficiency is achieved with a quarter wavelength matching layer between the transducer and the load, as discussed by Kossoff (1966, 1971). The signal duration is usually of the order of 1 cycle at a frequency of 1 MHz.

The transducer diameter is normally at least the equivalent of 10 wavelengths in order to produce a reasonably narrow beam. To narrow the beam further, weak focusing may be employed either by shaping a ceramic transducer or adding an acoustic lens made of Perspex or epoxy resin.

The basic pulse–echo method, often called the **A-scan method**, is useful when a particular measurement is required, for instance, in determining the position of the midline echo in the brain or measuring the anterior and posterior chambers in the eye. The transducer is moved across the body in a line whilst it transmits and receives ultrasonic pulses. The received echoes arise from changes in the acoustic impedance at the tissue interfaces through which the ultrasonic beam passes. For each position of the transducer a pulse–echo pattern is recorded. The vertical deflection gives the amplitude of the reflected ultrasonic pulses, while the time scale can be calibrated in terms of the depth of penetration from the surface.

In the **B-scan method**, the direction of motion of an oscilloscope time-base is linked with the direction of the ultrasonic beam in the patient. Length along the time-base direction is made proportional to the depth of penetration. The rectified amplitude of the received pulses is used to modulate the intensity of the oscilloscope beam. A cross-sectional picture may therefore be built up for a plane parallel to the direction of propagation of the ultrasonic beam, as shown in figure 8.12. While B-scan displays are in frequent use, their interpretation can be difficult when compared with an x-ray image which represents a cross-sectional view perpendicular to the direction of the x-ray beam. A further problem arises since the ultrasonic reflections are essentially specular. Each

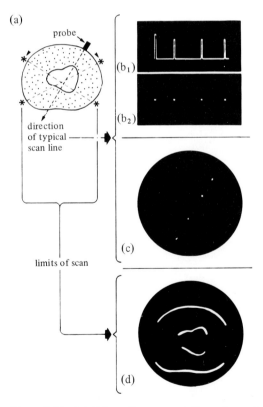

Figure 8.12. (a) Schematic representation of a section through a patient; (b_1) A-scope presentation of a typical scan line; (b_2) B-scope presentation of the same scan line; (c) B-scan as in (b_2), but with the direction of the time base linked to the direction of the ultrasonic beam; (d) compound B-scan, integrated from many individual scans, each one similar to (c) (after Wells, 1969).

interface segment must therefore be viewed at right angles. To achieve
this condition, compound scanning may be used, as shown in figure 8.12d,
in which a series of scans from a number of directions around the body
are superimposed (Kossoff, 1971; Wells, 1969). Figure 8.13 shows two
examples of a compound B-scan together with corresponding anatomical
sectional drawings.

The **C-scan method** produces a display showing a greater similarity to
the x-ray picture. In this method the horizontal and vertical deflections
of the oscilloscope beam are made proportional to the horizontal and
vertical coordinates of the transducer as it maps out an area on the surface
of the body. The light intensity of the beam is modulated by the
ultrasonic signal which is subjected to range gating. In this way depth
information is lost, but the display now corresponds to a given transverse
section with a thickness of a few wavelengths. There is still a difference
when compared with a transverse x-ray picture in that the latter represents
a compression of all the depth information into one plane whereas both
B-scan and C-scan pictures are two-dimensional cross-sections.

A modification of the A-scan method is the **M-scan** or **time-motion
method.** The echo pattern is allowed to modulate the oscilloscope beam

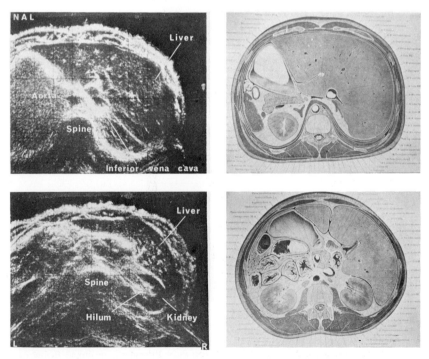

Figure 8.13. Examples of compound B-scans together with corresponding anatomical
sectional drawings (by courtesy of G. Kossoff).

and the trace so obtained is slowly moved along the y axis of the oscilloscope. The transducer is held stationary so that, if the echo producing structure is also stationary, vertical lines appear on the oscilloscope wherever echoes appear. If the structure is moving, curved lines will appear on the screen as the trace follows the movement.

Instead of moving a single transducer to make a scan, a faster system would involve an array of fixed transducers with electronic switching. Such a system, based on the A-scan, has been developed by Bom *et al.* (1971) for the examination of moving cardiac structures. An array of 20 transducer elements allows an almost instantaneous display of the set of A-scans. A method for producing fast B-scans, using a 21 element transducer array with electronic steering, has been described by Somer (1968). A further advantage of an array system is that the focusing of the array may be controlled to allow examination of specific small regions.

A number of computer-based methods have been devised which store the digitised signals arising from B- and C-type scans and thereby enable the display of data corresponding to any arbitrary section or the display of compounded signals from different sections to provide the equivalent of a three-dimensional display (Onoe *et al.*, 1972; Robinson, 1972). A three-dimensional optical system, designed to present a series of B-scan images in such a way as to form a three-dimensional image, has been described by Szilard (1974). A wide angle of view is possible and colour is used to discriminate between different depths.

A detailed discussion of the clinical applications of the above methods is presented by Wells (1969).

Acoustic holography techniques have been applied to biological problems (Kock, 1973). Some success has been achieved with the real-time liquid surface method and with the ultrasonic camera in which a piezoelectric crystal or array is scanned electronically. Attempts are being made to use DKDP crystals and liquid crystals as part of a real-time imaging system (Greguss, 1973).

The combination of pulse–echo methods with holographic processing can lead to significant improvements in resolution. With pulse–echo methods, object points can only be resolved if their range separation is greater than $\frac{1}{2}c\tau$, where c is the sound velocity and τ is the pulse duration. The angular resolution is determined by the beam spread due to diffraction effects at the transducer, that is by the ratio λ/d, where d is the diameter of the transducer. Holographic resolution is determined by the ratio λ/L, where L is the aperture. An increase in L will therefore improve the resolution of the system.

One method for increasing L is to adapt the synthetic-aperture method used in radar (Kock, 1973). A relatively wide-angle ultrasonic beam is used to scan the field as the transducer moves along a given path. The received echoes are caused to interfere with a reference beam derived from the transmitted signal. In this way a one-dimensional hologram is formed for each sweep of the transducer. The effective aperture of the system is equal to the length of the sweep assuming the source to remain coherent during this time. The holograms may be recorded on film or digitised and stored in a computer for subsequent reconstruction. The main disadvantage of the system is that it is not a real-time system.

A further use for holography is in multiplexing B-scan records. A number of B-scans are made at closely spaced intervals in a number of different directions to form a set of perspective views (Sollish and Glaser, 1972). An optical hologram is synthesised from these scans in which each B-scan is used to form its own subhologram. Thus, a three-dimensional image is built up step by step and may be viewed by optical reconstruction. An advantage of this system is that there is no distortion introduced because of wavelength conversion. The acoustic–optic conversion is nonholographic, while the holographic steps are entirely optical. The result is a full-size, undistorted, three-dimensional image.

Experiment E8.4 Rapid B-scanner for heart examination
(McDicken, W. N., Bruff, K., Paton, J., 1974, *Ultrasonics*, **12**, 269.)

Several ultrasonic techniques have been developed to record cardiac movements, including the M-scan method and stop-action scanning. In the latter method a B-scan image is recorded in one specific phase in the heart cycle by using an ECG signal to select the times at which echoes are recorded. The design of the present B-scanner was determined by the fact that access to the heart with an ultrasonic beam is limited by the strongly attenuating properties of bone and lung. Maximum use must therefore be made of the access points between rib spaces at the front of the chest.

With a well-balanced mechanical system it was found feasible to make a standard transducer oscillate at 8 Hz while pivoted at the front face of the crystal. The face of the crystal can then be placed at a rib space so that rapid sector scans are made of a

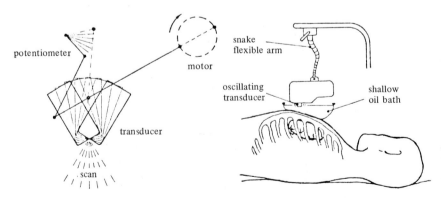

Figure E8.5. Oscillatory motion performed by the ultrasonic transducer.

Figure E8.6. Method used to couple the ultrasonic instrument to the patient. The shallow oil bath is convenient and provides a high degree of flexibility.

plane through the heart. With this instrument it is possible to stop the oscillation at a particular beam position and then record a high-quality M-scan. The system used is shown in figure E8.5.

With the transducer oscillating at 8 Hz, 16 sector B-scans with an angular sweep of $60°$ can be made each second. The angular position of the transducer was measured with a sine-cosine potentiometer. A shallow bath of castor oil was used to couple the oscillating transducer to the patient, as shown in figure E8.6. The castor oil was supported by a thin polythene membrane which was coupled to the chest of the patient by a layer of oil or gel. The ultrasonic transducer operated at a frequency of $2\cdot5$ MHz and had a diameter of 15 mm. The pulse repetition frequency of the transmitter was 2000 Hz. With this rapid scan system the ultrasound beam sweeps through an angle of $0\cdot5°$ between transmissions, providing a total of 120 lines of scan in each $60°$ sweep. This number of lines was found sufficient to produce an acceptable image of the cardiac structures. An example of the image produced by this technique is shown in figure E8.7.

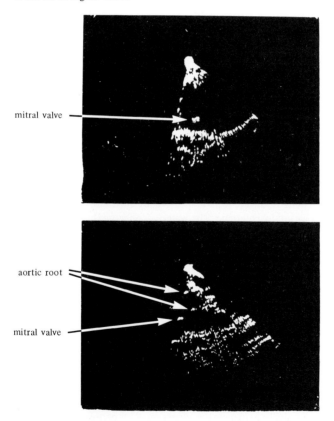

Figure E8.7. Two images produced by rapid single sweeps of the ultrasonic beam through the heart. The bottom image is a plane section longitudinal to the heart and shows the aortic root and mitral valve. The top image is a section transverse to the heart at the level of the tip of the mitral valve.

Experiment E8.5 The Octoson—a multi-transducer echoscope
(Kossoff, G., Carpenter, D. A., Radovanovich, R., Robinson, D. E.,
Garrett, W. J., 1975, in *Proceedings of the Second European Congress on
Ultrasonics in Medicine, Munich.*)

The time taken to form a high-quality compound B-scan echogram with current
instruments is between 10 and 20 s. The quality of the image is determined by the
number of resolvable elements it contains, and the quality of a scanned echogram is
considered to be high if it contains over 300 lines of information. Echograms are
usually displayed on a 10 cm × 10 cm oscilloscope screen and to display 300 lines of
information the spot size of the oscilloscope must be less than $0 \cdot 3$ mm. To achieve
the improvements made possible with compound scanning, it is necessary to examine
the tissues from many different directions and a compound scan echogram formed by
the superposition of eight simple scans contains 2500 lines of sight.

There are a number of problems that arise if fast scanning is attempted with a
single transducer. The need for rapid mechanical scanning with a single transducer
may be obviated by using a number of transducers suitably positioned on an arm
which extends the length of the examined tissue. A compound scanned echogram is
then obtained by simultaneously oscillating all the transducers and sequentially energising
and displaying the echoes received by each transducer. The sequence rate is chosen so
that each transducer is energised when the trace representing its line of sight moves to
its next discernable position. In this way a compound B-scan may be obtained with
one sweep of the transducers. For instance, suppose the transducers sweep through an
angle of $30°$ in 1 s, then if the transducers are energised once every 500 μs, they move
through an angle of $0 \cdot 015°$ between transmitted pulses.

The multi-transducer scanner, shown in principle in figure E8.8, is immersed in a
water bath, the top of which forms a couch on which a patient lies. The ultrasonic
beam is transmitted to the patient through a coupling polythene membrane. A quick
interchange mechanism allows the use of a variety of different sizes and shapes of
membrane for coupling to different parts of the body such as the uterus, abdomen,
thyroid and brain of young children. The water coupling method was chosen as it
allows the use of large-diameter and therefore high-resolution transducers for the
examination. These can be either of the fixed focus or of the electronically variable
focus annular phased array construction. Large aperture focused transducers are not as
critically dependent on the inclination of interfaces.

The scanner consists of eight transducers mounted on an arm with the oscillating
movement of the transducers in the plane of the length of the arm which defines the
plane of the echogram. The transducers oscillate through an angle of $50°$ and an
echogram is acquired in a little over 2 s. The arm may be rotated through $180°$ to
give transverse, longitudinal, or oblique echograms and it may be tilted through an
angle of $\pm30°$ to give inclined sections.

The arm may also be translated along the x and y axes in increments ranging from
1 mm to 2 cm to give automatic acquisition of a number of parallel echograms
separated by the selected distance. A routine examination, consisting of a set of 30
echograms in two mutually perpendicular planes, may be performed in under 3 min.
With the water coupling method all of these sections are obtained without changing
the coupling to the patient, thus allowing reliable cross-correlation of detail in
echograms obtained in different planes. The arm may also be moved up and down
along the z axis to position it at a selected level relative to the patient.

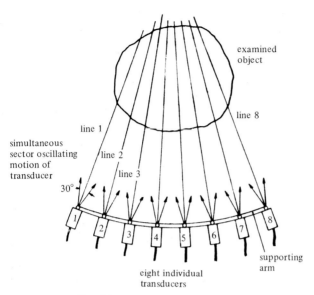

examined
object

line 8

line 1

simultaneous
sector oscillating   line 2
motion of
transducer   line 3

30°

1 2 3 4 5 6 7 8

supporting
arm

eight individual
transducers

Time (ms)	Energising sequence
0	line 1
0·5	line 2
1·0	line 3
.	.
.	.
3·5	line 8
4·0	line 1
4·5	line 2

Figure E8.8. Principle of operation of Octoson echoscope.

The rapid rate of acquisition of high-quality echograms has obvious advantages. It reduces blurring of detail caused by movement of structures during the echogram forming period and allows quasi-real-time viewing of the same section. The instrument also has the advantage that it utilises the minimum number of pulses to acquire a compound scan echogram and no point in the examined tissue is irradiated with more than eight pulses. Two echograms taken with this instrument are shown in figure 8.13.

References

Benedek, G. B., Lastovka, J. B., Fritsch, K., Greytak, T., 1964, *J. Opt. Soc. Am.,* **54,** 1284.

Bom, N., Lancee, C. T., Honkoop, J., Hugenholtz, P. G., 1971, *BioMed. Eng.,* **6,** 500.

Bragg, W. L., 1913, *Proc. Cambridge Philos. Soc.,* **17,** 43.

Brenden, B. B., 1972, in *Optical and Acoustical Holography,* Ed. E. Camatini (Plenum Press, New York).

Brillouin, L., 1922, *Ann. Phys. (Paris),* **17,** 88.

Gabor, D., 1948, *Nature (London),* **161,** 777.

Gabor, D., 1949, *Proc. R. Soc. A,* **197,** 454.

Gabor, D., 1969, *Rep. Prog. Phys.,* **32,** 395.

Gabor, D., Stroke, G. W., 1969, *Endeavour,* **28,** 40.

Greguss, P., 1973, *Acustica,* **29,** 52.

Haines, K. A., Hildebrand, B. P., 1966, *Appl. Opt.,* **5**, 595.

Kock, W. E., 1971, *Seeing Sound* (Wiley-Interscience, New York).

Kock, W. E., 1973, in *Physical Acoustics,* Eds W. P. Mason, R. N. Thurston, Volume 11, chapter 5 (Academic Press, New York).

Kock, W. E., Harvey, F. K., 1951, *Vell Syst. Tech. J.,* **20**, 564.

Kossoff, G., 1966, *IEEE Trans. Sonics Ultrason.,* **SU-13**, 20.

Kossoff, G., 1971, *Aust. Phys.,* **8**, 100.

Landry, J., Powers, J. P., Wade, G., 1969, *Appl. Phys. Lett.,* **15**, 186.

Mezrich, R., Etzold, K. F., Vilkomerson, D., 1974, *Proc. Soc. Photo-Opt. Instrum. Eng.,* **47**, 69.

Morse, P. M., Ingard, K. U., 1968, *Theoretical Acoustics* (McGraw-Hill, New York).

Mueller, R. K., Sheridon, N. K., 1966, *Appl. Phys. Lett.,* **9**, 328.

Onoe, M., Takagi, M., Masumoto, T., Hamano, N., 1972, in *Acoustical Holography,* Ed. G. Wade, Volume 4 (Plenum Press, New York).

Pohlman, R., 1939, *Z. Physik,* **113**, 697.

Powell, R. L., Stetson, K. A., 1965, *J. Opt. Soc. Am.,* **55**, 1593.

Powers, J. P., Wang, K. Y., Wade, G., 1972, *J. Acoust. Soc. Am.,* **51**, 1593.

Robinson, D. E., 1972, *J. Acoust. Soc. Am.,* **52**, 673.

Sokolov, S. I., 1941, *J. Tech. Phys. (USSR),* **11**, 160.

Sollish, B. D., Glaser, I., 1972, in *Acoustical Holography,* Ed. G. Wade, Volume 4 (Plenum Press, New York).

Somer, J. C., 1968, *Ultrasonics,* **6**, 153.

Stroke, G. W., 1965, *Int. Sci. Technol.,* May, p.52.

Szilard, J., 1974, *Ultrasonics,* **12**, 273.

Wells, P. N. T., 1969, *Physical Principles of Ultrasonic Diagnosis* (Academic Press, London).

REVIEW QUESTIONS

8.1   What is the principle of the Bragg diffraction effect that occurs between a light wave and a sound wave? (§8.2)

8.2   How may Bragg diffraction be used to produce an image of an acoustic field? (§8.2) (E8.1)

8.3   What is meant by Brillouin scattering? (§8.3)

8.4   Show how both amplitude and phase of an optical field may be recorded using the principle of optical holography. (§8.4)

8.5   How may an optical hologram be reconstructed? (§8.4.1)

8.6   What are the physical principles that govern the formation of an acoustic hologram on a liquid surface? (§8.5.1)

8.7   What is the principle of operation of a Pohlman cell? How could such a cell be used to produce a real-time acoustic hologram? (E8.2)

8.8   How would you proceed to make a time-average hologram of a vibrating plate? (E8.3)

8.9   What type of reflection phenomenon from an object of diameter d would be observed if (a) $\lambda = 5d$; (b) $\lambda = 0 \cdot 2d$? (§8.6.1)

8.10 What are the essential differences between a B-scan and a C-scan ultrasonic pulse display? (§8.6.2)

8.11 What are the main problems involved in making a compound B-scan echogram? (E8.5)

PROBLEMS

8.12 Brillouin scattering from a liquid is observed in a direction perpendicular to the incident laser beam (frequency 5×10^{14} Hz). In this direction the frequency shift for the two Brillouin lines is found to be 3×10^9 Hz and the half-intensity width of a line is 10^8 Hz. If the index of refraction of the liquid is $n = 1 \cdot 4$, determine the frequency, phase velocity, and spatial attenuation coefficient of the phonons involved.

8.13 A sensitive system for visualising ultrasonic wave fields and for measuring displacement amplitudes may be based on the principle of the Michelson interferometer (Mezrich et al., 1974). A reference mirror is mounted external to the sound field while a second flexible mirror (in the form of a thin metallised plastic film) is immersed in the fluid through which the ultrasonic wave passes.
(a) Find an expression for the displacement reflection coefficient of a thin film.
(b) Show that, if the film is thin enough, the displacement of the film is equal to that of the ultrasonic wave passing through it.
(c) When the mirrors are illuminated by coherent light (via a beam splitter) and the two reflected light beams are combined, find an expression for the total light intensity.
(d) Assuming that the relative phase shift between the two beams is 90° and that a photocell measures the intensity of the combined beam, show that the photocurrent is proportional to the amplitude of the ultrasonic wave.

8.14 Show that, in the formation of a liquid surface hologram, the displacement of the surface can be expressed as

$$\zeta(x, y) = 2A \cos[2k_y y + \phi(x, y)] + B$$

where A and B are given by equations (8.20) and (8.21) in the text.

M

Solutions to problems

Chapter 1

1.7 The general expression for the elastic free energy, $F = \frac{1}{2}c_{iklm}\epsilon_{ik}\epsilon_{lm}$, is unchanged if i and k are interchanged ($\epsilon_{ik} = \epsilon_{ki}$) or if l and m are interchanged ($\epsilon_{lm} = \epsilon_{ml}$). In addition, the strain terms may be interchanged, that is $\epsilon_{lm}\epsilon_{ik}$ for $\epsilon_{ik}\epsilon_{lm}$, and then the pairs of suffixes exchanged as above. The following symmetry relations then hold for the c's:

$$c_{iklm} = c_{kilm} = c_{ikml} = c_{kiml} ,$$
$$c_{lmik} = c_{mlik} = c_{lmki} = c_{mlki} .$$

In computing the free energy it is necessary to include all finite constants, not only the independent constants. For a cubic crystal there are only three independent constants:

$$c_{11}^{11} = c_{22}^{22} = c_{33}^{33} ,$$
$$c_{22}^{11} = c_{33}^{22} = c_{33}^{11} ,$$
$$c_{23}^{23} = c_{13}^{13} = c_{12}^{12} .$$

However, because of the symmetry relations, there are altogether twenty-one finite constants. The additional equivalent constants are:

$$c_{22}^{11} = c_{11}^{22} , \quad c_{33}^{22} = c_{22}^{33} , \quad c_{33}^{11} = c_{11}^{33} ;$$
$$c_{23}^{23} = c_{32}^{23} = c_{23}^{32} = c_{32}^{32} ;$$
$$c_{13}^{13} = c_{31}^{13} = c_{13}^{31} = c_{31}^{31} ;$$
$$c_{12}^{12} = c_{21}^{12} = c_{12}^{21} = c_{21}^{21} .$$

For energy considerations it is necessary to use the complete 9×9 elastic constant matrix. Hence

$$F = \frac{1}{2}c_{11}^{11}(\epsilon_{11}^2 + \epsilon_{22}^2 + \epsilon_{33}^2) + \frac{1}{2}c_{22}^{11}(\epsilon_{11}\epsilon_{22} + \epsilon_{22}\epsilon_{11} + \epsilon_{22}\epsilon_{33} + \epsilon_{33}\epsilon_{22} + \epsilon_{11}\epsilon_{33} + \epsilon_{33}\epsilon_{11})$$
$$+ \frac{1}{2}c_{23}^{23}(\epsilon_{23}^2 + \epsilon_{32}^2 + \epsilon_{23}\epsilon_{32} + \epsilon_{32}\epsilon_{23} + \epsilon_{13}^2 + \epsilon_{31}^2 + \epsilon_{13}\epsilon_{31} + \epsilon_{31}\epsilon_{13} + \epsilon_{12}^2 + \epsilon_{21}^2$$
$$+ \epsilon_{12}\epsilon_{21} + \epsilon_{21}\epsilon_{12}) .$$

Collecting equivalent terms we obtain

$$F = \frac{1}{2}c_{11}^{11}(\epsilon_{11}^2 + \epsilon_{22}^2 + \epsilon_{33}^2) + c_{22}^{11}(\epsilon_{11}\epsilon_{22} + \epsilon_{22}\epsilon_{33} + \epsilon_{11}\epsilon_{33}) + 2c_{23}^{23}(\epsilon_{23}^2 + \epsilon_{13}^2 + \epsilon_{12}^2) .$$

Addendum: Write out the complete elastic constant matrix for this problem using the following scheme for the suffixes for both rows and columns:

$$(1111), \ (2211), \ (3311), \ (2311), \ (3211), \ (1311), \ (3111), \ (1211), \ (2111).$$

1.8 In the orthorhombic system there are nine independent elastic constants: c_{11}^{11}, c_{22}^{22}, c_{33}^{33}, c_{22}^{11}, c_{33}^{22}, c_{33}^{11}, c_{23}^{23}, c_{13}^{13}, c_{12}^{12}. From the symmetry relations, additional finite constants are c_{11}^{22}, c_{22}^{33}, and three further constants for each of c_{23}^{23}, c_{13}^{13}, and c_{12}^{12} as in Problem 1.1. Hence

$$F = \frac{1}{2}c_{iklm}\epsilon_{ik}\epsilon_{lm} = \frac{1}{2}c_{11}^{11}\epsilon_{11}^2 + \frac{1}{2}c_{22}^{22}\epsilon_{22}^2 + \frac{1}{2}c_{33}^{33}\epsilon_{33}^2 + c_{22}^{11}\epsilon_{11}\epsilon_{22} + c_{33}^{22}\epsilon_{22}\epsilon_{33} + c_{33}^{11}\epsilon_{11}\epsilon_{33}$$
$$+ 2c_{23}^{23}\epsilon_{23}^2 + 2c_{13}^{13}\epsilon_{13}^2 + 2c_{12}^{12}\epsilon_{12}^2 .$$

1.9 As discussed in Appendix 1.1, a fourth-rank tensor, such as c_{iklm}, can be rendered isotropic by the following equation

$$c_{iklm} = \lambda\delta_{ik}\delta_{lm} + \mu(\delta_{il}\delta_{km} + \delta_{im}\delta_{kl}) + \nu(\delta_{il}\delta_{km} - \delta_{im}\delta_{kl}) .$$

[An alternative geometrical method is given in Long (1961), section 5.2.] We investigate the possible combinations of i, k, l, and m. If $i = k = l = m$, then

$$c_{1111} = c_{2222} = c_{3333} = \lambda + 2G .$$

If $i = k, l = m$,

$$c_{1122} = c_{1133} = c_{2211} = c_{2233} = c_{3311} = c_{3322} = \lambda .$$

If $i = l, k = m$, or $i = m, k = l$,

$$c_{iklm} = \tfrac{1}{2}(c_{1111} - c_{1122}) .$$

All other combinations, such as $i = k, l \neq m$ lead to zero values.

1.10 (a) Consider a hydrostatic compression applied to an isotropic body. Then, Hooke's law [equation (1.36)] is

$$\sigma_{ik} = -p\delta_{ik} = 2G\epsilon_{ik} + \lambda\theta\delta_{ik} ;$$

in incremental form:

$$-\Delta p\delta_{ik} = 2G\Delta\epsilon_{ik} + \lambda\Delta\theta\delta_{ik} .$$

Contracting, that is putting $i = k$ and summing, gives

$$-3\Delta p = 2G\Delta\theta + 3\lambda\Delta\theta ,$$

since $\theta = \epsilon_{ii}$. Hence, from the definition of K [equation (1.38)]

$$K = -\frac{\Delta p}{\Delta\theta} = \lambda + \tfrac{2}{3}G .$$

Therefore Hooke's law may be written

$$\sigma_{ik} = 2G\epsilon_{ik} + (K - \tfrac{2}{3}G)\epsilon_{ll}\delta_{ik} .$$

[Note that the suffixes on the dilatation must be different from i and k in this equation to avoid incorrect summation.]

(b) Contracting the last equation, we obtain

$$\sigma_{ii} = 2G\epsilon_{ii} + 3(K - \tfrac{2}{3}G)\epsilon_{ii} = 3K\epsilon_{ii}$$

(ϵ_{ll} has been written as ϵ_{ii} since there is now no confusion concerning summation). Thus,

$$\epsilon_{ii} = \frac{\sigma_{ii}}{3K} .$$

1.11 Combining equations (1.36), (1.41) and (1.42), namely

$$\sigma_{ik} = 2G\epsilon_{ik} + \lambda\theta\delta_{ik} ,$$

$$\lambda = \frac{E\mu}{(1 + \mu)(1 - 2\mu)} ,$$

$$G = \frac{E}{2(1 + \mu)} ,$$

gives the required result:

$$\sigma_{ik} = \frac{E}{1 + \mu}\left(\epsilon_{ik} + \frac{\mu}{1 - 2\mu}\theta\delta_{ik}\right) .$$

The equations of equilibrium for an isotropic body are

$$\partial_k \sigma_{ik} = 0 ,$$

$$\partial_k \left[\frac{E}{1+\mu} \left(\epsilon_{ik} + \frac{\mu}{1-2\mu} \theta \delta_{ik} \right) \right] = 0 ,$$

where $\partial_k = \partial/\partial x_k$. Using the definition of strain, equation (1.9), we obtain

$$\frac{E}{1+\mu} \left[\tfrac{1}{2} \partial_k (\partial_k u_i + \partial_i u_k) + \frac{\mu}{1-2\mu} \partial_k \theta \delta_{ik} \right] = 0 ;$$

$$\frac{E}{1+\mu} \left[\tfrac{1}{2} \partial_k \partial_k u_i + \tfrac{1}{2} \partial_i \theta + \frac{\mu}{1-2\mu} \partial_i \theta \right] = 0 ,$$

since $\theta = \partial_k u_k$;

$$\frac{E}{2(1+\mu)} \left[\partial_k \partial_k u_i + \frac{1}{(1-2\mu)} \partial_i \partial_l u_l \right] = 0 .$$

1.12 Young's modulus—cubic crystal

Suppose a cylindrical rod is cut from the crystal with its long axis in the direction of the unit vector n. The boundary conditions are such that the longitudinal stress in the rod must act perpendicular to the end faces, while the resultant stress at the side faces must be zero. Let p be the external force per unit area applied along the axis of the cylinder. Then, the stress components within the crystal must satisfy the conditions: (i) when σ_{ik} is multiplied by n_i, the resulting extension force must be parallel to n; (ii) when σ_{ik} is multiplied by a vector perpendicular to n, the result must be zero. To satisfy these conditions we may write

$$\sigma_{ik} = p n_i n_k . \tag{S1.1}$$

We shall now find expressions for the stress and strain components within the crystal, using the free energy equation computed in Problem 1.7.

The elastic free energy for a cubic crystal is given by

$$F = \tfrac{1}{2} c_1 (\epsilon_{11}^2 + \epsilon_{22}^2 + \epsilon_{33}^2) + c_2 (\epsilon_{11}\epsilon_{22} + \epsilon_{22}\epsilon_{33} + \epsilon_{11}\epsilon_{33}) + 2c_3 (\epsilon_{23}^2 + \epsilon_{13}^2 + \epsilon_{12}^2) , \tag{S1.2}$$

where, for conciseness, we have written

$$c_1 = c_{11}^{11} , \quad c_2 = c_{22}^{11} , \quad c_3 = c_{23}^{23} .$$

The stress components are found with the aid of equation (1.26):

$$\sigma_{ik} = \frac{\partial F}{\partial \epsilon_{ik}} . \tag{S1.3}$$

For instance, for a typical shear stress component:

$$2\sigma_{12} = \frac{\partial F}{\partial \epsilon_{12}} = 4c_3 \epsilon_{12} .$$

Note that a factor 2 appears for shear stress components. The reason for this is explained by Landau and Lifshitz (1970) as follows: the expression $\sigma_{ik} = \partial F/\partial \epsilon_{ik}$ is meaningful only as indicating that $dF = \sigma_{ik} d\epsilon_{ik}$; in the sum $\sigma_{ik} d\epsilon_{ik}$, however, the term in the differential $d\epsilon_{ik}$ for each component with $i \neq k$ of the symmetrical tensor ϵ_{ik} appears twice. Hence,

$$\sigma_{12} = 2c_3 \epsilon_{12}$$

and

$$\epsilon_{12} = \frac{\sigma_{12}}{2c_3} = \frac{pn_1n_2}{2c_3} , \tag{S1.4}$$

with similar expressions for the other shear strain components.

For the principal components of stress we have, from equation (S1.1),

$$\sigma_{11} + \sigma_{22} + \sigma_{33} = p(n_1^2 + n_2^2 + n_3^2) = p .$$

Hence, from equations (S1.2) and (S1.3), we obtain

$$c_1\epsilon_{11} + 2c_2\epsilon_{22} + c_1\epsilon_{22} + c_2(\epsilon_{11} + \epsilon_{33}) + c_1\epsilon_{33} + c_2(\epsilon_{11} + \epsilon_{22}) = p ,$$

$$\epsilon_{11}(c_1 + 2c_2) + 2(c_1 + 2c_2)\epsilon_{22} = p ,$$

putting $\epsilon_{22} = \epsilon_{33}$ because of transverse isotropy. But $\sigma_{11} = pn_1^2 = c_1\epsilon_{11} + 2c_2\epsilon_{22}$; therefore

$$\epsilon_{11}(c_1 + 2c_2) + 2(c_1 + 2c_2)\frac{pn_1^2 - c_1\epsilon_{11}}{2c_2} = p .$$

Thus,

$$\epsilon_{11} = p\frac{(c_1 + 2c_2)n_1^2 - c_2}{(c_1 - c_2)(c_1 + 2c_2)} , \tag{S1.5}$$

with similar expressions for the other principal components.

From figure 1.4 in the text, the distance between two points of a deformed body after deformation, dl', is related to the corresponding distance before deformation, dl, by

$$dl'^2 = dl^2 + 2\epsilon_{ik}\, dx_i\, dx_k .$$

The relative longitudinal extension of the rod is

$$\epsilon = \frac{dl' - dl}{dl} = \epsilon_{ik}n_in_k$$

where $n_i = dx_i/dl$. Thus,

$$\epsilon = \epsilon_{11}n_1^2 + \epsilon_{22}n_2^2 + \epsilon_{33}n_3^2 + \epsilon_{12}n_1n_2 + \epsilon_{23}n_2n_3 + \epsilon_{13}n_1n_3 .$$

Substitution from equations such as (S1.4) and (S1.5) yields Young's modulus from

$$E = \frac{p}{\epsilon} = \left\{ \frac{c_1 + c_2}{(c_1 + 2c_2)(c_1 - c_2)} + \left[\frac{1}{c_3} - \frac{2}{(c_1 - c_2)} \right](n_1^2n_2^2 + n_1^2n_3^2 + n_2^2n_3^2) \right\}^{-1} .$$

Chapter 2

2.6 Plane waves in [111] direction in cubic crystal

The independent elastic constants are c_{11}, c_{12}, and c_{44}. For this problem, $n_1 = n_2 = n_3 = 3^{-\frac{1}{2}}$. Equation (2.23) may then be written

$$\begin{vmatrix} A & B & B \\ B & A & B \\ B & B & A \end{vmatrix} = 0 ,$$

where

$$A = \tfrac{1}{3}(c_{11} + 2c_{44}) - \rho c^2 ,$$

$$B = \tfrac{1}{3}(c_{12} + c_{44}) .$$

Hence

$$(A - B)[A(A - B) + B(A - B)] + B(A - B)^2 = 0 ,$$

$$A + 2B = 0 , \quad \text{or} \quad (A - B)^2 = 0 .$$

That is

$$\tfrac{1}{3}(c_{11} + 2c_{12} + 4c_{44}) - \rho c^2 = 0 ,$$

or

$$[\tfrac{1}{3}(c_{11} - c_{12} + c_{44}) - \rho c^2]^2 = 0 .$$

Therefore

$$c_1 = \left(\frac{c_{11} + 2c_{12} + 4c_{44}}{3\rho} \right)^{1/2} ,$$

$$c_2 = c_3 = \left(\frac{c_{11} - c_{12} + c_{44}}{3\rho} \right)^{1/2} .$$

2.7   Plane waves in [100] direction in hexagonal crystal

The independent elastic constants are $c_{11}, c_{12}, c_{13}, c_{33}, c_{44}, c_{66} = \tfrac{1}{2}(c_{11} - c_{12})$. For this case, $n_1 = 1, n_2 = n_3 = 0$. The determinantal equation (2.23) reduces to

$$\begin{vmatrix} c_{11} - \rho c^2 & 0 & 0 \\ 0 & c_{66} - \rho c^2 & 0 \\ 0 & 0 & c_{44} - \rho c^2 \end{vmatrix} = 0$$

giving

$$(c_{11} - \rho c^2)(c_{44} - \rho c^2)(c_{66} - \rho c^2) = 0 ,$$

whence

$$c_1 = \left(\frac{c_{11}}{\rho} \right)^{1/2} , \quad c_2 = \left(\frac{c_{44}}{\rho} \right)^{1/2} , \quad c_3 = \left(\frac{c_{66}}{\rho} \right)^{1/2} = \left(\frac{c_{11} - c_{12}}{2\rho} \right)^{1/2} .$$

Chapter 3

3.8   Method of separation of variables

The wave equation to be solved is

$$\nabla^2 \phi = \frac{1}{c_\ell^2} \frac{\partial^2 \phi}{\partial t^2} . \tag{S3.1}$$

Solution by the method of separation of variables assumes that ϕ can be written as the product of functions of one variable only. Hence put

$$\phi = f_1(x_1) f_2(x_2) f_3(x_3) g(t) . \tag{S3.2}$$

Substitution of (S3.2) in (S3.1) and division by $f_1 f_2 f_3 g$ gives

$$\frac{1}{f_1} \frac{d^2 f_1}{dx_1^2} + \frac{1}{f_2} \frac{d^2 f_2}{dx_2^2} + \frac{1}{f_3} \frac{d^2 f_3}{dx_3^2} = \frac{1}{g c_\ell^2} \frac{d^2 g}{dt^2} . \tag{S3.3}$$

Introduction of suitable separation constants now yields the set of ordinary differential equations

$$\frac{d^2 f_1}{dx_1^2} + k_1^2 f_1 = 0 , \qquad \frac{d^2 f_3}{dx_3^2} + k_3^2 f_3 = 0 ,$$

$$\frac{d^2 f_2}{dx_2^2} + k_2^2 f_2 = 0 , \qquad \frac{d^2 g}{dt^2} + \omega^2 g = 0 . \tag{S3.4}$$

The solutions of equations (S3.4) may be written in the form

$$f_1(x_1) = C_1 \exp(ik_1x_1) + D_1 \exp(-ik_1x_1),$$
$$f_2(x_2) = C_2 \exp(ik_2x_2) + D_2 \exp(-ik_2x_2),$$
$$f_3(x_3) = C_3 \exp(ik_3x_3) + D_3 \exp(-ik_3x_3),$$
$$g(t) = C \exp(i\omega t) + D \exp(-i\omega t).$$

For this problem there is no motion in the x_2 direction or in the $-x_3$ direction. We also put $C = 0$ to give the usual time dependence. Hence the solution of the wave equation is

$$\phi = [C_1 \exp(ik_1x_1) + D_1 \exp(-ik_1x_1)]C_3 \exp(ik_3x_3)D \exp(-i\omega t).$$

At the boundaries the resultant pressure is zero, that is

$$p_{res}\big|_{x_1 = \pm a} = -\rho\frac{\partial^2\phi}{\partial t^2} = 0.$$

Hence

$$\rho\omega^2[C_1 \exp(ik_1a) + D_1 \exp(-ik_1a)]C_3 \exp(ik_3x_3) = 0.$$

This condition will hold for all values of x_3 if

$$C_1 \exp(ik_1a) + D_1 \exp(-ik_1a) = 0.$$

On expanding, we obtain

$$C_1 \cos k_1a + D_1 \cos k_1a + i[C_1 \sin k_1a - D_1 \sin k_1a] = 0.$$

Equating real and imaginary parts yields

$$C_1 = -D_1, \qquad \sin k_1a = 0.$$

Therefore the solution for the resultant field pattern in the waveguide is

$$\phi = C_1[\exp(ik_1x_1) - \exp(-ik_1x_1)]C_3 \exp(ik_3x_3)D \exp(-i\omega t)$$
$$= A[\exp(ik_1x_1) - \exp(-ik_1x_1)] \exp i(k_3x_3 - \omega t),$$

which is the same as equation (3.13) in the text with $k_1 = k\cos\alpha$, $k_3 = k\sin\alpha$ and the time dependence included.

3.9   Consider a plane acoustic wave travelling at an angle α to the x_1 axis. The displacement potential, ϕ, for the wave can be written

$$\phi = A \exp i(k_jx_j - \omega t).$$

The particle displacement in the x_1 direction is then

$$u_1 = \mathrm{grad}\phi = \frac{\partial\phi}{\partial x_1} = ik_1\phi.$$

The particle velocity in the x_1 direction is

$$\dot{u}_1 = \frac{\partial^2\phi}{\partial x_1\partial t} = \omega k_1\phi = \omega k\cos\alpha\phi.$$

The acoustic pressure is

$$p = -\frac{\rho\partial^2\phi}{\partial t^2} = \rho\omega^2\phi.$$

The wave impedance is defined as the ratio of the acoustic pressure to the particle velocity:

$$Z \equiv \frac{p}{\dot{u}_1} = \frac{\rho\omega^2}{\omega k \cos\alpha} = \frac{\rho c}{\cos\alpha} \ .$$

3.10 From the equation for the reflection coefficient [equation (3.25)] we have, putting $m = \rho'/\rho$ and $n = c_\ell/c'_\ell$,

$$R = \frac{m\cos\alpha - (n^2 - \sin^2\alpha)^{\frac{1}{2}}}{m\cos\alpha + (n^2 - \sin^2\alpha)^{\frac{1}{2}}} \ ,$$

since $n\cos\beta' = (n^2 - \sin^2\alpha)^{\frac{1}{2}}$. Therefore when $Z = Z'$, $R = 0$,

$$m\cos\alpha = (n^2 - \sin^2\alpha)^{\frac{1}{2}} \ ,$$

whence

$$\sin^2\alpha = \frac{m^2 - n^2}{m^2 - 1} \ .$$

3.11 From equations (3.10) and (3.17), we have

$$|p_{\text{inc}}| = \rho\omega^2 A \ , \qquad |p_{\text{refl}}| = \rho\omega^2 RA \ , \qquad |p_{\text{trans}}| = \rho'\omega^2 TA \ .$$

Since

$$I = \frac{|p|^2}{2\rho c} \ ,$$

$$I_{\text{inc}} = \frac{(\rho\omega^2 A)^2}{2\rho c} \ ,$$

$$I_{\text{refl}} = \frac{(\rho\omega^2 RA)^2}{2\rho c} = R^2 I_{\text{inc}} \ ,$$

$$I_{\text{trans}} = \frac{(\rho'\omega^2 TA)^2}{2\rho'c'} = \frac{\rho c}{\rho'c'}(1+R)^2 I_{\text{inc}} \ .$$

3.12 If a transverse wave only is present, then equation (3.47) must be zero, that is

$$\left(\frac{c_t}{c_\ell}\right)^2 \sin 2\beta \sin 2\gamma - \cos^2 2\gamma = 0 \ .$$

Since the angles of incidence and reflection are equal for the longitudinal waves ($\alpha = \beta$)

$$\sin 2\alpha = \sin 2\beta = \frac{\cos^2 2\gamma}{(c_t/c_\ell)^2 \sin 2\gamma} \ .$$

Chapter 4
4.11 Wave reflection method
The relevant wave equation is

$$\nabla^2\phi = \frac{1}{c_\ell^2}\frac{\partial^2\phi}{\partial t^2} \ .$$

The displacement potential for the incident wave is (see figure 4.1)

$$\phi_{\text{inc}} = A\exp i(k_j x_j - \omega t) = A\exp i(k_1 x_1 + k_3 x_3 - \omega t) \ .$$

Since the impedance ratio at each boundary is very high, the reflection coefficient R has a value 1 or -1 depending on the relative phase of the incident and reflected waves.

The displacement potential for the reflected wave is

$$\phi_{\text{refl}} = A \exp i(-k_1 x_1 + k_3 x_3 - \omega t) \quad \text{for } R = 1 ,$$

$$\phi_{\text{refl}} = -A \exp i(-k_1 x_1 + k_3 x_3 - \omega t) \quad \text{for } R = -1 .$$

At the boundaries,

$$p_{\text{res}}|_{x_1 = \pm a} = -\rho \frac{\partial^2 \phi}{\partial t^2} = 0 ,$$

where $\phi = \phi_{\text{inc}} + \phi_{\text{refl}}$.
When $R = 1$:

$$p_{\text{res}} = \rho \omega^2 A \exp(ik_3 x_3)[\exp(ik_1 x_1) + \exp(-ik_1 x_1)] .$$

Thus $p_{\text{res}}|_{x_1 = \pm a} = 0$ if $\cos k_1 a = 0$, i.e. when $k_1 a = \frac{1}{2} m\pi$ ($m = 1, 3, 5, \ldots$).
When $R = -1$:

$$p_{\text{res}} = \rho \omega^2 A \exp(ik_3 x_3)[\exp(ik_1 x_1) - \exp(-ik_1 x_1)] .$$

Now $p_{\text{res}}|_{x_1 = \pm a} = 0$ if $\sin k_1 a = 0$, i.e. when $k_1 a = \frac{1}{2} m\pi$ ($m = 2, 4, 6, \ldots$).
The case $R = 1$ corresponds to symmetric modes, and the case $R = -1$ to asymmetric modes, as illustrated in figure 4.2.

Since

$$k_1^2 + k_3^2 = k^2 = \left(\frac{\omega}{c_\varrho}\right)^2 ,$$

$$k_3^2 = \left(\frac{\omega}{c_\varrho}\right)^2 - k_1^2 = \left(\frac{\omega}{c_\varrho}\right)^2 - \left(\frac{m\pi}{2a}\right)^2 .$$

4.12 Cylindrical rod—Rayleigh correction
It is required to prove that

$$c_L' = \frac{\omega}{k_1} = c_L \left[1 - \mu^2 \pi^2 \left(\frac{a}{\lambda}\right)^2 \right] ,$$

where μ is Poisson's ratio and a is the radius of the rod.

When a rod is deformed longitudinally, there is also transverse (rotational) motion which involves additional kinetic energy. Potential energy is not increased since there is no opposing force; the radial stress vanishes at the side boundaries. Take the x axis in the direction of the axis of the rod. Then the longitudinal kinetic energy of the vibrating rod is

$$\mathcal{E}_{\text{long}} = \int_0^l \tfrac{1}{2} \rho A \dot{u}_x^2 \, dx = \tfrac{1}{4} \rho A V_x^2 l ,$$

assuming $\dot{u}_x = V_x \cos k_1 x$ for a free-free bar. Here ρ is the density, A the cross-section, l the length of the rod, and V_x the velocity amplitude.

The kinetic energy due to transverse motion is evaluated by finding the kinetic energy of a disc of thickness dx and then integrating along the axis. Thus, the transverse kinetic energy is

$$\mathcal{E}_{\text{trans}} = \rho \int_0^l dx \int_0^a \tfrac{1}{2} \dot{u}_r^2 2\pi a \, da$$

$$= \pi \rho \mu^2 \int_0^l dx \int_0^a \left(\frac{\partial \dot{u}_x}{\partial x}\right)^2 a^3 \, da$$

$$= \frac{\pi \rho \mu^2 a^4}{4} \int_0^l \left(\frac{\partial \dot{u}_x}{\partial x}\right)^2 dx ,$$

where the radial velocity, $\dot{u}_r$, is related to the axial velocity by $\dot{u}_r = \mu a(\partial \dot{u}_x/\partial x)$. But

$$\frac{\partial \dot{u}_x}{\partial x} = -k_1 V_x \sin k_1 x ,$$

so

$$\mathcal{E}_{\text{trans}} = \tfrac{1}{8}\pi\rho\mu^2 a^4 k_1^2 V_x^2 l .$$

Thus

$$\mathcal{E}_{\text{total}} = \mathcal{E}_{\text{long}}(1+\tfrac{1}{2}\mu^2 a^2 k_1^2)$$

assuming circular cross-section ($A = \pi a^2$).

As shown in a theorem of Rayleigh (Theory of Sound, Vol.1, §88), if the kinetic energy is increased, the period of vibration must increase. The ratio of periods will be the same as the square root of the ratio of the kinetic energies.

Alternatively, the increase in kinetic energy may be regarded as an increase in the effective mass. When the increased mass is substituted into the wave equation for the rod, the velocity of propagation is reduced. The original velocity is $c_{\text{L}} = (E/\rho)^{\frac{1}{2}} = (EAl/m)^{\frac{1}{2}}$, where m is the mass of the rod. The new velocity is $c_{\text{L}}' = [EAl/(m+\Delta m)]^{\frac{1}{2}}$, where Δm is the effective increase in mass due to transverse motion. Hence,

$$c_{\text{L}}' = c_{\text{L}}[1-\tfrac{1}{2}a^2\mu^2 k_1^2]^{\frac{1}{2}} = c_{\text{L}}\left[1-\mu^2\pi^2\left(\frac{a}{\lambda}\right)^2\right].$$

Alternative derivation from the Pochhammer–Chree equations
Expanding the Bessel's functions of the Pochhammer–Chree solution [equation (4.29)] we obtain

$$J_0(ka) = 1 - \tfrac{1}{4}(ka)^2 + \tfrac{1}{64}(ka)^4 - \dots ,$$
$$J_1(ka) = \tfrac{1}{2}(ka) - \tfrac{1}{16}(ka)^3 + \dots .$$

Taking only the first-order terms gives the simple classical solution:

$$\frac{\omega}{k_1} = c_{\text{L}} = \left(\frac{E}{\rho}\right)^{\frac{1}{2}} .$$

When second-order terms are included we obtain

$$\frac{\omega}{k_1} = \left(\frac{E}{\rho}\right)^{\frac{1}{2}}(1-\tfrac{1}{2}a^2\mu^2 k_1^2) = c_{\text{L}}\left[1-\mu^2\pi^2\left(\frac{a}{\lambda}\right)^2\right].$$

A graph of Rayleigh's equation is shown in figure 4.6a.

4.13 Force–current analog

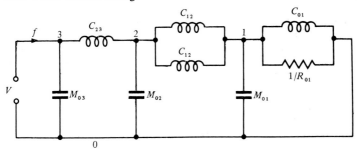

Force-voltage analog

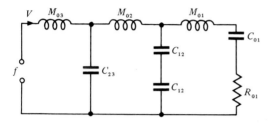

4.14 Force (pressure)-voltage analog

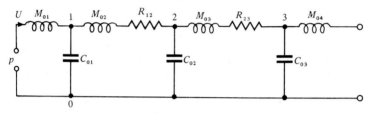

Pressure-current analog

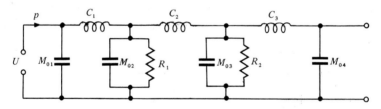

4.15 (a) *First method.* The displacement at any point in a rod due to a wave pulse is given by d'Alembert's solution [equation (4.39)]:

$$u(x, t) = f(ct - x) + g(ct + x) \ .$$

The corresponding velocity is given by

$$\dot{u}(x, t) = c\dot{f}(ct - x) + c\dot{g}(ct + x) \ .$$

If $x = 0$ at a free end, the resultant velocity there is

$$\dot{u}_{res} = c[\dot{f}(ct) + \dot{g}(ct)] \ .$$

But (see section 4.5.2), $f(ct) = g(ct)$ for a velocity pulse. Hence, $\dot{u}_{res} = 2c\dot{f}(ct)$; that is, at a free end the velocity is doubled.

Second method. The velocity (current) for a resonant transmission line may be written as

$$v(x, t) = [A \exp(-ikx) + B \exp(ikx)] \exp(i\omega t) \ .$$

At a free end (short-circuit), $Z_T = 0$ and, from equation (4.46), $|B| = |A|$. Hence,

$$v(x, t) = A[\exp(-ikx) + \exp(ikx)] \ ,$$

neglecting the time variation. Thus,

$$v(x) = 2A \cosh ikx = 2A \cos kx ,$$

and, if the free end occurs at $x = 0$,

$$v(0) = 2A .$$

(b) *First method.* At any point in a finite rod, the resultant stress will be the sum of the stress due to the incident pulse plus that due to the reflected pulse, that is $\sigma_{res} = \sigma_i + \sigma_r$.

Since $\sigma_{xx} = E\partial u/\partial x$, where E is Young's modulus and the axis of the rod is assumed to be in the x direction,

$$\sigma_{res} = E[g'(ct + x) - f'(ct - x)] .$$

Put $x = 0$ at the clamped end. For this case $f(ct) = -g(ct)$; hence,

$$\sigma_{res} = 2Ef'(ct) .$$

Second method. The stress in a resonant transmission line has the form [equation (4.43)]:

$$\sigma_{xx} = -A \exp(-ikx) + B \exp(ikx) .$$

At a clamped end (open-circuit), $Z_T = \infty$ and, from equation (4.46), $|B| = -|A|$. Then

$$\sigma_{xx} = -A[\exp(-ikx) + \exp(ikx)] = -2A \cosh ikx = -2A \cos kx .$$

If $x = 0$ at the clamped end,

$$\sigma(0) = -2A .$$

4.16 The input impedance of a finite rod follows from equation (4.44):

$$\hat{Z}_m = \hat{Z}_0 \frac{\hat{A} \exp(-i\hat{k}x) - \hat{B} \exp(i\hat{k}x)}{\hat{A} \exp(-i\hat{k}x) + \hat{B} \exp(i\hat{k}x)} .$$

The input impedance, $\hat{Z}_{11}$, at $x = 0$ is

$$\hat{Z}_{11} = \hat{Z}_0 \frac{\hat{A} - \hat{B}}{\hat{A} + \hat{B}} = \hat{Z}_0 \frac{1 - \hat{B}/\hat{A}}{1 + \hat{B}/\hat{A}}$$

$$= \hat{Z}_0 \frac{1 - [(\hat{Z}_0 - \hat{Z}_T)/(\hat{Z}_0 + \hat{Z}_T)] \exp(-2i\hat{k}l)}{1 + [(\hat{Z}_0 - \hat{Z}_T)/(\hat{Z}_0 + \hat{Z}_T)] \exp(-2i\hat{k}l)} \quad \text{[from equation (4.46)]}$$

$$= \hat{Z}_0 \frac{\hat{Z}_0[1 - \exp(-2i\hat{k}l)] + \hat{Z}_T[1 + \exp(-2i\hat{k}l)]}{\hat{Z}_0[1 + \exp(-2i\hat{k}l)] + \hat{Z}_T[1 - \exp(-2i\hat{k}l)]}$$

$$= \hat{Z}_0 \frac{\hat{Z}_0 \sinh i\hat{k}l + \hat{Z}_T \cosh i\hat{k}l}{\hat{Z}_0 \cosh i\hat{k}l + \hat{Z}_T \sinh i\hat{k}l} .$$

4.17

$$Y_{11} = \frac{1}{Z_0} \coth(\alpha + ik)l$$

where

$$\coth(\alpha + ik)l = \frac{\sinh \alpha l \cosh \alpha l - i \sin kl \cos kl}{\sinh^2 \alpha l \cos^2 kl + \cosh^2 \alpha l \sin^2 kl} .$$

Hence

$$|Y| = [(\text{Re } Y)^2 + (\text{Im } Y)^2]^{1/2}$$

$$= \frac{1}{Z_0}\left[\frac{\sinh^2\alpha l \cosh^2\alpha l + \sin^2 kl \cos^2 kl}{(\sinh^2\alpha l \cos^2 kl + \cosh^2\alpha l \sin^2 kl)^2}\right]^{1/2}.$$

$$Y_{12} = \frac{1}{Z_{12}} = \frac{1}{Z_0 \sinh ikl} = \frac{1}{Z_0 \sinh(\alpha + ik)l}$$

where

$$\sinh(\alpha + ik)l = \sinh\alpha l \cos kl + i \cosh\alpha l \sin kl$$
$$\equiv A + iB .$$

Then

$$Y_{12} = \frac{1}{Z_0}\frac{A - iB}{(A + iB)(A - iB)} = \frac{A - iB}{Z_0(A^2 + B^2)}$$

$$= \frac{1}{Z_0}\left[\frac{\sinh^2\alpha l \cos^2 kl + \cosh^2\alpha l \sin^2 kl}{(\sinh^2\alpha l \cos^2 kl + \cosh^2\alpha l \sin^2 kl)^2}\right]^{1/2}.$$

The maxima in figure 4.12 represent the maximum response to unit input force and correspond to resonant frequencies of the system whenever $kl = n\pi$. The minima for Y_{11} show pronounced dips or antiresonances since the contributions of adjacent modes are out of phase. For Y_{12} adjacent modes are in phase and there are no sharp antiresonances. The background level corresponds to an impedance value of Z_0.

4.18 As shown in section 4.5.5.2, the input impedance of a short-circuited transmission line may be written

$$\hat{Z}_{11} = Z_0 \tanh ikl = Z_0 \tanh(\alpha + ik)l = R_{11} + iX_{11}$$

where R_{11} and X_{11} are given by equations (4.56) and (4.57). Assume that near resonance the damping is small, then

$$\sinh\alpha l \rightarrow \alpha l , \qquad \cosh\alpha l \rightarrow 1 .$$

Also, if l is a multiple of $\frac{1}{4}\lambda$, $l = \frac{1}{4}m\lambda$, then $kl = \frac{1}{2}m\pi$, where m is an integer.
From equation (4.57), $X_{11} = 0$ when m is both even and odd. When m is odd:

$$\cos kl = 0 , \qquad \sin kl = 1$$

and equation (4.56) becomes

$$R_{11} = \frac{Z_0}{\alpha_n l} = \frac{4Q_n Z_0}{m\pi}$$

which is equation (4.60). When m is even:

$$\cos kl = 1 , \qquad \sin kl = 0 ,$$

and

$$R_{11} = Z_0\alpha_n l = \frac{m\pi Z_0}{4Q_n}$$

which is equation (4.61).

The transfer impedance of a short-circuited line is [equation (4.51)]

$$\hat{Z}_{12} = \hat{Z}_0 \sinh i\hat{k}l$$

$$= Z_0[\sinh\alpha l \cos kl + i \cosh\alpha l \sin kl] = R_{12} + iX_{12} .$$

Then,

$$R_{12} = Z_0 \sinh \alpha l \cos k l , \qquad X_{12} = Z_0 \cosh \alpha l \sin k l .$$

For small damping,

$$\sinh \alpha l \to \alpha l , \qquad \cosh \alpha l \to 1 .$$

Therefore

$$R_{12} \approx Z_0 \alpha l \cos k l , \qquad X_{12} \approx Z_0 \sin k l .$$

If l is a multiple of $\frac{1}{4}\lambda$, $l = \frac{1}{4}m\lambda$, and $kl = \frac{1}{2}m\pi$.
When m is odd:

$$\cos k l = 0 , \qquad \sin k l = 1 ,$$

and

$$R_{12} = 0 , \qquad X_{12} = Z_0$$

[equation (4.62)]. When m is even:

$$\cos k l = 1 , \qquad \sin k l = 0 ,$$

and

$$R_{12} = Z_0 \alpha_n l = \frac{m\pi Z_0}{4 Q_n} , \qquad X_{12} = 0$$

[equation (4.63)].

4.19 Consider three transmission lines in tandem with line 3 short-circuited, as shown in the figure.

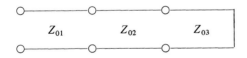

Let the characteristic impedance and length of each line be denoted by Z_{01}, Z_{02}, Z_{03}, and l_1, l_2, l_3 respectively. For a terminated transmission line it may be shown (e.g., R. A. Chipman *Transmission Lines*, Schaum, 1968, chapter 7) that

$$e_1 = e_T \left(\cosh i k_1 l_1 + \frac{Z_0}{Z_T} \sinh i k_1 l_1 \right) ,$$

where e_1 is the input voltage to line 1, e_T is the voltage across the termination, and Z_T is the terminating impedance.
For lines 1 and 2, then,

$$e_2 = \frac{e_1}{\cosh i k_1 l_1 + (Z_{01}/Z_{22}) \sinh i k_1 l_1} ,$$

where Z_{22} is the input impedance to line 2 and is given by

$$Z_{22} = Z_{02} \frac{Z_{02} \tanh i k_2 l_2 + Z_{03} \tanh i k_3 l_3}{Z_{02} + Z_{03} \tanh i k_2 l_2 \tanh i k_3 l_3} .$$

e_2 is now taken as the input voltage to line 2 which is terminated by Z_{33}; thus

$$e_3 = \frac{e_2}{\cosh i k_2 l_2 + (Z_{02}/Z_{33}) \sinh i k_2 l_2} ,$$

where $Z_{33} = Z_{03} \tanh i k_3 l_3$.

The required transfer impedance, Z_{13}, is given by

$$Z_{13} = \frac{e_1}{i_{T,3}} = \frac{e_1 Z_{33}}{e_3} \; ,$$

where $i_{T,3}$ is the current through the termination to line 3. Thus,

$$Z_{13} = \left[\cosh i k_1 l_1 + \frac{Z_{01}}{Z_{22}} \sinh i k_1 l_1 \right] \left[\cosh i k_2 l_2 + \frac{Z_{02}}{Z_{33}} \sinh i k_2 l_2 \right] Z_{03} \sinh i k_3 l_3 \; .$$

Again, damping will be assumed small and the length of each line an integral multiple of $\frac{1}{4}\lambda$. Neglecting higher order terms, we obtain

$$Z_{13} = -i \frac{Z_{01} Z_{03}}{Z_{02}}$$

when l_1, l_2, l_3 are odd multiples of $\frac{1}{4}\lambda$, and

$$Z_{13} = Z_{01}\alpha_1 l_1 + Z_{02}\alpha_2 l_2 + Z_{03}\alpha_3 l_3$$

when l_1, l_2, l_3 are even multiples of $\frac{1}{4}\lambda$.

4.20 In the frequency equation (4.95), as $\omega \to \infty$ the tanh functions $\to 1$. This follows since

$$\tanh x = \frac{\sinh x}{\cosh x} = \frac{\exp x - \exp -x}{\exp x + \exp -x} = \frac{1 - \exp(-2x)}{1 + \exp(-2x)} \; .$$

As $x \to \infty$, $\tanh x \to 1$. If we put $r = c/c_T$ and $s = c_T/c_L$, equation (4.95) becomes

$$(1 - \tfrac{1}{2}r^2)^2 = [1 - (rs)^2]^{\frac{1}{2}}[1 - r^2]^{\frac{1}{2}} \; .$$

Squaring and rearranging leads to

$$r^6 - 8r^4 + 8r^2(3 - 2s^2) - 16(1 - s^2) = 0$$

which is equation (3.64) for Rayleigh waves with $c = c_R$.

4.21 The function to be analysed is shown in the figure.

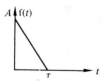

The Fourier transform of $f(t)$ is

$$F(i\omega) = \int_0^\tau f(t) \exp(-i\omega t) \, dt = \int_0^\tau A\left(1 - \frac{t}{\tau}\right) \exp(-i\omega t) \, dt$$

$$= \frac{A}{\omega^2 \tau}(1 - \cos\omega\tau) + i\frac{A}{\omega^2 \tau}(\sin\omega\tau - \omega\tau) \; .$$

Hence,

$$|F(i\omega)| = \frac{A}{\omega^2 \tau}[(1 - \cos\omega\tau)^2 + (\sin\omega\tau - \omega\tau)^2]^{\frac{1}{2}} \; .$$

In the limit, as $\tau \to 0$, $|F(i\omega)| \to \frac{1}{2}A\tau$ which is the area of the pulse.

Chapter 6

6.11 See section 6.3.2 for the derivation of the equation of motion:

$$\rho \ddot{u}_1 = (\lambda + 2G)\frac{\partial^2 u_1}{\partial x_1^2} + (\chi + 2\eta)\frac{\partial^3 u_1}{\partial x_1^2 \partial t} .$$

Assume a solution of the form

$$u_1 = A \exp i(\hat{k}x_1 - \omega t) ,$$

where $\hat{k} = k_1 + i\alpha$. Then,

$$\ddot{u}_1 = -\omega^2 u_1 ,$$

$$\frac{\partial^2 u_1}{\partial x_1^2} = -\hat{k}^2 u_1 \approx -(k_1^2 + 2i\alpha k_1)u_1$$

(neglecting the term in α^2),

$$\frac{\partial^3 u_1}{\partial x_1^2 \partial t} = i\omega(k_1^2 + 2i\alpha k_1)u_1 .$$

Substituting in the equation of motion and collecting real and imaginary parts yields the two equations:

Re: $-\omega^2\rho + k_1^2(\lambda + 2G) + 2\alpha\omega k_1(\chi + 2\eta) = 0 ,$

Im: $2\alpha k_1(\lambda + 2G) - (\chi + 2\eta)\omega k_1^2 = 0 .$

From the real part,

$$c^2 = \frac{\omega^2}{k_1^2} = \frac{(\lambda + 2G)}{\rho} + \frac{2\alpha c(\chi + 2\eta)}{\rho} ;$$

from the imaginary part,

$$\alpha = \frac{(\chi + 2\eta)\omega k_1}{2(\lambda + 2G)} \approx \frac{\omega^2(\chi + 2\eta)}{2\rho c^3} .$$

If this value of α is substituted into the velocity equation, the second term on the right hand side becomes $[k_1^2(\chi + 2\eta)^2]/\rho^2$, which is a second order quantity and may be neglected. The velocity of longitudinal waves is therefore given by

$$c_\varrho = \frac{\omega}{k_1} \approx \left(\frac{\lambda + 2G}{\rho}\right)^{\frac{1}{2}} .$$

This equation can be obtained directly if attenuation is neglected, that is if k is assumed real.

6.12 If the particle displacement is assumed to be in the x_2 direction and the direction of propagation is the x_1 direction, we obtain from equation (6.15):

$$\rho \ddot{u}_2 = G\frac{\partial^2 u_2}{\partial x_1^2} + \eta\frac{\partial^3 u_2}{\partial x_1^2 \partial t} .$$

Assume that

$$u_2 = B \exp i(\hat{k}x_1 - \omega t)$$

where $\hat{k} = k_1 + i\alpha$. Substitution shows that

Re: $-\omega^2\rho + Gk_1^2 + 2\alpha\omega k_1\eta = 0 ,$

Im: $2\alpha k_1 G - \omega k_1^2 \eta = 0 .$

From the imaginary part

$$\alpha \approx \frac{\omega^2 \eta}{2\rho c^3} ,$$

and from the real part, on neglecting the second order term,

$$c_t = \frac{\omega}{k_1} \approx \left[\frac{G}{\rho}\right]^{1/2} .$$

6.13 Starting from the wave equation (6.40) and assuming a solution of the form

$$\phi = C \exp i(\hat{k}x_1 - \omega t) ,$$

where $\hat{k} = k_1 + i\alpha$, we have for the real and imaginary parts of the solution:

$$\text{Re:} \quad k_1^2 \left[\frac{M_R}{\rho} + \frac{2\alpha\omega\tau_2 M_R}{\rho k_1}\right] - \omega^2 = 0 ,$$

$$\text{Im:} \quad \frac{2\alpha k_1 M_R}{\rho\omega} - \frac{k_1^2 \tau_2 M_R}{\rho} + \omega^2 \tau_1 = 0 .$$

Thus,

$$c_\varrho = \frac{\omega}{k_1} = \left[\frac{M_R}{\rho}\left(1 + \frac{2\alpha\omega\tau_2}{k_1}\right)\right]^{1/2} ,$$

$$\alpha = \frac{\omega^2}{2c}\left(\tau_2 - \tau_1 c^2 \frac{\rho}{M_R}\right) = \frac{\omega^2}{2c}\left[\tau_2 - \tau_1 c^2(1 + 2\alpha c\tau_2)\frac{k_1^2}{\omega^2}\right]$$

$$= \frac{\omega^2}{2c}\frac{(\tau_2 - \tau_1)}{(1 + \omega^2 \tau_1 \tau_2)} .$$

6.14 Wave propagation in thermoelastic solid

The equation of motion for an isotropic thermoelastic solid for the x direction is

$$(\lambda + 2G)\frac{\partial^2 u}{\partial x^2} - \gamma\frac{\partial\phi}{\partial x} = \rho\frac{\partial^2 u}{\partial t^2} , \tag{1}$$

where the displacement is

$$u = A \exp\{i\hat{k}(x - ct)\} ,$$

$\hat{k} = k + i\alpha$, and $\phi = T - T_0$ is the temperature deviation from the equilibrium temperature T_0. A further equation is required for ϕ which may be found as follows (see Sneddon, I.N., and Berry, D.S. in *Encyclopaedia of Physics*, volume 6).

The entropy per unit volume of a solid is

$$S = C_\epsilon \rho \ln\left(1 + \frac{\phi}{T}\right) + \gamma\theta ,$$

where C_ϵ is the specific heat at constant strain, θ is the dilatation (for total strain). If $\phi \ll T$,

$$S = \frac{\rho C_\epsilon \phi}{T} + \gamma\theta . \tag{2}$$

The quantity of heat absorbed per unit volume, h, is given by

$$h = TS = \rho C_\epsilon \phi + \gamma T\theta . \tag{3}$$

Variations in temperature within the solid are assumed to obey the thermal conductivity equation

$$\frac{\partial h}{\partial t} = K\nabla^2\phi + q , \tag{4}$$

where K is the coefficient of thermal conductivity and q is the quantity of heat per unit volume generated in the solid. Combining equations (2)-(4) and assuming that $q = 0$ gives the required temperature equation, we obtain

$$\frac{\partial^2 \phi}{\partial x^2} = \frac{\rho C_\epsilon}{K}\frac{\partial \phi}{\partial t} + \frac{\gamma T_0}{K}\frac{\partial^2 u}{\partial x \partial t} . \tag{5}$$

If ϕ is assumed to have the same form as u so that

$$\phi = T - T_0 = B \exp\{i\hat{k}(x - ct)\},$$

then substitution in equations (1) and (5) gives

$$(\lambda + 2G)(k^2 + 2i\alpha k)u + i(k + i\alpha)\gamma\phi - \rho(k^2 + 2i\alpha k)c^2 u = 0 , \tag{6}$$

$$(k^2 + 2i\alpha k)\phi - \frac{i\rho C_\epsilon}{K}(k + i\alpha)c\phi + \frac{\gamma T_0}{K}(k^2 + 2i\alpha k)cu = 0 . \tag{7}$$

From equation (7) we have

$$\phi = -\frac{\gamma T_0(k^2 + 2i\alpha k)cu}{Kk^2 + \rho C_\epsilon \alpha c + i(2\alpha k K - \rho C_\epsilon kc)} . \tag{8}$$

Substituting equation (8) in equation (6), we obtain

$$(k^2 + 2i\alpha k)[(\lambda + 2G) - \rho c^2]u + i(k + i\alpha)\gamma\left[-\frac{\gamma T_0 kc(k + i2\alpha)}{Kk^2 + \rho C_\epsilon \alpha c + i(2\alpha k K - \rho C_\epsilon kc)}\right]u = 0 . \tag{9}$$

Separating real and imaginary parts of equation (9) leads to the required solutions:

$$c^2 = \frac{\lambda + 2G}{\rho} + \frac{\gamma^2 T_0 k^2 c(C_\epsilon c + \alpha T_0/\rho)}{K^2 k^2 + \rho^2 C_\epsilon^2 c^2} , \tag{10}$$

$$\alpha = \frac{\gamma^2 T_0 k^2 K}{2(K^2 k^2 + \rho^2 C_\epsilon^2 c^2 + \gamma^2 T_0 \rho C_\epsilon c)} . \tag{11}$$

Chapter 7

7.13 Equation (7.23) for the decrement can be written

$$\Delta = \Delta_0 \Omega \Lambda L^2 \frac{\omega\tau}{(1 + \omega^2\tau^2)} , \tag{1}$$

where $\tau = BL^2/\pi^2 C$ and it is assumed that $\omega^2/\omega_0^2 \ll 1$ (true for most practical cases). The condition for maximum decrement is that

$$\frac{d\Delta}{d\omega} = 0 ,$$

whence it follows that

$$2\omega^2\tau^2 = (1 + \omega^2\tau^2) .$$

Thus, the frequency, ω_m, for which Δ is a maximum is

$$\omega_m = \frac{1}{\tau} = \frac{\pi^2 C}{BL^2} .$$

Substituting in the first equation shows that

$$\Delta_m = \tfrac{1}{2}\Delta_0 \Omega \Lambda L^2 .$$

7.14 Substituting in equation (1) of problem 7.13 for the case $\omega\tau \gg 1$, we obtain

$$\Delta \to \frac{\Delta_0 \Omega \Lambda L^2}{\omega\tau} = \frac{8Eb^2}{\pi^3 C}\frac{\Omega \Lambda L^2}{\omega}\frac{\pi^2 C}{BL^2} .$$

Since $\alpha = \omega\Delta/2\pi$,

$$\alpha_\infty = \frac{4\Omega\Lambda Eb^2}{\pi^2 B} .$$

It is noteworthy that α_∞ depends inversely on B but does not depend at all on the loop length.

7.15 Assuming $\omega \ll \omega_0$, we obtain from equation (7.21),

$$\frac{\Delta c}{c_0} \to \frac{\Delta_0}{2\pi}\omega_0^2 \Lambda\Omega L^2 \frac{\omega_0^2}{\omega_0^4 + \omega^2 d^2} = \frac{\Delta_0 \Lambda\Omega L^2}{2\pi} = \frac{4Eb^2 \Lambda\Omega L^2}{\pi^2 C}$$

which is equation (7.30). From equation (7.23),

$$\Delta \to \Delta_0 \omega_0^2 \Lambda\Omega L^2 \frac{\omega d}{\omega_0^4 + \omega^2 d^2} = \Delta_0 \Lambda\Omega L^2 \frac{\omega d}{\omega_0^2} = \frac{8Eb^2 \Lambda\Omega L^4 \omega B}{\pi^5 C^2} = \frac{8Eb^2 \omega B\Lambda\Omega L^4}{\pi^5 C^2}$$

which is equation (7.31).

7.16 As may be observed from figure 7.24, the frequency dependence of the decrement is considerably broadened when a random distribution of loop lengths is assumed. As shown by Granato and Lücke (1956a), if Koehler's distribution [equation (7.27) in the text] is adopted, the decrement may be computed from

$$\Delta = \int_0^\infty \Delta(l)N(l)\,\mathrm{d}l ,$$

by expanding in terms of exponential integral functions of complex arguments. The result is the curve shown in figure 7.24 with a maximum given by equation (7.28). A simpler argument leading to a somewhat different expression is given by Koehler (1952).

7.17 From equation (7.35), taking logarithms we obtain

$$\ln(\Delta_H\epsilon_0) = \ln c_1 - \frac{c_2}{\epsilon_0} .$$

If a plot is made of $\ln(\Delta_H\epsilon_0)$ versus $1/\epsilon_0$, then the slope is

$$-c_2 = -\frac{K\eta b}{L_c} .$$

The pinning point density is defined as

$$n_d = \frac{L(0)}{L(t)} - 1 .$$

Thus, if the *slope* is measured at two different times (following deformation or irradiation), we find

$$\frac{c_2(t)}{c_2(0)} = \frac{L_c(0)}{L_c(t)} ,$$

and hence n_d may be evaluated.

The *intercept* of the graph is

$$\ln c_1 = \frac{c_2}{\epsilon_0} = -\frac{\text{slope}}{\epsilon_0} .$$

This yields c_1; Λ may be found from the relation

$$c_1 = \frac{\Delta_0 \Lambda \Omega L_N^3 c_2}{\pi L_c}$$

provided the other terms are known.

Chapter 8

8.12 (a) The phonon frequency is equal to the Doppler frequency shift, that is 3×10^9 Hz.

(b) The phase velocity is found from equation (8.7):

$$c^* = \frac{c_0 \Delta \nu}{2n\nu \sin\theta} = \frac{(3 \times 10^8)(3 \times 10^9)2^{\frac{1}{2}}}{2(1 \cdot 4)(5 \times 10^{14})} = 909 \text{ m s}^{-1} .$$

(c)
$$\alpha = \frac{\pi w}{c} ,$$

where w is the bandwidth. Hence

$$\alpha = \frac{10^8 \pi}{909} = 3 \cdot 46 \times 10^5 \text{ m}^{-1} .$$

8.13 Ultrasonovision

(a) Consider a thin film of thickness d. Let

$$u_i = A \exp i(kx - \omega t)$$

be the displacement of a wave incident on the first surface. Let R_{12} be the displacement reflection coefficient between media 1 and 2, and T_{12} the corresponding transmission coefficient. For convenience take the origin as coincident with the first surface. Then a series of reflections occur as shown in the diagram. (The time dependence has been omitted.)

Allowing for multiple internal reflections, the overall displacement reflection coefficient, R, is

$$R = R_{12} + T_{21} R_{21} T_{12} \exp(2ikl) \times (1 + r + r^2 + ...)$$

$$= R_{12} + \frac{T_{21} R_{21} T_{12} \exp(2ikl)}{1 - R_{21}^2 \exp(2ikl)} ,$$

where $r = R_{21}^2 \exp(2ikl)$.

(b) If $l \to 0$,

$$R \to R_{12} + \frac{T_{21} R_{21} T_{12}}{1 - R_{21}^2} \ .$$

Since (see section 3.3)

$$R_{12} = \frac{Z_2 - Z_1}{Z_2 + Z_1} = -R_{21} \ ,$$

$$T_{12} = \frac{\rho_1}{\rho_2}(1 + R_{12}) = \frac{\rho_1}{\rho_2}(1 - R_{21}) \ ,$$

and

$$T_{21} = \frac{\rho_2}{\rho_1}(1 + R_{21}) \ ,$$

we have

$$R \to R_{12} + \frac{R_{21}(1 - R_{21}^2)}{(1 - R_{21}^2)} = 0 \ .$$

If $R \to 0$, the total transmission coefficient $\to 1$. Hence, for a very thin film (compared with the wavelength), the displacement of the film will be the same as that of the transmitted ultrasonic wave.

(c) The basic arrangement of the apparatus is shown in the sketch. The amplitude of light reflected from the reference mirror M_1 may be written

$$A_r = A_{r0} \exp(i\phi_r)$$

where A_{r0} is the amplitude of the light incident on the reference mirror and ϕ_r is a constant phase factor related to the distance between the beam splitter B and the reference mirror.

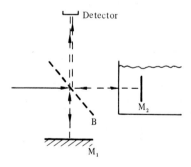

In the case of the thin film mirror M_2, the amplitude of the reflected light is

$$A_f = A_{f0} \exp[i(\phi_f + \phi_s)]$$

where ϕ_f depends on the distance between the beam splitter and the film, and ϕ_s is the additional phase shift caused by the motion of the film. Since the displacement of the film is assumed to be equal to that of the ultrasonic wave passing through it, then

$$\text{film displacement} = u_0 \cos\omega_s t$$

where u_0 is the amplitude of the ultrasonic displacement and ω_s is the ultrasonic frequency. Since

$$\text{phase difference} = \frac{2\pi}{\lambda} \times \text{path difference} \, ,$$

$$\phi_s = 2\frac{2\pi}{\lambda} u_0 \cos\omega_s t \, .$$

The additional factor 2 arises from the fact that the relative phase shift is doubled on reflection.

The total light amplitude is therefore

$$A_t = A_r + A_f \, ,$$

and the total light intensity is

$$I = A_t A_t^* = I_r + I_f + 2A_{r0}A_{f0} \cos\left(\phi_f - \phi_r + \frac{4\pi}{\lambda} u_0 \cos\omega_s t\right)$$

where $I_r = |A_{r0}|^2$ is the intensity of the light reflected from the reference mirror, and $I_f = |A_{f0}|^2$ is the intensity of the light reflected from the thin film mirror.

(d) Assume that $\phi_f - \phi_r = 90°$, then

$$I = I_r + I_f + 2I_r^{\frac{1}{2}}I_f^{\frac{1}{2}} \cos\left(90° + \frac{4\pi}{\lambda} u_0 \cos\omega_s t\right)$$

$$= I_r + I_f + 2I_r^{\frac{1}{2}}I_f^{\frac{1}{2}} \sin\left(\frac{4\pi}{\lambda} u_0 \cos\omega_s t\right) \, .$$

Since $u_0 \ll \lambda$ except for very high acoustic intensities,

$$I = I_r + I_f + 2I_r^{\frac{1}{2}}I_f^{\frac{1}{2}}\frac{4\pi}{\lambda} u_0 \cos\omega_s t \, .$$

The photocurrent, i, will be a function of the total intensity of the light. The ultrasonic contribution may be extracted using a low-pass filter; thus

$$i \approx \frac{8\pi}{\lambda} I_r^{\frac{1}{2}}I_f^{\frac{1}{2}} u_0 \cos\omega_s t \, .$$

The filtered photocurrent is proportional to the displacement amplitude, u_0, of the ultrasonic wave.

Note from the original paper: "The ultimate sensitivity of the system to acoustic waves is determined by the available laser light. With a 5 mW laser, acoustic wave intensities of $0·25 \, \mu W \, cm^{-2}$, equivalent to a displacement of $0·05$ Å at $1·5$ MHz, can be measured. As the light level is increased, the sensitivity (and the signal-to-noise ratio) increases." A given ultrasonic field may be scanned by moving the laser beam over the surface of the thin film mirror.

8.14 Following the analysis given in section 8.5.1.1, the intensity distribution on the surface of the liquid is given by [equation (8.16)]

$$I(x,y) = \frac{|p_A + p_B|^2}{2\rho c} \, ,$$

where

$$p_A = P_A \exp(ik_y y) \qquad \text{(reference wave)},$$

$$p_B = P_B \exp(-ik_y y - i\phi) \quad \text{(object wave)}.$$

The radiation pressure at the surface is [equation (8.17)]

$$\Pi = \frac{2I}{c} \ .$$

The preceding equations are now combined with the use of the relation

$$|p_A + p_B|^2 = (p_A + p_B)(p_A^* + p_B^*) = p_A p_A^* + p_A p_B^* + p_A^* p_B + p_B p_B^*$$
$$= P_A^2 + P_B^2 + 2P_A P_B \cos(2k_y y + \phi) \ .$$

Thus,

$$\Pi = \frac{P_A^2 + P_B^2}{\rho c^2} + \frac{2P_A P_B}{\rho c^2} \cos(2k_y y + \phi) \ .$$

The equation of motion is [equation (8.18)]

$$\Pi(x,y) = \rho g \zeta - \gamma \left(\frac{\partial^2 \zeta}{\partial y^2} + \frac{\partial^2 \zeta}{\partial x^2} \right) .$$

Its solution is

$$\zeta = 2A \cos(2k_y y + \phi) + B$$

provided

$$A = \frac{P_A P_B}{\rho c^2 (\rho g + 4\gamma k_y^2)} \ ; \qquad B = \frac{P_A^2 + P_B^2}{\rho^2 c^2 g} \ .$$

Author index

Subject index